Lecture Notes in Computer Science

Commenced Publication in 1973
Founding and Former Series Editors:
Gerhard Goos, Juris Hartmanis, and Jan van Leeuwen

Editorial Board

Luís Carriço Nelson Baloian
Benjamim Fonseca (Eds.)

Groupware: Design, Implementation, and Use

15th International Workshop, CRIWG 2009
Peso da Régua, Douro, Portugal
September 13-17, 2009
Proceedings

 Springer

Volume Editors

Luís Carriço
University of Lisbon, Faculty of Sciences
Department of Informatics, Campo Grande
Edifício C6, Piso 3, Sala 6.3.25, 1749-016 Lisboa, Portugal
E-mail: lmc@di.fc.ul.pt

Nelson Baloian
University of Chile, Blanco Encalada 2120, Santiago, Chile
E-mail: nbaloian@dcc.uchile.cl

Benjamim Fonseca
University of Trás-os-Montes e Alto Douro
School of Science and Technology, Engineering Department
Apartado 1013, 5001-801 Vila Real, Portugal
E-mail: benjaf@utad.pt

Library of Congress Control Number: 2009933476

CR Subject Classification (1998): H.5, K.3, K.4, C.2.4, H.5.3, K.4.3

LNCS Sublibrary: SL 3 – Information Systems and Application, incl. Internet/Web and HCI

ISSN 0302-9743

ISBN 978-3-642-04215-7 Springer Berlin Heidelberg New York

springer.com

© Springer-Verlag Berlin Heidelberg 2009

Typesetting: Camera-ready by author, data conversion by Scientific Publishing Services, Chennai, India
Printed on acid-free paper   SPIN: 12749676   06/3180   5 4 3 2 1 0

Preface

This volume presents the proceedings of the 15th International Workshop of Groupware (CRIWG 2009). The conference was previously held in USA, (Omaha) in 2008, Argentina (Bariloche) in 2007, Spain (Medina del Campo) in 2006, Brazil (Porto de Galinhas) in 2005, Costa Rica (San Carlos) in 2004, France (Autrans) in 2003, Chile (La Serena) in 2002, Germany (Darmstadt) in 2001, Portugal (Madeira Island) in 2000, Mexico (Cancun) in 1999, Brazil (Buzios) in 1998, Spain (El Escorial) in 1997, Chile (Puerto Varas) in 1996, and Portugal (Lisbon) in 1995.

The CRIWG workshops seek to advance theoretical, experimental, and applied technical knowledge of computer supported collaboration. In the CRIWG workshops, researchers and professionals report findings, exchange experiences, and explore concepts for improving the success of people making a joint effort toward a group goal. Topics of discussion are wide ranging, encompassing all aspects of design development, deployment, and use of groupware.

CRIWG embraces both mature works that are nearly ready for publication in peer review journals, and new, cutting-edge works in progress. A total of 30 papers were accepted for presentation this year—16 full papers and 14 works in progress. Papers were subjected to double-blind review by at least three members of the Program Committee. The papers are organized into nine sessions, on eight different themes: Mobile Collaboration, Social Aspects of Collaboration I & II, Technologies for CSCW, Groupware Evaluation, CSCW Design, Geo Collaboration, Collaborative Learning and Modeling CSCW.

CRIWG 2009 would not have been possible without the work and support of a great number of people. We thank the members of the Program Committee for their valuable reviews, the CRIWG Steering Committee for its timely and sagacious advice and support. We owe a special debt of gratitude to our Local Organizing Committee, who worked long hours to produce a fine workshop. Finally, we honor the authors and attendees for their substantial contributions that made CRIWG 2009 a valuable experience for all involved.

September 2009

Nelson Baloian
Luís Carriço

Organization

Conference Chair

Benjamim Fonseca   Universidade de Trás-os-Montes e Alto Douro, Portugal

Program Chairs

Luís Carriço   University of Lisbon, Portugal
Nelson Baloian   University of Chile, Santiago, Chile

Steering Committee

Carolina Salgado   Universidade Federal de Pernambuco, Brazil
Gert-Jan de Vreede   University of Nebraska at Omaha, USA
Jesus Favela   CICESE, Mexico
Jörg M. Haake   FernUniversität in Hagen, Germany
José A. Pino   Universidad de Chile, Chile
Stephan Lukosch   Delft University of Technology, The Netherlands
Pedro Antunes   Universidade de Lisboa, Portugal
Marcos Borges   Federal University of Rio de Janeiro, Brazil

Organizing Committee

Hugo Paredes   Universidade de Trás-os-Montes e Alto Douro, Portugal
Leonel Morgado   Universidade de Trás-os-Montes e Alto Douro, Portugal
Paulo Martins   Universidade de Trás-os-Montes e Alto Douro, Portugal
Vasco Amorim   Universidade de Trás-os-Montes e Alto Douro, Portugal

Program Committee

Alberto Morán   UABC, Mexico
Alberto Raposo   Catholic University of Rio de Janeiro, Brazil
Alejandra Martínez   Universidad de Valladolid, Spain
Alejandro Fernández   Universidad Nacional de La Plata, Argentina
Alicia Díaz   Universidad Nacional de La Plata, Argentina
Álvaro Ortigoza   Universidad Autónoma de Madrid, Spain

Atanasi Daradoumis	Open University of Catalonia, Spain
Atul Prakash	University of Michigan, USA
Aurora Vizcaíno-Barceló	Universidad de Castilla-La Mancha, Spain
Benjamim Fonseca	Universidade de Trás-os-Montes e Alto Douro, Portugal
Bertrand David	Ecole Centrale de Lyon, France
Carlos Duarte	Universidade de Lisboa, Portugal
César Collazos	Universidad del Cauca, Colombia
Choon Ling Sia	University of Hong Kong, Hong Kong
Christine Ferraris	Université de Savoie, France
Christoph Rensing	Technische Universität Darmstadt, Germany
Dominique Decouchant	LSR-IMAG, Grenoble, France
Eduardo Gómez-Sánchez	Universidad de Valladolid, Spain
Filippo Lanubile	University of Bari, Italy
Flávia Santoro	Universidade Federal do Estado do Rio de Janeiro, Brazil
Gert-Jan de Vreede	University of Nebraska at Omaha, USA
Guillermo Simari	Universidad Nacional del Sur, Argentina
Gustavo Zurita	Universidad de Chile, Chile
Gwendolyn Kolfschoten	Delft University of Technology, The Netherlands
Hugo Fuks	Pontifícia Universidade Católica do Rio de Janeiro, Brazil
Hugo Paredes	Universidade de Trás-os-Montes e Alto Douro, Portugal
Jesus Favela	CICESE, Mexico
Joey F. George	Florida State University, USA
José A. Pino	Universidad de Chile, Chile
Julita Vassileva	University of Saskatchewan, Canada
Luis A. Guerrero	Universidad de Chile, Chile
Marcos Borges	Universidade Federal do Rio de Janeiro, Brazil
Martin Wessner	Fraunhofer IPSI, Germany
Miguel Nussbaum	Pontificia Universidad Católica de Chile, Chile
Niels Pinkwart	Clausthal University of Technology, Germany
Nelson Baloian	Universidad de Chile, Chile
Nuno Preguiça	Universidade Nova de Lisboa, Portugal
Pedro Antunes	Universidade de Lisboa, Portugal
Ralf Steinmetz	Technische Universität Darmstadt, Germany
Richard Anderson	University of Washington, USA
Robert O. Briggs	University of Nebraska at Omaha, USA
Sergio F. Ochoa	Universidad de Chile, Chile
Stephan Lukosch	Delft University of Technology, The Netherlands
Steven Poltrock	Boeing, USA
Till Schümmer	FernUniversität in Hagen, Germany

Table of Contents

Mobile Collaboration

Social Aspects of Collaboration I

Social Aspects of Collaboration II

Technology for CSCW

Groupware Evaluation

CSCW Design

Geo Collaboration

Collaborative Learning

Modeling CSCW

Building Real-World Ad-Hoc Networks to Support Mobile Collaborative Applications: Lessons Learned

Roc Messeguer[1], Sergio F. Ochoa[2], José A. Pino[2], Esunly Medina[1], Leandro Navarro[1], Dolors Royo[1], and Andrés Neyem[3]

[1] Department of Computer Architecture, Universitat Politècnica de Catalunya, Spain
{messeguer,esunlyma,leandro,dolors}@ac.upc.edu
[2] Department of Computer Science, Universidad de Chile, Chile
{sochoa,jpino}@dcc.uchile.cl
[3] Department of Computer Science, Pontificia Universidad Católica de Chile, Chile
aneyem@ing.puc.cl

Abstract. Mobile collaboration is required in several work scenarios, i.e. education, healthcare, business and disaster relief. The features and capabilities of the communication infrastructure used by mobile collaborative applications will influence the type of coordination and collaboration that can be supported in real work scenarios. Developers of these applications are typically unaware of the constraints the communication infrastructure imposes on the collaborative system. Therefore, this paper presents an experimental study of how ad-hoc networks can effectively support mobile collaborative work. The article analyzes several networking issues and it determines how they influence the collaborative work. The paper also presents the lessons learned and it provides recommendations to deal with the networking issues intrinsic to ad-hoc networks.

Keywords: Mobile Collaboration, Communication Support, Wireless Networks.

1 Introduction

Collaborative applications are intended to support the work performed by a group of collaborators who pursue a common goal. The focus in recent years has been station-ary collaboration; however, advances in mobile computing and communication have made mobile collaboration a real possibility. Medical applications [12, 2], collabora-tive learning [20], emergency management [14], and productive activities [15] are some of the application areas for mobile collaborative solutions.

The idea behind this new CSCW paradigm is to support collaboration among mo-bile users, regardless of their physical location. The physical location of each partici-pant can be a precious source of contextual information for collaboration support applications, and it should not be a limitation to collaborate.

Two decades ago, Ellis et al. showed the coordination and collaboration depended on communication [6]. Therefore, if we want to enable coordination and collaboration in several mobile work scenarios, then we have to improve the communication capabilities

L. Carriço, N. Baloian, and B. Fonseca (Eds.): CRIWG 2009, LNCS 5784, pp. 1–16, 2009.

among the mobile users. Today, the wireless communication technologies partially match the mobile users' communication needs. That means mobile groupware designers must take into account the capabilities and limitations of the communication infrastructure, when they are designing a new collaborative application.

This article presents a study showing when and how an ad-hoc network can be used to support mobile collaborative work, and also which are the limitations and the considerations to be applied for mobile groupware application design. The study involved several experiments carried out in representative work scenarios, using real-world ad-hoc network implementations.

The next section presents the communication requirements to support mobile collaborative work. Section 3 describes the test bed used in this study. Section 4 describes the experimental evaluation. Section 5 presents the lessons learned and section 6 lists the recommendations to deal with ad-hoc networking issues. Finally, section 7 presents the conclusions and future work.

2   Requirements on Ad-Hoc Communication for Mobile Collaborative Applications

We define an ad-hoc network as an autonomous and decentralized system formed by a collection of cooperating nodes which are connected by wireless links. They can dynamically self-organize and communicate among them, in order to make up a network without necessarily using any pre-existing infrastructure. Some of the key features of these networks are the following: self-organizing network, multi-hop routing, wireless links with a dynamic topology (joining/leaving nodes and wireless links changes).

Mobile collaboration supported by these networks must consider several communication requirements in order to enable group work in ad-hoc settings. These requirements are briefly explained below.

1.   *Adequate network performance.* Mobile collaboration requires a stable and sufficient network performance. Whenever two persons decide to interact, the communication link must be able to support it. Network performance problems are frequent as users move and the network topology changes. The most critical parameters are latency (i.e. time required to transport the information between two locations), jitter (i.e. the variance of the latency) and insufficient throughput (i.e. data transfer rate) [4].

2.   *Reliability.* Communication reliability is related to the trustworthiness of the communication link to transfer information between two points. Reliability design software techniques are used to deal with typical data transfer problems such as packet loss and ordering [4].

3.   *Dynamic network architecture.* Typically, the network architecture controls the data distribution strategy used to transport the information between two points. Since ad-hoc networks are dynamic in terms of topology, performance and reliability, the groupware system has to adapt itself to determine the distribution schemes which are best suited for a particular application and collaboration scenario [4].

4.   *Interoperability.* Mobile users must be allowed to interact with anyone else on a casual or opportunistic collaboration. As a consequence their collaborative mobile applications should offer interoperability of communication, data and services [13].

5.   *Awareness of users' reachability.* Mobile users need to know if a particular user is reachable when they intend to start a collaborative activity [13].

Several studies have been published about how these issues affect groupware applications when they run over Internet. A study presented by Gutwin et al. shows performance/usability of real-time distributed groupware applications depends on network parameters such as latency, jitter, packet loss, bandwidth and type of traffic (UDP/TCP) [7]. Network delays due to latency and jitter have serious effects on users' work, causing difficulties in coordination and forecasting [7]. In extreme situations they cause communication break downs; this occurs, e.g., when latency > 300 ms [18], or jitter > 500 ms [5]. It has been also reported that insufficient bandwidth increases latency and packet loss [7]. It is clear at least some of these issues will also affect collaboration on ad-hoc networks. That motivates this paper which tries to determine how networking issues affect mobile collaboration supported by ad-hoc networks.

The study presented in this article evaluates whether the networking issues in ad-hoc networks are comparable to those obtained on extreme situations. It intends to provide mobile application designers with a range of values that can be found in real-world work scenarios, for a set of key networking parameters such us throughput (i.e. real bandwidth), packet loss, latency and jitter. The study also provides advice on how to keep the network performance and reliability within acceptable ranges.

3   Test Bed Description

All the tests involved in this study used real-world ad-hoc networks. The traffic was emulated to ensure the repeatability of the results. In order to provide realism to the simulated traffic, the study followed the recommendations by Kiess and Mauve [11]. Next sections present the hypotheses and the experimentation scenarios involved in this study. We also describe the routing protocols, hardware and software used in the experimentation process.

3.1   Work Hypotheses

The work hypotheses involved in this study are the following ones:

Hypothesis 1. The network bandwidth and reliability decrease when increasing number of hops between the sender and receiver nodes.

Hypothesis 2. The network bandwidth and reliability decrease with increasing mobility of the nodes.

Hypothesis 3. The network bandwidth and reliability decrease due to increasing interference generated by mobile devices from other users.

Hypothesis 4. Routing protocols based on number of hops (such as BATMAN [9]) have better reliability and bandwidth than protocols based on statistics (such as OLSR [8]).

Hypothesis 5. In ad-hoc networks, communication on UDP has better performance than communication on TCP.

3.2 Experimentation Scenarios

In order to validate the hypotheses, the study considered three experimentation scenarios: static, mobile and working group scenarios. These three scenarios represent key situations in collaborative activities. Each one is briefly described below.

Static work scenario: Communication can occur between any pair of users independently of the distance between them. There was no mobility in this scenario (each node was stationary), and the network topology was chain. Five nodes were located at 11 to 14 meters of distance between them; therefore, a 4-hops ad-hoc network was established. Since this scenario allows monitoring and reproducing the network multi-hop behavior, it was used to validate *hypothesis 1*.

Mobile work scenario: A single person moves around the room while the other users are still working. This test scenario is similar to the first one, because it uses the same hardware and network topology. However, this new one introduces nodes mobility, implemented through a mobile user who continually walked between both ending points of the network. In this case, an extra node (i.e. *node 6:* the node used by the mobile user) was included in the test scenario. The test began with node 6 located close to node 1 (network ending point), and the data transfer is always done between the mobile user and node 5 (the other network ending point). Packet filtering was also used in order to force the communication between these nodes always goes through at least 2 hops. This scenario was used to validate *hypothesis 2* because it is possible to isolate the effect produced by a mobile user.

Working groups scenario: People work in groups and wish to exchange information within their group or between groups. We prepared two different sub-scenarios. The first one involved two groups composed of three mobile nodes each, and one group consisting of two mobile nodes. These tests were carried out in a laboratory of 146 m^2 and the distance between groups was about 12 to 18 meters. The second sub-scenario involved two groups of four mobile nodes each. The distance between groups was about 8 meters. In both cases the data transfer must be performed among group members or between groups. This test scenario was used to validate *hypothesis 3*.

The validation of *hypotheses 4* and *5* does not require specific settings; therefore they can be evaluated in each of the described test scenarios.

3.3 Routing Protocols of Mobile Ad-Hoc Networks

The routing protocol usually has an important role in network reliability and performance. Thus, we reviewed the most widespread protocols and we selected the following ones:

Better Approach To Mobile Ad-hoc Networking (BATMAN) [9]. The routing metric used by this protocol is the number of hops involved in the communication. This is a proactive routing protocol using a distance vector approach to determine the best route between sender and receiver. The routing metric represents the main criterion to determine the best path for data transmission, from a given source to a given destination. The protocol implementation used during the tests was BATMANd for Linux.

Optimized Link State Routing Protocol (OLSR) [8]. The routing metric used by this protocol is Expected Transmission Count (ETX) [3]. This is a proactive routing protocol using a link state approach to select the optimal route. The routing metric represents the main criterion to determine the best path for data transmission, from a given source to a given destination. The protocol implementation used in the test was OLSRd for Linux and Windows.

3.4 Metrics for Link Quality Evaluation

The metrics used to assess link communication quality during the tests were those relevant for mobile collaborative work (described in section 2) and also those designed to determine the Mobile Ad-hoc Networks (MANETs) protocols performance [3, 11]. The metrics provided by the traffic generator itself were also taken into account [16]. These metrics were divided into three groups, depending on the type of traffic used for measuring them: (1) ICMP traffic metrics using Round-Trip Time (RTT), (2) UDP traffic metrics considering throughput, packet loss and jitter, and (3) TCP traffic metrics involving throughput, handshake time, out of order packets and number of re-transmissions.

The UDP/TCP traffic was generated using the Iperf tool [16]. The metrics were measured by conducting a 60 seconds test, on which UDP/TCP packets were transferred between a given source-destination pair. For UDP, packets were generated at different bit rates; consequently, several UDP traffic loads were offered to the network. Unlike with UDP, in the TCP experiment we tested the maximum achievable throughput; therefore, no given bit rates were specified.

The RTT was measured by conducting a 60 seconds test, on which ICMP packets were transferred between a source-destination pair, using the regular ping service. Those experiments were carried out for packet sizes of 64 or 1024 bytes.

3.5 Hardware

The experiments were carried out using eight laptops. Six of them were HP NX6310 with an IBM Intel Core 2 T5500 of 1.66 GHz processor and 1 GB of RAM. Each of these computers had an internal Intel PRO/Wireless 3945ABG Network Connection card for IEEE 802.11b/g wireless connectivity. In addition, two HP NX6110 laptops were also used in the experiments. These computers are almost equal to the previous ones; the only difference is the model of the internal wireless card (Intel PRO/Wireless 2200BG card for IEEE 802.11b/g). During the experiments, the wireless cards on the laptops were set to channel 1 at the 802.11b/g band, using auto rate, transmission power 1 dBm and RTC/CTS off, following recommendations in [10].

3.6 Test Bed Supporting Software

All laptops were equipped with Linux Operating System (Ubuntu 8.04 Linux distribution with the 2.6.24-19-generic kernel) and also MS Windows XP. The traffic generator used in the test was Iperf (version 2.4) and the regular ping service provided by the operating system.

The traffic analyzers were Wireshark, and tcpdump (similar to Wireshark but with a command line interface). A MAC filter was also used to classify packets on the MAC layer and force a multi-hop behavior avoiding direct communication between two nodes [11].

In order to avoid the human intervention as much as possible, a LiveCD was prepared. This LiveCD adds an extension to the operating system facilitating the test bed implementation, use and data gathering [17].

4   Experimentation / Evaluation Ad-Hoc Networks

This section presents the obtained results in the tests performed in the three described work scenarios. The results allow mobile application designers to see the range of values that can be found in real work scenarios, for each key networking issue. These results allow also understanding the degree of validity of the hypotheses.

4.1   Static Work Scenario

The tests in all work scenarios involve at least 10 repetitions in order to get representative values. In the static work scenario it is possible to see that the RTT (Round-Trip Time) increases with the number of required hops and also with the packet size. For small packets the behavior of OLSR and BATMAN is similar (Fig. 1) and the RTT seems to be comparatively better with large size packets.

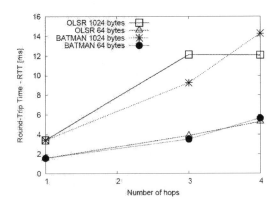

Fig. 1. RTT of the ping service, using different routing protocols and packet sizes

Figure 2 and 3 show the throughput decreases with the number of hops using both, UDP and TCP transport protocols. Once again, for both types of traffic, the behavior of OLSR and BATMAN was similar. In case of TCP, the packet out of order, retransmission and handshake time (about 0.5 ms) are zero or negligible numbers. In both routing protocols the values are almost equal (Fig. 3).

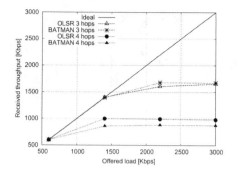

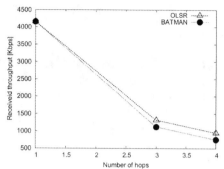

Fig. 2. Throughput on UDP, considering various routing protocols and number of hops

Fig. 3. Throughput on TCP, considering various routing protocols and number of hops

4.2 Mobile Work Scenario

For all considered TCP metrics, BATMAN has a better behavior than OLSR (Table 1). It can be due to the fact that OLSR uses the ETX metric for the selection of the routes, and the ETX metric utilizes statistical information of the 10 last probes to compute its current value. Since the mobile user location is constantly changing, the computed best route becomes out-dated soon when using OLSR. Therefore, it will involve a major rate of out-of-order packets and a higher number of retransmissions.

Table 1. Routing protocols comparison using TCP

	BATMAN	OLSR
Received Throughput (kbps)	2110	2035
Number of packets out of order	0.05	2.25
Number of re-transmissions	0.00	296.75
Handshake time (ms)	0.00	0.04
RTT (ms)	6.59	7.37

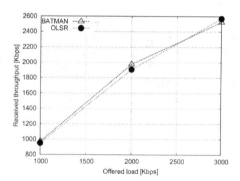

Fig. 4. Throughput on UDP, considering various offered loads

The RTT was similar for both protocols. However, BATMAN showed a slightly better UDP throughput than OLSR for medium offered loads. For higher loads the behavior of the routing protocols shows reversed results (Fig. 4).

4.3 Working Groups' Scenario

As mentioned in section 3.2, this experimentation scenario has two sub-scenarios to be implemented. The first one is composed of three groups of nodes, and the second one has two groups of mobile users. The experimentation results belonging to each sub-scenario are presented below.

4.3.1 First Experimentation Sub-scenario

This experimentation process involved two groups composed of three mobile nodes each one, and a third one composed of two mobile nodes. In this case the mobile nodes belonging to the three groups were transmitting and receiving information during all the observed period.

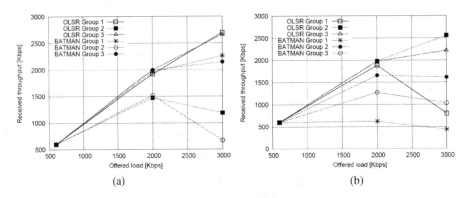

(a)   (b)

Fig. 5. UDP throughput intra/inter-group transmissions, using two protocols and various offered loads

Figure 5a shows the UDP throughput obtained by three groups of mobile users, when they communicate inside the group (intra-group interactions) using both routing protocols. The experiment considered mainly offered loads of 2000 and 3000 kbps which are typical values to be found in mobile collaboration scenarios on wireless networks. Figure 5b shows the results of the same experiment, but now involving mobile users communicating with users belonging to other groups (inter-group interactions).

These results show both routing protocols are similar when the offered load is 2000 kbps or less. After that limit, the throughput obtained with OLSR is superior to BATMAN. This situation is a consequence of the high mobility of the nodes. In this case, the routing tables of both protocols become rapidly out-dated; however, OLSR has a better capability to adapt itself to the new network characteristics. In other words, that protocol reacts faster and better to the changes in the networking context. This capability is fundamental to support collaboration among users with high mobility, such as emergency management.

4.3.2 Second Experimentation Sub-scenario

This experimentation scenario involved two groups; each one composed of four mobile nodes. Intra and inter-group communication were evaluated in this scenario. Figure 6 shows results of UDP throughput considering intra-group communication. In this case both protocols show a similar behavior when the data transfer is between group members and the offered load is below 2000 kbps. After that limit OLSR is able to obtain up to 500 kbps over BATMAN.

However there is an important difference in the network behavior when the communication is among groups. In that case, BATMAN performs better than OLSR for loads over 600 Kbps. In those settings BATMAN is able to reach up to 800 Kbps of "extra" received throughput (Fig. 7). This situation could be explained by considering the ETX link quality metric. OLSR uses 2-hops routes, which means a lower received throughput in comparison with the 1-hop route used by BATMAN.

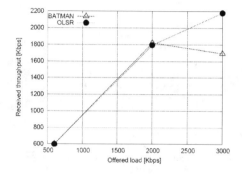
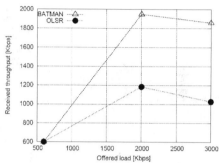

Fig. 6. Intra-group communication involving mobile nodes

Fig. 7. Inter-group communication involving mobile nodes

Table 2 shows the network throughput decreases with an increase in the number of transmitting nodes (1 tx: one node transmitting per group, 2 tx: two nodes transmitting per group, and so on). The rest of the networking issues are consistent with this situation because, e.g., the % of packet loss and the jitter increase with the number of transmitting nodes. These results could be showing hypothesis 3 is valid.

Table 2. Key networking issues vs. routing protocols and offered load

UDP Offered load (Kbps):	2000				3000			
Routing Protocol	BATMAN		OLSR		BATMAN		OLSR	
Number of Transmiting Nodes	Tx 1	Tx 2	Tx 1	Tx 2	Tx 1	Tx 2	Tx 1	Tx 2
Received throughput (Kbps)	2000	1620	1477	1327	2133	476	2737	369
% loss	0	0	0,53	0,93	0,18	28,67	6,59	63
Jitter (ms)	1,86	16,5	1,62	5,45	136,55	1085,78	140,23	195,15
Number of datagrams out of order	0	1	0	0	6	1	125	38

These experiments were reproduced, but using TCP as transport protocol. The obtained results were similar to those obtained on UDP. In other words, OLSR performs better than BATMAN when the communication is intra-group, but BATMAN has a better performance when communication is inter-group.

If we analyze the network behavior when UDP was used, it is possibly to notice the performance is affected (in terms of jitter, % packet loss and received throughput) by the number of nodes which are transmitting inside each group.

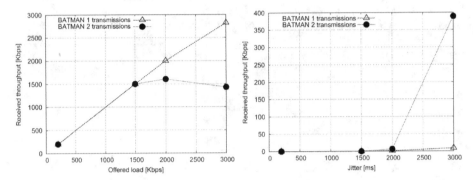

Fig. 8. UDP throughput vs. number of transmitting nodes

Fig. 9. UDP jitter vs. number of transmitting nodes

Table 3. Key networking issues vs. number of transmitting nodes (per group)

	UDP Received throughput (Kbps):		
UDP Offered load (Kbps):	**2 groups (2 tx in total)**	**3 groups (3 tx in total)**	**2 groups (4 tx in total)**
2000	2000	1822	1606
3000	2818	1694	1429

	UDP Jitter (ms):		
UDP Offered load (Kbps):	**2 groups (2 tx in total)**	**3 groups (3 tx in total)**	**2 groups (4 tx in total)**
2000	1,21	4,56	6,36
3000	8,81	30,37	389,43

	UDP % packet loss		
UDP Offered load (Kbps):	**2 groups (2 tx in total)**	**3 groups (3 tx in total)**	**2 groups (4 tx in total)**
2000	0	0	0,37
3000	0,28	3,49	24,61

	UDP # of out of order datagrams		
UDP Offered load (Kbps):	**2 groups (2 tx in total)**	**3 groups (3 tx in total)**	**2 groups (4 tx in total)**
2000	0,5	0,5	0,33
3000	0,67	3,61	42,67

In order to understand this influence, we used just one routing protocol: BATMAN. On the one hand, Fig. 8 shows the throughput increases with the number of transmitting nodes. On the other hand, Fig. 9 shows the network jitter increases with the number of transmitting nodes. However, the difference is relevant just over

2000 kbps of offered load. Table 3 summarizes the results obtained for the key networking issues, when two, three and four transmitting nodes are involved.

4.4 Validation of the Hypotheses

Next a brief validity analysis of the hypotheses is presented based on the results shown in the previous sections.

Hypothesis 1 - *The network bandwidth and reliability decrease when increasing the number of hops between the sender and receiver nodes.* Figure 2 and 3 show how the throughput decreases with the number of hops for both, UDP and TCP transport protocols. Therefore, hypothesis 1 is validated by the experiments.

Hypothesis 2 - *The network bandwidth and reliability decrease with the mobility of the nodes.* Unfortunately the study does not show significant information related to the validity of this hypothesis. However comparing the results obtained in mobile work (section 4.2) and working group scenarios (section 4.3) it seems the mobility of the users affects negatively the network throughput, if we consider the same number of transmitting nodes. If the mobile users are close to each other, network cards produce interference with each other and therefore the network performance degrades (Table 3).

Hypothesis 3 - *The network bandwidth and reliability decrease due to the interference from other mobile users.* Based on the results presented in Tables 2 and 3, it is highly probable hypothesis 3 be true.

Hypothesis 4 - *Routing protocols based on number of hops (such as BATMAN [9]) have better reliability and bandwidth than protocols based on statistics (such as OLSR [9]).* The study does not show results able to validate this hypothesis. In static multi-hop scenarios the results seem to be similar. However, the use of the ETX metric degrades the network performance when OLSR is used. In mobile scenarios the results depend on the level of mobility of the users. In high mobility, OLSR seems to be slightly better than BATMAN.

Hypothesis 5 - *In ad-hoc networks, communication on UDP has better performance than communication on TCP.* In the work scenarios involving mobility, the obtained results (Figs. 2 and 3; Tables 1 and 2) are showing the UDP communication degrades slower than TCP communication. Therefore, the throughput on UDP is higher than on TCP. These results are showing this hypothesis could be valid.

5   Lessons Learned

After the tests performed in the described scenarios, there is much empirical information that must be considered by mobile groupware application designers. This information will be relevant to understand the capabilities and limitations of a collaborative application, when it supports mobile work in a real work scenario. For example, the communication support degrades when:

- *The number of hops required to transport a message increases.* Evidence of it is a reduction in throughput, a greater latency, and major packet loss. It affects the mobile collaborative work when the users are disperse, for example on activities taking place in a hospital.
- *The mobility of users increases.* Typically the latency increases and the throughput decreases because the routing protocols are not fast enough to react to the changes in the network topology and composition. It affects the collaboration in scenarios with high mobility; for example in emergency situations, such as big fires or urban search-and-rescue.
- *Collaborators are very close to each other.* Typically the throughput decreases and the packet loss increases. Although the throughput in these scenarios is good enough to support collaboration at moderate rates, network demanding applications (such as audio/videoconference systems) can be seriously affected.

Routing protocols enable communication beyond the immediate neighbors, i.e. at more than one hop of network distance. It means that a user can collaborate with persons located in other buildings, just because of using a routing protocol. Otherwise the user is restricted to collaborate with persons that are physically sufficiently close.

The behavior of the routing protocols is diverse. Therefore, the designer must select that protocol depending on the users' mobility, and the features of the work physical scenario. Something similar happens with the transport protocols. Typically connection-less transport (e.g. UDP) provides a better throughput than connection oriented (e.g. TCP). Path redundancy could help to increase the communication throughput.

6   How Applications Can Deal with the Ad-Hoc Networking Issues

This section presents a list of recommendations to deal with the communication issues in mobile collaboration, which were presented in section 2: *network performance* (measured in terms of throughput, latency, and jitter), *communication reliability* (measured in terms of packet ordering and loss), *interoperability* (measured in terms of hardware, and data and services exchange), *awareness of users' reachability* and the *dynamic property of the network architecture*. The following sections present possible solutions on the application layer, which help to deal with the challenges imposed by the communication limitations of MANET networks.

6.1   Dealing with Network Performance and Reliability

Dealing with network performance and reliability can be faced by applications in several ways. Some of them are the following ones:

Data compression. Compression is applied to the volume of data required to represent information. It helps to increase the apparent throughput [4] and it also results in smaller packets on the network, which reduces the packet loss during the interactions among mobile users.

Rate control. In order to improve performance, network transmission should be decoupled from the system's event model, and the transmission rates should be carefully regulated. Because MANETs have a limited bandwidth which has to be shared among all nodes which are transmitting concurrently, the regulation of the transmission rate on each node could be a solution to avoid the collapse of the network when it is overloaded (system-wide rate regulation systems may also help to maintain a quality of service level). Therefore, applications might adapt by decoupling communication from user interaction or other tasks and by considering alternative data encodings selected depending on current network conditions.

Adaptive user interfaces. Applications that dynamically change their interaction paradigm can better accommodate to changing network conditions. Thus, applications can deliver a limited or extended level of service to the end-user depending on the network conditions (e.g. moving from high to low video, to audio only or text communication according to the network conditions). This type of self-adaptation mechanism allows applications to adapt gracefully   reducing the impact of the network changes on end-users.

Revealing network problems. Revealing network problems can be implemented as awareness components, which inform users about networking problems. Thus, users can take some action to try to solve or mitigate the problem. For example, when a mobile user becomes isolated, the awareness mechanism can inform him/her about that. Therefore, the user can change his/her location to be able to interact with other users.

Multicasting. Multicast, i.e., sending a message to multiple destinations at once, could help to eliminate data redundancy freeing network capacity, or allowing additional interactions among mobile users. However, multicast can become another issue when network links are diverse in capacity and for applications requiring reliable group communication.

6.2 Dealing with Interoperability, Awareness of Users' Reachability, and Dynamic Network Architectures

There are several ways of dealing with interoperability, awareness of users' reachability, and various network architectures. Some of them are the following ones:

Cross layer interaction. The cross layer pattern recommends the separation of concern in different layers, and the sharing of information among them [13]. Although sharing information between network layers adds complexity and breaks layered models, it provides a very useful mechanism to deal with dynamic network architectures. For example, the routing mechanisms can self-adapt based on information from the network layer and help to improve the throughput on the network.

Standardized solutions. Beyond typical data and service interoperability required for effective interaction among continuously changing groups of participants, mobile work scenarios will benefit from specific transport and routing protocols optimized for a large range of mobile devices (from cellular to laptops), supporting bi-lateral, multi-lateral, synchronous and asynchronous communication.

Gossip propagation mechanisms. Sometimes two collaborators are unreachable because there is no link between them. Then, it is possible to deliver a gossip, which is message travelling through the network during a certain time period, looking for the destination user. Typically, these messages (if they are received by the destination) try to promote an encounter. For example, *"URGENT: try to be at ... after lunch"*. Although this mechanism may not always succeed, it can contribute to improve the reachability of users in disperse work scenarios or in situations in which few persons work together at the same time.

Pushing notifications. Typically, after a couple of unsuccessful tries to interact with somebody, a mobile user may no longer be aware of such person. However, if the system notifies the user when the destination becomes reachable again, it enables additional collaborative interaction. Notification mechanisms must be autonomous, proactive and non-invasive to reduce the burden on the participants.

7 Conclusions and Further Work

This study tries to understand the challenges and it suggests how applications can effectively use ad-hoc networks in the CSCW domain. Particularly, this study addresses the communication issues affecting mobile collaborative work. Although there are several studies on real-world ad-hoc networks, most of them are just focused on data transfer and technologies analysis [11].

Designers of mobile groupware applications can take advantage of the results of this study when designing the functionality of applications in order to benefit from the opportunities and avoid the obstacles of these dynamic networks. The design of contextualized applications is always a challenge, but even greater for mobile groupware systems due to the intrinsic diversity and the dynamic work scenarios [1].

Three experimentation scenarios were considered in the study: static, low mobility and high mobility. In addition, intra-group and inter-group communications were analyzed to understand what network support can be required (and obtained) in real work scenarios when two or more mobile users decide to work together. Two typical setups for routing and data transportation were used in the tests. The results show a number of challenges which the designer has to face.

The paper also presents a list of lessons learned and a set of recommendations to consider in the design of applications; particularly in its communication support infrastructure. These recommendations can help mobile groupware designers to define a predictable behavior for their systems under a dynamic real-world scenario, according to the design objectives and the users' needs.

Next steps in this study involve extending the number of experimentation scenarios to cover an ample spectrum of mobile collaboration styles. In addition, the authors want to include additional hardware diversity in the tests to understand its impact on the collaboration capabilities provided by applications.

Acknowledgements

This work was partially supported by Fondecyt (Chile), grant Nos 11060467 and 1080352, by LACCIR grants R0308LAC002 and R0308LAC005, and by the Spanish MEC project P2PGrid TIN2007-68050-C03-01.

References

[1] Alarcón, R., Guerrero, L., Ochoa, S., Pino, J.: Analysis and Design of Mobile Collabora-tive Applications Using Contextual Elements. Computing and Informatics 25(6), 469–496 (2006)

[2] Antunes, P., Bandeira, R., Carriço, L., Zurita, G., Baloian, N., Vogt, R.: Risk Assessment in Healthcare Collaborative Settings: A Case Study Using SHELL. In: Briggs, R.O., Antunes, P., de Vreede, G.-J., Read, A.S. (eds.) CRIWG 2008. LNCS, vol. 5411, pp. 65–73. Springer, Heidelberg (2008)

[3] Draves, R., Padhye, J., Zill, B.: Comparison of Routing Metrics for Static Multi-hop Wireless Networks. In: Proc. of the 2004 Conference on Applications, Technologies, Architectures, and Protocols for Computer Communications (SIGCOMM 2004), pp. 133–144. ACM Press, New York (2004)

[4] Dyck, J.: A Survey of Application-Layer Networking Techniques for Real-time Distributed Groupware. Technical Report HCI-TR-06-06, University of Saskatchewan (2006)

[5] Dyck, J., Gutwin, C., Graham, T.C.N., Pinelle, D.: Beyond the LAN: Techniques from Network Games for Improving Groupware Performance. In: Proc. of Group 2007, pp. 291–300. ACM Press, New York (2007)

[6] Ellis, C.A., Gibbs, S.J., Rein, G.L.: Groupware: Some Issues and Experiences. Communications of the ACM 34(1), 38–58 (1991)

[7] Gutwin, C.: The Effects of Network Delays on Group Work in Real-Time Groupware. In: Proc. of the Seventh European Conference on Computer Supported Cooperative Work (ECSCW 2001), pp. 299–318. Kluwer Academic Publishers, Norwell (2001)

[8] RFC3626 - Optimized Link State Routing Protocol (OLSR)(2003),
     http://www.ietf.org/rfc/rfc3626.txt

[9] Better Approach To Mobile Ad-hoc Networking, B.A.T.M.A.N. (2008),
     http://tools.ietf.org/html/
     draft-wunderlich-openmesh-manet-routing-00

[10] Kanchanasut, K., Tunpan, A., Awal, M.A., Das, D.K., Wongsaardsakul, T., Tsuchimoto, Y., Dumbonet, M.: A Multimedia Communication System for Collaborative Emergency Response Operations in Disaster-affected Areas. International Journal of Emergency Management 4, 670–681 (2007)

[11] Kiess, W., Mauve, M.: A Survey on Real-World Implementations of Mobile Ad-hoc Networks. Ad Hoc Networks 5(3), 324–339 (2007)

[12] Mejia, D., Moran, A., Favela, J.: Supporting Informal Co-located Collaboration in Hospital Work. In: Haake, J.M., Ochoa, S.F., Cechich, A. (eds.) CRIWG 2007. LNCS, vol. 4715, pp. 255–270. Springer, Heidelberg (2007)

[13] Messeguer, R., Ochoa, S.F., Pino, J.A., Navarro, L., Neyem, A.: Communication and Coordination Patterns to Support Mobile Collaboration. In: Proc. of the 12th International Conference on Computer Supported Cooperative Work in Design (CSCWD 2008), pp. 565–570. IEEE CS Press, Los Alamitos (2008)

[14] Monares, A., Ochoa, S.F., Pino, J.A., Herskovic, V., Neyem, A.: MobileMap: A Collaborative Application to Support Emergency Situations in Urban Areas. In: Proc. of 13th International Conference on Computer Supported Cooperative Work in Design (CSCWD 2009), pp. 565–570. IEEE Press, Los Alamitos (2009)

[15] Rodríguez-Covili, J., Ochoa, S.f., Pino, J., Favela, J., Mejía, D., Moran, A.: Designing Mobile Shared Workspaces by Instantiation. In: Proc. of 13th International Conference on Computer Supported Cooperative Work in Design (CSCWD 2009), April 2009, pp. 402–407. IEEE Press, Los Alamitos (2009)

[16] Tirumala, A., Gates, M., Qin, F., Dugan, J., Ferguson, J.: Iperf: The TCP/UDP Bandwidth Measurement Tool, http://dast.nlanr.net/Projects/Iperf (Last visit, March 2009)

[17] Tschudin, C., Gunningberg, P., Lundgren, H., Nordstrom, E.: Lessons from Experimental MANET Research. Ad Hoc Networks 3(2), 221–233 (2005)

[18] Vaghi, I., Greenhalgh, C., Benford, S.: Coping with Inconsistency due to Network Delays in Collaborative Virtual Environments. In: Proc. of the ACM Workshop on Virtual Reality and Software Technology, pp. 42–49 (1999)

[19] Whalen, T., Black, J.P.: Adaptive Groupware for Wireless Networks. In: Proc. of the 2nd IEEE Workshop on Mobile Computer systems and Applications (WMCA 1999), p. 20. IEEE Press, Los Alamitos (1999)

[20] Zurita, G., Baloian, N., Baytelman, F.: Using mobile devices to foster social interactions in the classroom. In: Proc. of the 12th International Conference on Computer Supported Cooperative Work in Design (CSCWD 2008), pp. 1041–1046. IEEE Press, Los Alamitos (2008)

Preserving Interaction Threads through the Use of Smartphones in Hospitals

David A. Mejía[1], Jesús Favela[1], and Alberto L. Morán[2]

[1] Departamento de Ciencias de la Computacion, CICESE, Ensenada, Mexico
{mejiam,favela}@cicese.mx
[2] Facultad de Ciencias, UABC, Ensenada, Mexico
alberto_moran@uabc.mx

Abstract. Hospital workers need information to decide on the appropriate course of action for patient care; this information could be obtained from artifacts -such as medical records and lab results- or as a result of interactions with others. However, these exchanges could be a source of medical errors since this information is not usually preserved and could be lost –totally or partially- due to the volatility of human memory. This happens due to the verbal nature of the interaction or due to the lack of an infrastructure that facilitates the capture of information even when hospital workers are on the move. The capabilities increasingly found in Smartphones, such as WiFi, touch screen or a D-pad (directed pad), built-in camera, accelerometers, contact management software, among others, make it feasible to record significant information about the interactions that take place in the hospital and seamlessly retrieve it to support work activities. Thus, in this paper we propose a system to capture and manage collaboration outcomes in hospitals through the implementation of mobile collaboration spheres in Smartphones.

Keywords: Informal interactions, Capture of Interactions' Outcomes.

1 Introduction

Communication is an essential resource in hospital work; it is used to collaborate and coordinate the way in which work is performed, as well as to locate and gather the artifacts and human resources necessary for patient care [11]. Hospital workers frequently interact with their colleagues [10] due to their need to coordinate activities, exchange and integrate information from many devices or artifacts, and the mobility of workers, patients, and medical equipment.

Most communication among hospital workers is informal. By informal communication we mean those interactions that do not have a predefined schedule or place of encounter, are spontaneous, not planned and brief, and where the topic of the conversation can change during the course of the interaction [9].

The mobility experienced by hospital workers requires them to interact in places such as hallways, bed wards and consulting rooms [11]. In these interactions information is shared and/or exchanged. This information is crucial to hospital work, as hospital workers need to deal with events as they arise [3].

L. Carriço, N. Baloian, and B. Fonseca (Eds.): CRIWG 2009, LNCS 5784, pp. 17–31, 2009.

A recent study on informal communication in hospitals [11] shows that the information that hospital workers demand to achieve their goals while on the move includes: awareness of the occurrence of significant events related to their patients -such as the availability of lab results, changes on the patient's vital signs-, the diagnosis of a patient's condition –through the discussion of a medical case with other colleagues-, or information related to the coordination of activities or to patient care. Currently, most of these information exchanges occur during face to face encounters, through a verbal channel and without the mediation of technology [11]. Thus, one of the inconveniences with informal communication in hospitals is that this information is rarely recorded, because hospital workers do not have the tools to preserve this information when interacting with colleagues [10]. In addition, the purpose of these interactions is seldom achieved during one exchange [17]; thus, some information could be misplaced, confused or forgotten throughout the interaction thread.

This could cause that the information generated during an interaction might not be accurately recalled when hospital workers need it, due to communication problems during collaboration, or to the volatility of human memory [3]. Indeed, there is evidence that this could be the cause of deficient patient care that could even result in their death [8].

The proliferation of Smartphones with additional connectivity (e.g. WiFi and Bluetooth), processing and storage capabilities makes it feasible to capture and access information used or generated during informal interactions, most of which take place while hospital workers are on the move.

This paper proposes to support the capture and management of collaboration outcomes through the use of ubiquitous technology in Smartphones. We illustrate this kind of technology through the implementation of eMemoirs, a tool aimed at preserving the interaction threads by automatic recording of interactions and seamless access to interaction's outcomes. eMemoirs is an extension of SOLAR [10], a tool that implements pervasive services for supporting informal co-located interactions in hospitals.

2 Observational Study in a Hospital

Informal communication has been widely studied in traditional working environments, such as office work or those in which people or teams collaborate remotely. However, the working conditions of these environments are different to those faced by hospital workers. For this reason, we conducted a field study in a public mid-size hospital aimed at understanding informal communication in these settings.

In this study, we shadowed five physicians and five resident physicians during two complete work shifts. These roles were selected because they are in charge of providing medical care to patients, frequently collaborate with others and experience high mobility. Our goal was to have a good sample of interactions and capture all the communication in which those workers are involved as they conduct their work. We next report the finding of this field study and compare these results with those of other work environments.

2.1 Importance of Informal Communication in Hospitals

The importance of informal communication has been expressed in terms of their frequency and purpose:

Frequency of Informal Communication in Hospitals. Interactions can be classified based on the intentionality and location of the participants when the need for interacting emerges [3]. This includes: (a) conversations that were previously scheduled or arranged (scheduled), (b) those in which the initiator sets out specifically to visit another party (intended), (c) the ones in which the initiator had planned to talk to another participant and took advantage of a chance encounter to have a conversation (opportunistic), and (d) spontaneous interactions in which the initiator had not planned to talk to another participant (spontaneous).

Intended, opportunistic and spontaneous interactions are considered forms of informal communication [6]. We found that 95% of all the interactions observed in our study (45 interactions per day, per person, on average) were informal (39% are intended, 21% opportunistic and 35% spontaneous). This highlights the importance of informal communication in terms of frequency of use in hospitals.

Purposes of Informal Communication in Hospitals. The purposes of informal interactions have been analyzed as a way to understand the nature and characteristics of collaboration in a working environment [7]. We identified nine main purposes of informal communication in hospitals and measured them in terms of frequency as follows: (a) tracking people, which involves identifying the current whereabouts, activities, and future plans of intended collaborators (4%), (b) taking or leaving messages, which refers to contacting someone via a third party (2%), (c) making meeting arrangements, which refers to scheduling future interactions (1%), (d) delivering documents, which refers to handing off a document with actions attached to it and, in a more complex example, involves discussions of individual actions associated with different parts of a document (10%), (e) giving or getting help, which refers to joint problem solving for one person's benefit (16%), (f) reporting progress and news, which refers to updating people with relevant information (37%), g) delegating activities, which refers to asking someone to perform an activity or part of it, that can be performed by two or more hospital workers (11%), (h) tracking artifacts, which involves identifying the current location or state of specific artifacts, like medical records, medical equipment, or laboratory results among others (4%), and (i) personal conversations, which refer to social conversations (15%). We also observed that only 15% of informal interactions (personal conversations) were not directly related to working activities.

Summarizing, the exchange of information among hospital workers is frequent, usually unplanned and related to work topics.

2.2 Communication Problems in Hospitals

Despite the importance of informal communication in the execution of medical activities, disruptions arise during and after the exchange of information that could directly impact the care of patients. We show some of these problems through the next scenarios that illustrate real instances of interactions observed in the hospital. Later, we discuss the causes of these inconveniences through the analysis of some results of our field study.

Scenario 1. *The medical specialist and a resident physician are in the hallway, talking about the medical procedures that the resident must perform on two patients. The* *medical specialist starts explaining the procedure of each patient. Towards the end of the interaction, a physician arrives and asks the resident about the laboratory results of a patient. A few minutes later the resident resumes the interaction with the medical specialist, who completes the explanation. Then, the resident goes to the bed ward and reviews the medical record of each patient. At this time, the resident is confused about which procedure must be performed to each patient. Thus, the resident locates the medical specialist and initiates a new conversation.*

    [Resident physician]   *Excuse me, but I'm confused about the procedures.*
    [Medical specialist]   *Well, the (name of the medical procedure) is for the patient at (patient's room number) and the (name of the medical procedure) is for the patient at (patient's room number).*

In this scenario we observed that due to an interruption during the interaction, the resident physician could not remember the indications appropriately. This emphasizes the absence of mechanisms that ensure the persistency of verbal information.

Scenario 2. *The medical specialist is in his office discussing the care of a patient with leprosy with a fellow physician. During the interaction, they consult in an online pharmacy the cost of medication to treat this disease. The physician annotates the information on a piece of paper and says:*

    [Physician]   *I will ask the [hospital] pharmacy about the cost and availability of similar treatments.*

    The physician leaves the office. Five minutes later, the head physician asks the medical specialist about the patient. The medical specialist says that he and the physician in charge of the patient found the information on the treatment, but he doesn't have it, because the physician took the paper with the annotations with him.

    [Head physician]   *Please search the information again. We need it to acquire the medication immediately.*
    [Medical specialist]   *Ok.*

    Later, the medical specialist is consulting the information when the physician arrives again with information of this and other treatments. After this, the three workers meet to decide on the most appropriate treatment, based on this information and the patient's medical record..

In this scenario we observed that the information generated during one interaction is not always available to all participants, which could lead to additional tasks or interactions.

In the next subsection we discuss the probable causes of these problems.

2.3   Understanding Disruptions of Informal Communication in Hospitals

Hospital work is characterized by the need to manage multiple activities or collaborations simultaneously, constant local mobility, frequent interruptions, and intense collaboration and communication. These conditions impose important demands on users that need to frequently switch between tasks and/or interactions, contributing to a decrease in efficiency and becoming a source of errors and mishaps [1].

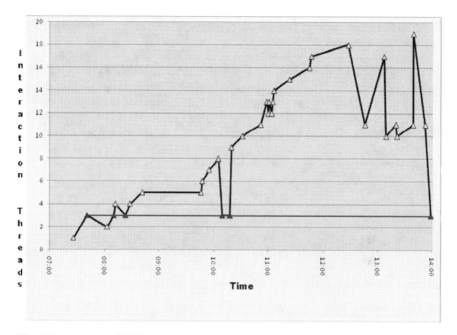

Fig. 1. Interactions of different conversation threads during one work shift of a physician

Interactions frequently involve the use of a diversity of information resources such as medical records, laboratory results, X-ray images, medical guides, books, or personal notes. The need to remember their location and gather all the information resources related to an activity or interaction is likely to involve certain effort and sometimes results in cognitive overload for the user [15].

Interactions in hospitals are frequently interrupted and resumed at a later time. Figure 1 shows the flow of interaction topic during one work shift of a physician as identified from the study. A good example to illustrate this phenomenon is conversation topic number 3, which in fact corresponds to the second scenario presented above related to the treatment of the patient with leprosy. This topic is initiated around 7:40 and resumed on 5 additional occasions, including the last interaction of the physician that day, which took place around 2pm.

In our field study we identified some characteristics of informal communication in hospitals that could lead to problems or inconveniences during and after an interaction.

- **Interactions are fragmented.** The purpose of an informal interaction is not always achieved in one single interaction. In our study, we found that 40% of the interactions were continuation of a previous interaction from this same day, and 13% were continuation of interactions from previous days. When an interaction is resumed, the participants need information relevant from the former exchange [4]. Another finding of our study is that hospital workers use information not only from the last interaction, but also from former interactions in the thread. Thus, we propose the use of mechanisms that facilitate the maintenance of the interaction thread, as well as the use of mechanisms that facilitate the capture of information, its organization and access.

- **Interactions are not documented.** During the interaction, information is exchanged, both verbally and through physical information artifacts, such as lab results or X-ray images. In our field study, we observed that in 74% of the interactions medical workers exchanged only verbal information, while in the remaining interactions they used or generated information in information artifacts (e.g. discuss a diagnosis based on the patient's medical record). Verbal information is not recorded. Due to the volatility of human memory, there is a risk of forgetting information that could be necessary for future activities or interactions.

  Information artifacts are often used to discuss diagnosis or give instructions about the treatment for a patient. However, since it is not always possible to make personal annotations in most of these artifacts, this information could be partially or completely lost. For this reason, we propose the use of mechanisms to preserve conversations as well as the information generated over information artifacts.

In the next section we describe eMemoires, a tool designed to address the issues mentioned above, through the capture and access of information related to hospital interactions.

3 eMemoirs: Capturing and Accessing Interaction's Outcomes in Hospitals

To assist mobile users in the management of their multiple collaborations, we designed and developed eMemoirs, a pervasive tool that provides seamless automatic capture of collaboration outcomes, organizes each interaction into interaction topics, and facilitates the retrieval or resumption of previous interactions. Each topic is considered a thread of interactions, thus, eMemoirs allows the user to access and manage their interaction resources around a topic of interaction.

Since most activities performed by hospital workers are centered on their patients, their care can be considered their main interaction topic. For this reason, by default, the system defines one interaction topic for each patient and associates to them information resources relevant to the interaction, such as the medical record, medical workers in charge of the patient, etc. Additionally, the system allows maintaining an interaction thread by saving the status of the corresponding interaction each time it occurs.

The system was developed for Smartphones with built-in accelerometer and WiFi. It is an extension of SOLAR [10], a collaborative tool that integrates the following services: i) location estimation of workers in the hospital displayed on the floor plan shown on the Smartphone, ii) awareness of relevant events and nearby colleagues and their social context, that allows them to select an appropriate moment to initiate an interaction with others, iii) transfer of files and URLs among heterogeneous devices just by clicking the right mouse button and choosing the target device, iv) screen sharing of any device in the vicinity (e.g. PDA, PC or public display) on a handheld computer, and remotely sharing the control of the device with its owner and/or other users, and v) notification of creation or modification of information, and storage of this information in a user's mobile device during a collaboration episode (capture of collaboration outcomes). Figure 2 shows the extended architecture of SOLAR including the eMemoirs components required for the capture of collaboration outcomes. A further description is presented in next subsections.

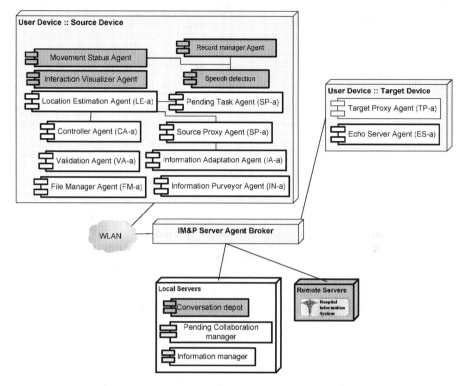

Fig. 2. Architecture of the integration of e-Memoirs (gray items) into SOLAR

3.1 Capturing Interaction Outcomes

As stated earlier, during an interaction hospital workers exchanges verbal and in-scribed information. cCapture is a component of eMemoirs that allows capturing both types of information.

Verbal Information. cCapture provides support for the seamless capture of verbal information. This means that based on location, movement status of potential partici-pants and on speech detection, the system decides when the recording of an interaction must start. To do this, the system uses three main components:

- **Location estimation (LE).** To introduce location awareness we use a location estimation component that estimates the position of users within a hospital [2]. Radiofrequency (RF) signals received by a mobile device carried by medical staff from at least three access points are measured to obtain their signal strength. A trained backpropagation neural network is embedded within the component and is used to estimate the approximate location of the users.
- **Movement status (MS).** Taking advantage of the built-in accelerometer, the system detects when a user is moving or not, as well as the direction or orientation of the movement.

- **Speech detection (SD).** Through a hands free microphone, the system detects when a conversation is initiated. For this purpose, we implemented a Voice Activity Detection (VAD) algorithm [5]. Thus, this component detects when the user or a colleague in the vicinity is talking, and automatically initiates the recording process.

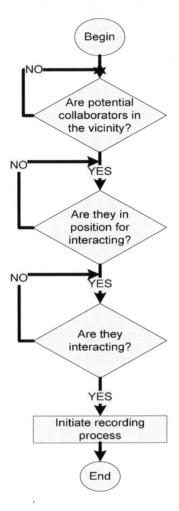

Fig. 3. Flow of activities for seamless capture of interactions

These three components work together to record verbal interactions (see Figure 3). The LE component estimates the hospital workers' location and whether a possible collaborator is in the vicinity, the MS component is activated. Then, the MS determines whether an interaction is close to occur if: i) all possible collaborators in the vicinity are standing still, ii) at least one of them is not moving and the remaining are moving to this location, iii) if two or more workers are moving towards the same place, iv) if two or more workers are moving together in the same direction. If some of these conditions occur, the SD starts monitoring for collaborators speech. If a verbal conversation begins, the SD activates the recording process. Through the VAD algorithm, we are differentiating between collaborators' voice and environmental noise in order to shorten the length of the conversation file.

The recording continues until the LE detects that the collaborators are not co-located or voice is not detected for several seconds.

Preserving Interaction Outcomes on Information Artifacts. Hospital workers use information artifacts in interactions or activities directly related to patient care. However, due to hospital policies, these artifacts cannot be modified with personal annotations to remember or recreate information exchanged during an interaction. To deal with this, we developed a component that allows users to save this information in a separate artifact, without modifying the original information. We classified this information in two main types: i) images and ii) annotations. With this component, dialog balloons with personal comments can be added over images or inscribed information.

To add personal comments, after selecting the "add comment" option from the main menu, the user selects the place in the artifact where he wants to insert the comment by tapping the screen into the desired position. Then, a dialog balloon appears and the user can type the information in it. In order to save this comment without modifying the information in the artifact, the component stores the position were the dialog balloon was inserted, as well as the information of this balloon in a separate file. Thus, when the user needs to access this information, the information in the artifact and the comment's file are presented in a single view (as it was presented during the interaction).

To add comments to images, the component saves the coordinates and color of the figures drawn in a separate file, and when the user requests this information, the component displays it again on the same image file. In both cases, the location where the interaction takes place is stored.

Storing Resources. eMemoirs provides medical workers with a component that allows them to the use the resources and storage capacity of their PC's. This component is an extension of SOLAR's seamless information transfer service [10], which provides seamless interaction between devices. Thus, when eMemoirs finishes recording interaction or saving modifications to information artifacts, these resources are sent and stored into the PC without user intervention, in order to avoid consuming the limited storage capacity of handheld devices. Then, when the user needs to access an interaction folder resource, the Smartphone makes a request to the user's PC, which sends a list with the location of resources to be displayed to the user on the Smartphone. When the user selects the required resources, the system in the Smartphone requests them to the user's PC which transfers the related files to be opened on the Smartphone.

3.2   Accessing Interactions Outcomes

Interaction can be characterized by their location, time of occurrence, topics, and artifacts used and generated. Some of the interactions in the hospital share some common characteristics (e.g occurs in the hallways, have the same participants or the same topic).

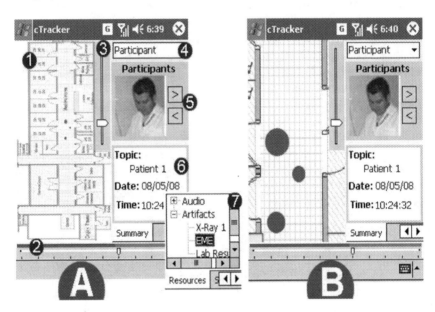

Fig. 4. Screen shoot of cTracker. A1) Map of interaction locations. A2) Time-line to look over the chronologically ordered interactions. A3) Control bar of interaction spots on the map. A4) Search option. A5) Participants in the interaction. A6) Summary of the interaction. A7) Resources of the interaction. B) Zoom of the cTracker map.

To ease resuming and/or accessing interaction outcomes, we are providing visual contextual information through a map (Figure 4-A1) that shows the location where every interaction occurred. These locations are indicated with a red or green spot on the map, where every spot represents an instance of interaction. Since not all information from interactions is missed or relevant for future activities, is not necessary hear all conversations. For this reason, cTracker allow users to search/filter interactions according to specific search options: by location, by participant, by topic, by time of occurrence or to view them all (Figure 4-A4). The user can decide how many circles he wants to see simultaneously (Figure 4-A3).

Using the time-line bar (Figure 4-A2), the user can look over the chronologically ordered interactions or zoom into a specific area in the map. According to the position of the time-line bar, interactions that occurred before the indicated time are colored in red and interactions that occurred after the indicated time are shown in green. In the same way, the time elapsed between the current position of the time-line bar and the time when the interaction took place is represented through the size of the spot. The nearer the occurrence of the interaction to the indicated time, the bigger the spot representation in the map, and vice versa, the farther in time the occurrence of the interaction the smaller the spot represention in the map.

By tapping on a spot, the system displays the participants in the interaction (Figure 4-A5), a summary of the interaction (Figure 4-A6), as well as the artifacts or elements generated or used during the interaction (Figure 4-A7). When a user switches between interaction spots, the user can decide to i) resume the workspace state or ii) see short-cuts to the interaction resources. In the first case (i), the interaction folder is enabled to quickly gather and retrieve its own workspace state (windows positions, status and overlay order of last interaction or activity) and context information such as opened documents, idle time, etc. in a seamless manner. On the second case (ii), the system presents the shortcuts to these resources and allows users to access them, through the shortcuts, when required. In both cases, the application hides the current workspace and shows the one related to the consulted instance of interaction. Additionally, cTracker allows zooming- in the map just by drawing a rectangle over the map of interactions (Figure 4B).

When the system is not able to identify the corresponding interaction's thread, due for instance to the location of participants, the user has to explicitly indicate this, either at the end of the interaction or at a later time. A problem that arises when the user decides to indicate this at a later time is that he/she could confuse which informa-tion belongs to a specific interaction folder, mainly due to the amount of interactions in which hospital workers are engaged in, during their work shift. Thus, cTracker (Figure 4) provides users with information to help them remember many of the ele-ments that shaped the interaction, such as participants, location, occurrence time, topics and artifacts used and generated during it.

3.3 Using eMemoirs at the Hospital

We illustrate the utility of eMemoirs describing how it would be used in the scenarios presented in Section 2.2:

A medical specialist and a resident physician meet in the hallway. When eMemoirs detects their mutual proximity, the recording component starts "listening" to them. When the medical workers begin to talk, eMemoirs starts recording the conversation. The medical specialist explains the procedures that the resident must perform on the two patients. After finishing the interaction with the medical specialist, the resident goes to the laboratory to get the results of another patient (unrelated to those of the previous interaction). When eMemoirs detects that the medical workers move away from each other, the recording process is terminated. Afterwards, on arrival to the bed ward the resident reviews the medical record of each patient, and initiates the proce-dure on the first patient.

However, the resident has forgotten some details of his discussion with the medical specialist and is confused about the procedure to perform to each patient. Thus, the resident opens cTracker to review his most recent interactions in order to retrieve the recording of the verbal conversation generated during the interaction with the medical specialist.

cTracker indicates the location of the most recent interactions by means of circles on the map. It recalls that the conversation took place a few minutes ago. Thus the most probable instances of the recorded interactions are the two large red spots on the hallway, since spot size relates to how recent the event was, considering the current time as the time marker. As illustrated in Figure 4 when the resident taps on one of the red spot a popup presents information related to the interaction, such as the time of day when the encounter took place and the actual participants. Thus, the resident confirms that this is indeed the interaction he wants to recall, plays it back, and then, performs the medical procedure to each patient without having to search for and consult the specialist.

3.4 Addressing Privacy Issues of the Recording Process

The automatic recording of conversations in hospitals could benefit medical workers in situations where they have an overload of verbal information. However, this process raises privacy issues to third-parties since the transient nature of verbal communication is made persistent. eMemoirs automatically initiates the recording process when it detects that two persons start interacting, but third persons could be recorded without being aware.

To address this issue we propose to introduce the concept of Quality of Privacy (QoP) [13] to the services provided by eMemoires. QoP is the probability that a pervasive environment meets a given privacy contract. This involves considering the inclusion of three components: (i) an event describing the need to execute an action, which is characterized by five contextual variables: location, identity, time, activity and artifacts, which change dynamically while the user's context varies; (ii) a condition defining rules that must be enforced to determine which action might need to be executed; and (iii) an action containing a set of functions that may be executed to enforce or relax privacy policies [14]. Based on this concept we propose to implement one or more of the following strategies:

- Notification of recording initiation. In this strategy eMemoirs notifies the third persons that their conversations could be recorded and later accessed by others.
- Automatically impeding the recording process. In this case, a conversation won't be record when a third person is close to them.
- Explicit permission to hear conversations. This strategy consists in notifying to users when somebody wants to hear conversations that were recorded in their presence. Thus, nobody could hear recorded conversations without the explicit permission of all those who were present when it was recorded.
- Allow recording interactions only at specific areas and/or time. Medical staff usually experiences information overload at specific time and/or areas. When this occurs medical staff is more prone to miss or forget information exchanged during collaboration. This strategy consists in recording interactions only at specific times and/or places according to the needs of users .
- Recording interactions only with specific colleagues. The most important information for a hospital worker is that related to the patients of whom he is in charge of. Thus, this strategy consists on recording information only from interactions among collaborators that have patients in common.
- Defining collaboration roles. This strategy consists on defining a group of privileges and grant them to others. These privileges relate to both, the recording and listening of conversations. Since all previous strategies have advantages and disadvantages, we propose to combine this one with the previous strategies and associate them to groups of users.

4 Related Work

The capture and retrieval of experiences has been a growing field of research within ubiquitous computing for two main reasons: i) these systems help to increase the capacity of human memory for remembering the events in which humans are involved, and ii) these systems allow to recreate an experience by replaying the documented events or activities [6]. Many research efforts have been focused on classroom or meeting room settings. Examples of these systems include TIVOLI and SMaRT.

TIVOLI [12] is an electronic whiteboard application designed to support informal meetings and targeted to run on a large pen-based interactive display. It documents the activity through the recording of audio and drawing sketches, facilitating the recreation of the meeting. SMaRT [16] provides meeting support services that do not require explicit human-computer interaction. Instead, by monitoring the activities in the meeting room using both video and audio analysis, the room reacts appropriately to users' needs, allowing them to focus on their goals.

Additionally, capture and retrieval system has been used in mobile settings. ButterflyNet [18] is a notebook augmented through digital media such as text, audio and pictures. It was designed to support fieldwork in biology. This system allows for the automatic transformation of the collected information into spreadsheets to facilitate access and customization of this information.

These tools require direct interaction between the user and the device. This could be problematic in hospitals where medical workers often require both hands to perform a procedure or to evaluate a patient. In these circumstances, having to operate a desktop or handheld computer to communicate with others is impractical. Another characteristic of these tools is that they consider that the purpose of the activity or interaction to document is achieved in a single session. This contrasts with what we observed in the hospital, where interactions are frequently fragmented.

These issues are partially addressed in ActiveTheatre [6] which is built to support collaboration in and around an operating theatre, to capture events instead of automatically capturing everything. An event can be a text note, a picture or a video clip. The system uses the palette metaphor, in which users can hold different types of data, put and take things from the palette at any point in time, and access them from the palette as needed. After the operation, a medical worker carries the palette to an office where the data on the palette is used to create different kinds of documents depending on who are the recipients. However, ActiveTheatre does not consider the use of information not included in the palette, and does not take into account the use of information generated during the surgical procedure for future activities.

This approach is not necessarily adequate for informal interactions, which are unplanned and where the setting and topic of interaction is not previously defined, because users are not able to carry information to resume previous interactions or enrich new ones. For this reason, eMemoirs deals with interruptions by providing automated access to information of previous interactions in order to resume pending interaction topics, and actually creating threads of such interactions. In addition, conversations, a rich source of information in eMemoires, are not considered as events in activeTheatre.

Perhaps the main difference between eMemoirs and these previous works, is that eMemoirs takes advantage of the features of the Smartphones in order to support collaboration in hospitals: i) WiFi connectivity, built-in accelerometer, long-life battery and media capabilities to support the seamless recording of interactions based on location, ii) powerful processor and storage capacity to support resuming an interaction's topic, and finally iii) the portability of the device, in order to support the high mobility of hospital workers.

5 Conclusions

The exchange of information during interactions could be a potential source of medical mistakes. In hospitals most of this information is verbal and it is mostly stored only in human memory; thus, this information is transient. Due to information overload, to communication problems and to the time critical nature of hospital activities, this information could be missed or forgotten, To date, there are some approaches for supporting the capture and management of events and experiences but they are not appropriate for hospitals since they are not designed for mobile workers and/or require direct manipulation from the users. For this reason, based on results of a field study that we conducted in a public hospital, we developed eMemoirs, a tool that seamlessly records conversations across the entire hospital area without human intervention, and provides contextual access to resources generated in previous interactions.

To gain contextual information of people, activities and/or events in the hospital, eMemoirs takes advantage of the capacities of current Smartphones, such as powerful processor, storage capacity and built-in-accelerometer, as well as the implicit support for mobility that this kind of devices provide.

Finally, one of the main concerns when recording information is privacy. We address this issue by proposing a set of strategies that could be implemented in order to minimize the impact of privacy issues in the recording process.

Acknowledgments

We thank Saul Cruz and Luz Lozano for their dedication in the development of eMemoirs. This work was partially supported by a research grant provided by the LACCIR grant No. RFP0012007.

References

1. Camacho, J., Favela, J., González, V.M.: Supporting the Management of Multiple Activities in Mobile Collaborative Working Environments. In: Dimitriadis, Y.A., Zigurs, I., Gómez-Sánchez, E. (eds.) CRIWG 2006. LNCS, vol. 4154, pp. 381–388. Springer, Heidelberg (2006)
2. Castro, L.A., Favela, J.: Contiuous Tracking of User Location in WLANs Using Recurrent Neural Networks. In: Proceedings of Sixth Mexican International Conference on Computer Science, Puebla, Mexico, pp. 174–181. IEEE Press, Los Alamitos (2005)
3. Coiera, E.: Clinical Communication - A New Informatics Paradigm. In: Proceedings of the American Medical Informatics Association, Washington, DC, pp. 17–21. J. Am. Med. Informatics Assoc. (1996)

4. González, V.M., Mark, G.: Constant, constant, multi-tasking craziness: managing multiple working spheres. In: Proceedings of the 2004 Conference on Human Factors in Computing Systems, CHI 2004, Vienna, Austria, pp. 113–120. ACM Press, New York (2004)
5. Górriz, J.M., Ramírez, J., Segura, J.C., Hornillo, S.: Voice Activity Detection Using Higher Order Statistics. In: Cabestany, J., Prieto, A.G., Sandoval, F. (eds.) IWANN 2005. LNCS, vol. 3512, pp. 837–844. Springer, Heidelberg (2005)
6. Hansen, T.R., Bardram, J.E.: ActiveTheatre – a Collaborative, Event-based Capture and Access System for the Operating Theatre. In: Beigl, M., Intille, S.S., Rekimoto, J., Tokuda, H. (eds.) UbiComp 2005. LNCS, vol. 3660, pp. 375–392. Springer, Heidelberg (2005)
7. Isaacs, E., Whittaker, S., Frohlich, D., O'Conaill, B.: Informal communication re-examined: New functions for video in supporting opportunistic encounters. In: Finn, K.E., Sellen, A.J., Wilbur, S.B. (eds.) Video-Mediated Communication, pp. 459–485. Lawrence Erlbaum, Mahwah (1997)
8. Kohn, L.T., Corrigan, J.M., Donaldson, M.S. (eds.): To Err is Human: Building a Safer Health System. Committee on Quality of Health Care in America
9. Kraut, R., Fish, R., Root, R., Chalfonte, B.: Informal communication in organizations: form, function and technology. In: The Claremont Symposium on Applied Social Psychology, California, United States, pp. 145–199. Sage Publications, Thousand Oaks (1990)
10. Mejia, D.A., Morán, A.L., Favela, J.: Supporting Informal Co-located Collaboration in Hospital Work. In: Haake, J.M., Ochoa, S.F., Cechich, A. (eds.) CRIWG 2007. LNCS, vol. 4715, pp. 255–270. Springer, Heidelberg (2007)
11. Morán, E.B., Tentori, M., Gonzalez, V.M., Favela, J., Martínez-Garcia, A.I.: Mobility in Hospital Work: Towards a Pervasive Computing Hospital Environment. International Journal of Electronic Healthcare (IJEH) 3(1), 72–89
12. Pedersen, E.R., McCall, K., Moran, T.P., Halasz, F.G.: Tivoli: an electronic whiteboard for informal workgroup meetings. In: Conference on Human Factors in Computing Systems, Amsterdam, The Netherlands, pp. 391–398. ACM Press, New York (1993)
13. Tentori, M., Favela, J., Rodríguez, M.D.: Privacy-Aware Autonomous Agents for Pervasive Healthcare. IEEE Intelligent Systems 21(6), 55–62 (2006)
14. Tentori, M., Favela, J., González, V.M.: Quality of Privacy (QoP) for the Design of Ubiquitous Healthcare Applications. J. UCS 12(3), 252–269 (2006)
15. Voida, S., Mynatt, E.D., MacIntyre, B., Corso, G.M.: Integrating virtual and physical context to support knowledge workers. IEEE Pervasive Computing 1(3), 73–79
16. Waibel, A., Schultz, T., Bett, M., Denecke, M., Malkin, R., Rogina, I., Stiefelhagen, R., Yang, J.: SMaRT: the Smart Meeting Room Task at ISL. In: Conference on Acoustics, Speech, and Signal Processing, pp. 752–755. IEEE, Los Alamitos (2003)
17. Whittaker, S., Swanson, G., Kucan, J., Sidner, C.: Telenotes: managing lightweight interactions in the desktop. ACM Transactions on Computer Human Interaction 4, 137–168
18. Yeh, R.B., Klemmer, S.R.: Guimbretière, ButterflyNet: Mobile Capture and Access for Biologists. In: Conference Supplement to UIST 2005: ACM Symposium on User Interface Software and Technology, Seattle, WA, p. 20 (2005)

Nomadic User Interaction/Cooperation within Autonomous Areas

Victor Gómez[1], Sonia Mendoza[1], Dominique Decouchant[2],
and José Rodríguez[1]

[1] Departamento de Computación, CINVESTAV-IPN, D. F., México
[2] C.N.R.S. - Laboratoire LIG de Grenoble, France
vgomez@computacion.cs.cinvestav.mx, smendoza@cs.cinvestav.mx,
Dominique.Decouchant@imag.fr, rodriguez@cs.cinvestav.mx

Abstract. Ubiquitous computing integrates Internet/Intranet small integrated sensors as well as powerful and dynamic devices into the people's working and domestic areas. An intelligent area contains many devices that provide information about the state of each artifact (e.g., power failure of the refrigerator) without user intervention. Service discovery systems are essential to achieve this sophistication as they allow services and users to discover, configure and communicate with other services and users. However, most of these systems only provide support for interaction between services and software clients. In order to cope with this limitation, the SEDINU system aims at supporting interactions between nomadic users and services provided by areas. As users may move within the organization from an area to another one in order to accomplish their tasks, this system also provides support for user-user interaction and collaboration under specific contexts (role, location and goals).

1 Introduction

For last years, we have witnessed important changes arising from the use of communication technology (mostly, wide-area network accessed by means of wired/wireless connections and mobile telephony) to satisfy some welfare needs (e.g., home-health care via Internet) and social needs (e.g., transmission of textual or multimedia messages via mobile telephony networks). The benefits of this technology become evident in working environments, where applications and users can easily access different services (e.g., database access, Web browsing, printing). In such environments, we can assume that connectivity provided by corporative networks is reliable and continuous, and offers high bandwidth.

Owing to the constant evolution of the communication technology and the fact that people can use mobile devices to interact as they move, people tends to be nomadic. However, moving from one place to another, people can find substantial changes concerning the: 1) interaction mobile devices, 2) available services, 3) communication technology, and 4) available bandwidth.

Wireless networks constitute the most suited interconnection technology for the mobile nature of new devices [16]. Thus, users of mobile phones can consult their email anywhere; travelers can surf the Web from airports; tourists can

L. Carriço, N. Baloian, and B. Fonseca (Eds.): CRIWG 2009, LNCS 5784, pp. 32–40, 2009.
© Springer-Verlag Berlin Heidelberg 2009

use GPS to locate museums, restaurants or streets; scientists can interchange documents during a conference; people at home can transfer and synchronize information between handheld devices and PCs. However, nomadic work requires efficient discovery systems to find available services within the network.

As service registration and cancellation are dynamic operations, most discovery systems asynchronously notify their clients of the service availability [4]. However, the clients of these services are mainly programs, e.g., a printing service can be registered as a proxy object in a lookup server, which acts as a remote control. Therefore, these systems do not support interactions between users and current environment services and even less user-user interactions.

In this paper, we describe the SEDINU (SErvice DIscovery for Nomadic Users) system, which provides support for these kinds of interaction within a specific context defined in terms of user's role, location and goals. More precisely, we are interested in defining a suited support for environments in which users are moving and collaborating within a quite large organization that includes several departments, services or units. These organizational units, called "autonomous areas", define and administrate specific tasks to be accomplished and offer contextual dedicated services. In such kind of environment, we aim at providing nomadic users with support to reach others to establish collaborative sessions and, within each specific context, to access services (e.g., slide projection) and perform special tasks (e.g., collaborative production of diagnosis). As highlighted, all services and tasks depend on the nomadic user's current context.

This document is organized as follows. After providing a synthesis of related work in section 2, we describe the SEDINU system is section 3. Particularly, we describe the concept of "autonomous area", as well as the SEDINU functional schema and a use scenario. We also provide some technical details about the implementation of the proposed system. Finally, in section 4, we present the current research directions of this work in progress.

2 Related Work

In ubiquitous computing environments, service discovery and notification systems must be deployed to deal with the dynamism of these environments as well as the growing number and diversity of devices that can provide and request services. Some proposed solutions have been defined depending on a network protocol (e.g., IP), whereas other solutions take part of a framework for the development of services within dynamic environments. However, their use depends on the way by which these solutions have been developed and provided [2].

During the last years, some representative efforts have been developed to provide service discovery systems [15]: in the academic field, we can mention Intentional Naming System (MIT) [1] and Ninja Service Discovery Service (UC Berkeley) [7]. In the same time, the main software companies developed service discovery systems integrated into their operating systems, e.g., Jini Network Technology (Sun Microsystems) [12], Universal Plug and Play (Microsoft) [9] and Bonjour (Apple) [13]. In the opposite way, DEAPspace (IBM Research)

[11], Salutation (Salutation Consortium) [8], Service Location Protocol (IETF) [14] and Bluetooth SDP (Bluetooth Special Interest Group) [3] propose service discovery systems that are independent of the operating system.

All these systems rely on the comparison of typical attributes (e.g., the type of service specified by an URL) and communication interface attributes (e.g., IP address and port number of the service host) to check the existence and availability of the services required by the user. Some systems are language-independent, e.g., Service Location Protocol (SLP), Salutation, and Universal Plug and Play (UPnP). In contrast, others are totally language-dependent, e.g., Jini depends on Java and thus a Java Virtual Machine is required on all devices. UPnP and Salutation use multicasting to find services, whereas Jini registers services (Java objects) in a directory JLS (Jini Lookup Services), i.e., clients and services discover each other by means of JLS.

However, all these service discovery systems only provide support in terms of localization or network topology [15]. Each one solves a different kind of problem (e.g., service access control, device recognition, network connection, or service identification), but most of them are only intended for domestic environments (e.g., localization of multimedia services to allow the user to receive a video call) and business environments (e.g., determination of available services to treat a printing request). Therefore, these discovery systems cannot be used in other contexts and specially in collaborative ubiquitous environments.

All analyzed systems explore different aspects of service discovery within (only) distributed environments. Thus, they do not offer a suited solution for problems arising from dynamic collaborative environments, e.g., the automatic detection of services provided by autonomous areas or the dynamic creation of ad-hoc networks to support both user-service and user-user interactions.

3   The SEDINU System

Nowadays, large organizations are divided into sub-organizations (e.g., departments, delegations and sections) in order to be more efficient at providing users with specialized services. Sub-organizations refer to autonomous areas, which make the global organization administration easier as resources (e.g., services, contextual information, workflows and roles) are handled and controlled in a distributed way. Thus, each autonomous area becomes an administration unit, which handles its own resources, while observing the englobing organization policies (e.g., security and invoicing). The different areas exchange information for coordinating themselves, in order to assist nomadic users to achieve their partial and global goals within the organization. An autonomous area can be mapped to several buildings, a building or a part of a building.

Areas are organized in a hierarchy of englobing/englobed organizational units where higher levels provide general services and policies, whereas lower levels offer specialized ones. Fig. 1 illustrates the principle of role attribution between two areas: A0 is an englobing area and A1 is an included sub-organization area. Each area defines a set of roles, e.g., area A0 defines roles R01 and R02, whereas

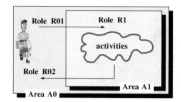

Fig. 1. Autonomous Area and Role Attribution Principles

area A1 defines role R1. Thus, when the nomadic user goes into area A0 for first time, role R01 is attributed to him, e.g., at the institution entrance, the role "Deliver_ComputerScience" is attributed to a deliveryman. Then, coming into area A1, role R1 is assigned to him, so that he can carry out some activities within area A1. When the nomadic user terminated his task, he comes back from area A1 to area A0, which assigns him role R02, i.e., he can obtain a different role he previously obtained coming into area A0 for first time.

3.1   SEDINU Functional Schema

The SEDINU system has been designed to allow the nomadic user located in an autonomous area to interact with: a) other users, e.g., for information cooperative production or b) the services available on such area, e.g., for information request. Interactions are directed by a context-aware workflow, which depends on the user's location, role and goals. Managed by the SEDINU system, each area has been designed as an autonomously administrated entity that relies on the RBAC-Soft system in order to communicate and coordinate itself with other areas. RBAC-Soft is a role-based access control system, which is responsible of managing the lists of roles, resources and permissions defined by each area.

Fig. 2 shows the steps followed by the SEDINU system: a) to provide the nomadic user with services related to his current interaction context and b) to allow him collaborating with other users, who are present within the same area. First, the Location Detector (see Fig. 2 ref. #1) determines the coordinates of the user's handheld within the current building. Then, the Location Detector transmits these coordinates to RBAC-Soft (see Fig. 2 ref. #2), which process them in order to identify the corresponding autonomous area. Depending on his current location, RBAC-Soft determines: a) the role attributed to the nomadic user and b) the services that are associated to this role (see Fig. 2 ref. #3).

Based on this contextual information, RBAC-Soft creates a workflow (or searches the already defined one) to guide the nomadic user towards the achievement of his goals (see Fig. 2 ref. #4). Each activity of the workflow is associated to a set of services that allows the nomadic user to carry out such an activity. In this way, RBAC-Soft uploads the corresponding workflow to the nomadic user's handheld (see Fig. 2 ref. #5) and asks the Service Manager to upload the client or peer applications to access the required services (see Fig. 2 ref. #6).

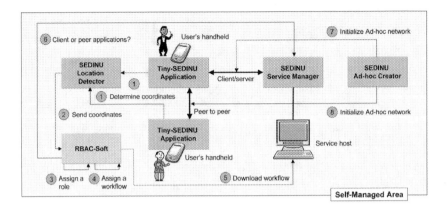

Fig. 2. Functional Schema of the SEDINU System

As soon as the nomadic user has access to the services, he can select one of them from his handheld. The set of services available at a given moment is controlled by the corresponding workflow, e.g., it can activate: a) one service available in the case of a single serialized task or b) several services available in the case of tasks whose execution order is irrelevant. Thus, the Ad-hoc Network Creator (see Fig. 2 ref. #7) dynamically establishes an ad-hoc network: a) between the Tiny-SEDINU Application (running on the user's handheld) and the Service Manager or b) between two Tiny-SEDINU Applications running on the handheld devices (see Fig. 2 ref. #8) of the users who can then start interacting/collaborating. This last kind of ad-hoc network facilitates direct information sharing (e.g., invoices or vouchers) and collaboration between users supported by a P2P communication, instead of passing by a central server.

The distribution architecture of the services is centered on organizing the institution in autonomous areas that manage their own roles, services and workflows. That comes from the decision to make: a) the area resources available and b) the system reliable and easily manageable. Thus, whenever an area corresponds to a whole building or a part of a building, its services are concentrated on a host. On the contrary, whenever an area corresponds to several distributed buildings (e.g., a department that is distributed on several cities) its services are replicated on different hosts that are distributed among different buildings.

3.2   Use Scenario: Management of a Conference Lecturer

The prototype of the SEDINU system has been deployed within our institution. Thus, we have divided the test bed organization into five nested areas (cf. Fig. 3) that are organized as follows: the first area (A1) corresponds to the whole institution (root of the area hierarchy). Three sub-areas are included: the Entrance (area A2), the Computer Science Dept. (area A3) and the Restaurant (area A4). In turn, the Computer Science Dept. includes the Auditorium (area A5).

Bluetooth ad-hoc networks can be established to support proximity communications, e.g., transfers from a user's PDA to an e-conference system or another

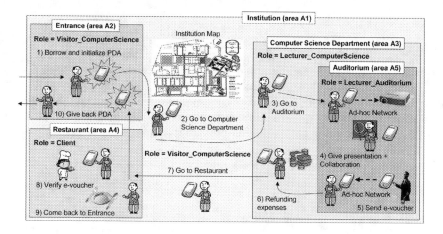

Fig. 3. Use Scenario of the SEDINU System

user's PDA. The institution area hierarchy is fully covered by a wireless network (WiFi) that provides communication supports between different areas. Within this environment, all users use mobile WiFi devices (PDA, smartphone or laptop) that allow them to interact/collaborate exchanging data in an easy way.

To illustrate the SEDINU system functionalities, let us suppose a computer scientist arrives at the institution (area A1) with the goal to give a lecture at the Computer Science Department (area A3). First, the lecturer is registered at the Entrance (area A2) where he receives a PDA: RBAC-Soft assigns him the role "Visitor_ComputerScience" and uploads a general workflow to his PDA to guide him and support his activities within the institution (see Fig. 3 ref. #1). By means of an organization map and GPS, the SEDINU system indicates him the easy way to reach the Computer Science Department (see Fig. 3 ref. #2).

When the lecturer arrives at the Computer Science Dept. (area A3), the SEDINU system determines the current coordinates of the user using two techniques: first, we designed and prototyped the RAMS face recognition system [5] which allows to dynamically identify users and to locate them within the multi-areas environment in which cameras have been installed at specific points (external public spaces, meeting rooms, corridors, secretary office). The goal is clearly to follow each user rather than to locate his mobile device, from which he may be regularly separated for a while. The second solution consists in locating the user's device using techniques of triangulation [8]. These two different location techniques are complementary and can be combined in a fruitful way. The RBAC-Soft system relates the nomadic user's coordinates to the autonomous area A3 and consequently attributes the role "Lecturer_ComputerScience" to him. In a complementary way, the RBAC-Soft system uploads a specific workflow to the lecturer's PDA to guide his activities within area A3. Relatively to the user's current Computer Science Dept. environment, the SEDINU system relies on a map and on the WiFi technology to help him to localize the Auditorium (area A5) where his lecture has to take place (see Fig. 3 ref. #3).

As soon as the lecturer arrives to the Auditorium, the SEDINU system determines his new location and sends it to the RBAC-Soft system, which associates his new location with the autonomous area A5 and then assigns him the role "Lecturer_Auditorium". The SEDINU system provides the lecturer with a dedicated "Slide Projector" service that allows him to present his slides (see Fig. 3 ref. #4). Thus, a bluetooth-based ad-hoc network is dynamically created between the lecturer's PDA and the projector service to allow him: a) to transfer his slides and b) to control his presentation using the PDA. During the lecturer's speech, all conference listeners can use the "Slide Projector" service ("Listener" role) and consult/annotate the slides. During the question session, all users (lecturer, chairman and listeners) can establish a fruitful question-answer collaborative session supported by the "Slide Projector" service and their personal devices.

When the lecture finished, the lecturer is invited to have a lunch at the institution Restaurant. The chairman transfers an e-voucher from his PDA to the lecturer's PDA by a Bluetooth-based ad-hoc network (see Fig. 3 ref. #5). Then, the SEDINU system provides the lecturer with relevant information to guide him to the restaurant (area A4), where he acts with the role "Client" (see Fig. 3 ref. #6 and #7). There, the cashier verifies and charges the lecturer's e-voucher (see Fig. 3 ref. #8). One hour later, the lecturer leaves the Restaurant (area A4) and comes back to the Entrance (area A2) with the role "Exiting_Visitor" always guided by the system (see Fig. 3 ref. #9). Finally, the lecturer gives back the PDA (see Fig. 3 ref. #10) and leaves the institution (area A1).

4   Conclusion and Work in Progress

In this paper, we described the SEDINU system, which facilitates interactions between nomadic users and services provided by autonomous areas, as well as interactions among nomadic users under specific contexts. This service discovery system is complemented by a recognition face system [5] which allows to detect users' face and then locate them within the organization (corridors, meeting room, offices, cafetaria, etc) better than trying to detect their wireless device. These two systems constitute the main components of a powerful collaborative organizational environment that is basically composed of hierarchical autonomous areas. This environment is designed to efficiently support collaborative ubiquitous work within large organizations (e.g. companies or universities).

These organizations are usually divided into departments or units that are structured following a well defined hierarchical relation. Our proposal defines a hierarchy of areas that follows the organization structure: each organizational unit may be associated to a single autonomous area or to several ones. In this way, each area defines specific and administrative policies for the management of their resources and tasks which require a sophisticated workflow system [6]. However, defining and managing a global workflow system constitutes a complex and almost unrealizable objective because of the specificity of the administrative policies of each organizational unit. Thus, among other considerations, it appears really interesting to take benefits from the area-based structuring to efficiently define and administrate the organization workflow system. In this way, the work-flow is logically distributed among the sub-areas: englobing areas provide general workflows whereas englobed ones offers specialized workflows.

As future extensions of the SEDINU system, we will develop modules to create ad-hoc networks through the ZigBee protocol [10] as it can be used by applications that do not require high data transmission. Moreover, a ZigBee device battery consumption is much lower than a Bluetooth or WiFi device's. The SED-INU system constitutes the basis for the definition of a powerful framework to support ubiquitous and nomadic collaboration.

References

1. Adjie-Winoto, W., Schwartz, E., Balakrishnan, H., Lilley, J.: The Design and Implementation of an Intentional Naming System. In: Proc. of SOSP 1999, The 17th ACM Symposium on Operating System Principles, December 12-15, pp. 186–201. ACM Press, Charleston (1999)
2. Campo, C., Almenarez, F., Diaz, D., Garcia-Rubio, C., Lopez, A.M.: Secure Service Discovery based on Trust Management for ad-hoc Network. Journal of Universal Computer Science 12(3), 340–356 (2006)
3. Chang, C.-Y., Sahoo, P.K., Lee, S.-C.: A Location-Aware Routing Protocol for the Bluetooth Scatternet. Wireless Personal Communications: An International Journal 40(1), 117–135 (2007)
4. Edwards, W.K.: Discovery Systems in Ubiquitous Computing. IEEE Pervasive Computing 5(2), 70–77 (2006)
5. Garcia, K., Mendoza, S., Decouchant, D., Olague, G., Rodriguez, J.: Shared Resource Availability within Ubiquitous Collaboration Environment. In: Briggs, R.O., Antunes, P., de Vreede, G.-J., Read, A.S. (eds.) CRIWG 2008. LNCS, vol. 5411, pp. 25–40. Springer, Heidelberg (2008)
6. Han, J., Cho, Y., Kim, E., Choi, J.: A Ubiquitous Workflow Service Framework. In: Gavrilova, M.L., Gervasi, O., Kumar, V., Tan, C.J.K., Taniar, D., Laganá, A., Mun, Y., Choo, H. (eds.) ICCSA 2006. LNCS, vol. 3983, pp. 30–39. Springer, Heidelberg (2006)
7. Herborn, S., Lopez, Y., Seneviratne, A.: A Distributed Scheme for Autonomous Service Composition. In: Proc. of the First ACM Int. Workshop on Multimedia Service Composition, November 6-11, pp. 21–30. ACM Press, Singapore (2005)
8. Jamalipour, A.: The Wireless Mobile Internet: Architectures, Protocols and Services. John Wiley & Sons, Inc., New York (2003)
9. Jeronimo, M., Weast, J.: UPnP Design by Example: A Software Developer's Guide to Universal Plug and Play. Intel Press, Hillsboro (2003)
10. Labiod, H., Afifi, H., De Santis, C.: Wi-Fi, Bluetooth, ZigBee and WiMax, 1st edn. Springer, Dordrecht (2007)
11. Nidd, M.: Service Discovery in DEAPspace. IEEE Personal Communications 8(4), 39–45 (2001)
12. Oaks, S., Wong, H.: Jini in a Nutshell: A Desktop Quick Reference. O'Reilly & Associates, Inc., Sebastopol (2000)
13. White, K.M.: Apple Training Series Mac OS x Support Essentials, 2nd edn. Peach-pit Press, Berkeley (2008)

14. Zhao, W., Schulzrinne, H.: Enhancing Service Location Protocol for Efficiency, Scalability and Advanced Discovery. Journal of Systems and Software 75(1-2), 193–204 (2005)
15. Zhu, F., Mutka, M.W., Ni, L.M.: Service Discovery in Pervasive Computing Environments. IEEE Pervasive Computing 4(4), 81–90 (2005)
16. Zhu, I.H., Chlamtac, I.: Admission Control and Bandwidth Reservation in Multihop ad hoc Networks. Computer Networks 50(11), 1653–1674 (2006)

Increasing Opportunities for Interaction in Time-Critical Mobile Collaborative Settings

Valeria Herskovic[1], David A. Mejía[2], Jesús Favela[2], Alberto L. Morán[3], Sergio F. Ochoa[1], and José A. Pino[1]

[1] Department of Computer Science, Universidad de Chile, Chile
{vherskov,jpino,sochoa}@dcc.uchile.cl
[2] Departamento de Ciencias de la Computación, CICESE, Ensenada, México
{mejiam,favela}@cicese.mx
[3] Facultad de Ciencias, UABC, Ensenada, México
alberto_moran@uabc.mx

Abstract. The critical nature of some working environments, such as hospitals or search and rescue operations, gives rise to the need for timely collaboration. However, interactions are not always possible since potential collaborators may be unreachable because of the lack of a communication channel to carry out the interaction or due to their involvement in other activities. The use of adequate interaction facilitators may allow users to collaborate even in these circumstances. This paper presents a characterization of this type of situation and then introduces a set of design suggestions that may help improve opportunities for user interaction in time critical mobile collaborative settings.

1 Introduction

Recent mobile technologies, such as smartphones and 3G, allow people to work collaboratively while on the move, and motivate developers to build new mobile groupware applications. However, we still need to increase our understanding of mobile environments to develop applications that better support collaborative work.

Communication is essential in mobile collaborative environments. It is the basis to make coordination and collaboration possible in any work scenario [7]. In mobile collaboration, people may need to move to interact with others, or they may need to interact with others while they are on the move. Collaboration among persons and resource sharing usually require a communication channel. Nevertheless, there are some situations in which interaction is not possible even when a communication channel is available, e.g., when one of the collaborators is unavailable. In non-critical activities, people are willing to wait for a natural opportunity for interaction when this situation occurs. However, people do not have time to wait for an interaction in time-critical scenarios. In these settings, finding a particular person or resource as soon as possible can be vital to the success of the collaborative effort. For this reason, improving interaction opportunities can benefit the results of the collaborative work.

Section 2 presents observations of two mobile collaborative work settings. Section 3 characterizes this type of work. Section 4 presents the proposal. Section 5 presents related work. Finally, section 6 presents the conclusions.

L. Carriço, N. Baloian, and B. Fonseca (Eds.): CRIWG 2009, LNCS 5784, pp. 41–48, 2009.

2 Time-Critical Mobile Collaborative Settings

The scenarios of emergency management and hospital work are similar in that the work is collaborative, mobile, and by nature urgent during certain periods of time. We define this type of scenario as a *time-critical mobile collaborative setting* (TMC). In these settings, mobile workers are autonomous most of the time, and interactions are motivated by the occurrence of triggering events. This means that at some point in time, which cannot be anticipated, the actors will need to interact in order to respond to the event, usually involving information exchange and/or work synchronization.

This section presents observations from studies in two TMC settings. The information was collected in a medium-size public hospital through shadowing and reported in [7], and in urban search and rescue exercises through interviews, observation, and the study of documents detailing work procedures [3].

2.1 Hospital Work

Hospital work is characterized by the need for coordination and collaboration among specialists with various areas of expertise, task switching, information exchange, data integration, and the mobility of staff, patients, documents and equipment [8]. At the same time, this working environment is highly critical and error prone [5].

In this TMC environment, informal communication is an essential resource for supporting coordinated medical tasks, to discuss diagnoses and treatments and to gather the artifacts and human resources necessary for patient care [7; 8]. Hospital staff spend 53% of their time outside their base location [8], which triggers spontaneous interactions. At other times, workers may be found in specific locations.

Face-to-face is not the only communication channel that medical workers use to interact. In some situations, when one of the collaborators is unavailable (e.g., if they work in different shifts), medical workers communicate asynchronously by leaving instructions or notes in information artifacts such as medical records and clipboards.

In addition, in some situations a medical worker may be in the hospital, but unavailable for interaction due to the activities he is performing (e.g., surgical procedures, meetings). We illustrate this situation through the following scenario:

The medical intern goes to the specialist's consulting room to discuss a patient.

> **[Medical Intern]**   *Hi, is Dr. (the medical specialist's last name) here?*
> **[Secretary]**   *No, she is in Internal Medicine.*

The intern then goes to Internal Medicine and asks for the location of the specialist.

> **[Medical Intern]**   *Have you seen the medical specialist here?*
> **[Nurse]**   *Yes, she is in room 235.*
> **[Medical Intern]**   *Is she busy now?*
> **[Nurse]**   *I guess so, she is performing a (clinical procedure).*
> **[Medical Intern]**   *Ok, I will have to look for her later.*

In this scenario, the medical intern cannot interact with the medical specialist because she is unavailable. If the intern required an interaction to solve an urgent situation, this could have caused a problem in the patient's treatment.

2.2 Search and Rescue Operations

Some catastrophic events, such as floods, earthquakes and chemical spills require specialized teams of firefighters to conduct search and rescue operations. In this TMC environment, firefighters use radio, marking systems and face-to-face communication to exchange information about resource allocation, location of victims, and dangerous and safe areas. Their goal is to find and rescue all victims as fast as possible because the number of diseases grows exponentially during the first 24hs [11].

Inside the emergency area, teams of firefighters move constantly to assess the emergency situation, search for victims, and communicate work progress. These teams are autonomous; interactions among them occur when there is a specific need (e.g., a stretcher is needed to carry a victim) or triggering event (e.g, an earthquake replica has trapped a firefighter). Firefighters' main communication channel is a radio system (2 to 3 channels per emergency). This system is used both for important and non-important messages (e.g. reporting the injuries a victim has sustained). Frequently, the messages are not heard (e.g., due to environmental noise); they often have to be repeated several times. Sometimes, urgent messages cannot be delivered. We illustrate these communication difficulties with the following observed scenario:

The incident commander (IC) gives directions through the radio to the team captain in charge of evaluating the stability of the affected civil infrastructure.

[IC] Please report on progress so a team can move into the building to search.
(The captain's team is cutting wood and the environmental noise is very loud)
[IC] Please report on progress so a team can move into the building to search.
(A couple of minutes later)
[IC] Please report on progress so a team can move into the building to search.
[Captain] OK... (He reports on progress)

In highly critical situations, firefighters use an audible signal that is loud enough to always be heard. There is a small set of these signals, which are delivered only in critical situations, e.g., when the collapse of a building is imminent.

2.3 Discussion of Communication Patterns in TMC Settings

The observations in two TMC settings allowed us to identify two important dimensions of communication in TMC settings: communication and time.

Communication is basic for many strategies mobile workers use. Communication problems strongly decrease the opportunities to carry out collaborative interactions, some of which could be urgent or critical. For example, a rescuer has found a group of trapped victims and he needs to get help on time. If the communication channel and the people involved are available, the interactions will be possible and the critical situations may evolve positively. Otherwise, the consequences will be unpredictable.

In the case of firefighters, most breakdowns in communication are due to the failure or unavailability of a communication channel. Communication through media such as radio and marking systems is often unnoticed. The only effective channel is the audible alarm; yet, it can only convey simple predefined messages. In hospitals, communication failures are usually due to unavailability of one of the collaborators.

Time is important when providing support to collaboration. In the case of TMC settings it is particularly so because communication should be carried out during particular time frames. After that, the communication goal may become obsolete.

3 Characterization of Interaction States in TMC Settings

The observations presented above highlight the importance of finding the required collaborators and communicating with them. We define two dimensions that are of interest to study mobile collaboration. They are *simultaneity* and *reachability*, which are strongly related to time and communication. When an actor requires an interaction with someone, their relationship will be in one of the quadrants shown in Fig. 1 [3].

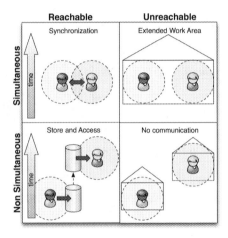

Fig. 1. The four possible interaction states in TMC settings

Whenever the actors need to interact, they will be in a particular quadrant of the classification. There are four possible states: simultaneous/reachable (SR), simultaneous/unreachable (SU), non-simultaneous/reachable (NR) and non-simultaneous/unreachable (NU). During their work, two actors may pass through several states. Fig. 2 presents a state diagram for one actor, illustrating the transitions between states that may happen during his workday in relation to a collaborator.

The SR situation represents a typical synchronous collaboration scenario. Several types of technological support for this scenario currently exist, such as IM, file transfer, etc. In the NR situation, some solutions (e.g., email, synchronization with a server) also exist. The NU scenario is infrequent, and may happen e.g., when collaboration is being established or between actors who communicate erratically.

The SU scenario occurs whenever the communication channel between two actors is unavailable, or when one of the actors is not available. Consequently, even if the collaboration could benefit from two actors' interaction, it is not possible. In non time-critical work environments, the change from state SU to SR or NR may occur naturally within some time; however, in time-critical situations, actors are reluctant to wait for the state change. In these cases, helping actors move from a SU state to SR or a NR state could enable collaboration or improve it to address the critical situation faster. The next section proposes strategies that can be incorporated into mobile collaborative systems to move collaboration scenarios towards favorable settings.

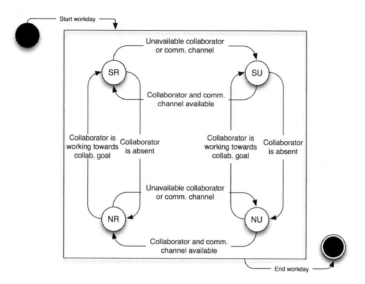

Fig. 2. State diagram for an actor with respect to a collaborator

4 Allowing Interactions in SU States

In TMC settings, unreachability is a synonym of no collaboration. However, unreachability may sometimes be transformed into reachability. The solutions proposed in this section use existing and new mechanisms to try to deal with the causes of unreachability, namely, lack of a communication channel and unavailability of collaborators. Even though not all instances of SU states may be solved, in some cases a user may obtain the results needed from the interaction and continue working.

4.1 Strategies to Increase Opportunities for Interaction

Uncoupled Management of Resources. Actors often need to interact to exchange resources, but when a user is unavailable his/her resources are also unavailable. If we uncouple user management from resource management it could be possible to move the interaction from the SU to the SR quadrant. For example, a user might anticipate that he will become unavailable for some time, and choose to make some of his resources accessible in the system for this time period and/or for certain users.

Use of Intermediaries for Synchronous Interactions. Typically, in nomadic work a person can easily interact with other persons or resources that are "one-hop" away [9]. In that scenario each person or resource located two or more hops away will be unreachable for potential requesters. If we take advantage of the intermediaries that are present in the work area (other mobile computing devices and access points) it could be possible to move the interaction from the SU to the SR quadrants. Therefore, if the intermediaries are used as part of the solution, reachability can be achieved.

Use of Equivalent Alternatives. Mobile workers usually need to exchange ideas with other people and have access to skills, and other knowledge. Thus, when a user needs to interact with an unavailable collaborator, he may be redirected to an appropriate intermediary, either referred by the original collaborator or through an automatic recommendation system. It is then possible to move a SU interaction to a SR one by replacing the original collaborator with an equivalent alternative one.

Use of Collaboration Promoting Awareness (CPA). We call CPA to those proactive mechanisms which provide awareness information to a user to promote collaborative interactions. These awareness mechanisms must be autonomous and use the context information [1] to induce a collaboration instance, once the user has expressed his/her need to interact with another one. In this way, a user may be aware of other users' availability. Awareness may also be provided to users when they are being required, to provide them with an opportunity to interact or to clarify their interrupt status.

4.2 Design Suggestions

Displaying user availability. When a user needs to interact with another user, he must be aware of his partner's availability to determine whether the interaction can occur. Therefore, the system may provide users with contextual information (e.g., user location, nearby users) that helps them determine whether a user is available or not.

Searching for a potential collaborator. The system should allow users to specify the identity of the user they are looking for, even if he/she is unreachable.

Identifying alternative collaborators. When a user needs to interact with an unavailable collaborator, he may contact another person who may have a similar set of skills or knowledge. In this case, the system may suggest these alternative collaborators by using a manual or automatic role-based recommendation system.

Contextual management of resources. When a user wants to interact with another one, he may actually need to access a resource containing ongoing work between them. Thus, contextual information about resource use (e.g., location) should be stored. This data may be used to automatically grant or deny access when the owner is not available. The system may also grant access to restricted resources in critical events.

Searching for an equivalent resource. A person may need a resource belonging to an unavailable information source (a person or an intermediary server). The application should allow searching for a similar resource among reachable information sources.

Establishing alternative routes for communication. Two users may need to interact while lacking a direct communication channel. In this case, the system may implement alternative communication routes using intermediary nodes to forward messages between requester and responder. Routing may provide a channel between collaborators or extend the communication threshold of the system.

Allowing gossip mechanisms. Sometimes collaborators are unable to interact even if the mechanisms described above are implemented (e.g., in disperse work scenarios). In that case, a gossip, which is a message traveling through the network searching for the destination user during a time period, may be used. These messages try to force a reachable situation, e.g., *"URGENT: be at X at 9AM"*.

Pushing notifications. After a few unsuccessful tries to interact with somebody, a user may desist trying to reach the person, who might become available a moment later. Thus, the system should notify the user when the destination becomes reachable.

Table 1. Matching between strategies and design guides

Strategies	Design Suggestion
Uncoupled Management of Resources	• Contextual management of resources • Searching for an equivalent resource • Displaying user availability
Use of Intermediaries for Synchronous Interactions	• Establishing alternative routes for communication • Allowing gossip mechanisms • Searching for potential collaborators
Use of Equivalent Alternatives	• Identifying alternative collaborators • Displaying user availability • Searching for potential collaborators
Use of Collaboration Promoting Awareness	• Pushing notifications • Searching for potential collaborators

5   Related Work

Time-critical settings include several areas of mobile work such as service technicians, police patrols and firemen [6], as well as hospitals. Several systems have been proposed to support firefighting work. Most identify the lack of a reliable communication channel and provide alternatives, effectively diminishing the time users spend in a SU state. For example, Siren [4] supports work in search and rescue operations using a peer-to-peer architecture, so intermediary nodes are used to transfer information between firefighters in the field. Unreachability is usually caused by lack of a reliable communication channel, but it may also be caused by unavailability, e.g., in an emergency in which all firefighters are engaged in urgent tasks. This case is seldom acknowledged in the emergency management literature.

Hospital settings provide conditions for reliable communication channels. In hospitals, workers become unreachable either because they cannot be located or because they are unavailable. To address user location, some mechanisms are in use (e.g., megaphones) or have been proposed, such as a neural network embedded within the component and used to estimate location of users and devices [2]. Another approach is [10], which presents an approach to estimate activities. This data could be used as a trigger to start interaction by advertising user availability. Thus, research in hospitals is aimed at showing where an actor is or what he/she is doing in order to approach him/her. However, these efforts do not deal with user unavailability.

6 Conclusions

Communication and time are essential resources in time-critical mobile environments. We identified four collaboration situations based on the dimensions of reachability and simultaneity: Simultaneous reachable (SR), non-simultaneous reachable (NR), simultaneous unreachable (SU), and non-simultaneous unreachable (SU). There are several strategies that may be used to facilitate communication in two of these situations (SR and NR). However, the SU state currently does not allow for interaction among actors, which may be crucial in time-critical settings. This paper presented strategies that assist moving from a SU setting to a SR or NR setting, and design suggestions that may help implement these strategies in a mobile application.

This paper identifies the existence of SU states and affords visibility to this problem, as well as allows designers and developers to consider it when developing systems in TMC settings. We believe that incorporating these strategies into the design of a system may help increase the number of interactions that occur during collaboration, improving the results of the collaborative process.

Acknowledgements. This work was partially supported by Fondecyt (Chile) grants 11060467 and 1080352, and LACCIR grants R0308LAC005 and S1108LAC002.

References

1. Brezillon, P., Borges, M., Pino, J.A., Pomerol, J.C.: Context and awareness in Group Work. Lessons learned from three case studies. J. of Decision Systems 17(1), 27–40 (2008)
2. Castro, L.A., Favela, J.: Continuous Tracking of User Location in WLANs Using Recurrent Neural Networks. In: Proc. of the 6th Mexican International Conf. on Comp. Science, pp. 174–181. IEEE Comp. Society, Los Alamitos (2005)
3. Herskovic, V., Ochoa, S.F., Pino, J.A.: Modeling Groupware for Mobile Collaborative Work. In: Proceedings of CSCWD 2009 (to be published, 2009)
4. Jiang, X., Chen, N., Hong, J.I., Wang, K., Takayama, L., Landay, J.: Siren: Context-Aware Computing for Firefighting. In: Ferscha, A., Mattern, F. (eds.) PERVASIVE 2004. LNCS, vol. 3001, pp. 87–105. Springer, Heidelberg (2004)
5. Kohn, L.T., Corrigan, J.M., Donaldson, M.S. (eds.): To Err is Human: Building a Safer Health System. National Academy Press, Washington (1999)
6. Landgren, J.: Making action visible in time-critical work. In: Proceedings of the SIGCHI conference on Human Factors in computing systems, pp. 201–210. ACM, New York (2006)
7. Mejia, D., Morán, A.L., Favela, J.: Supporting Informal Co-located Collaboration in Hospital Work. In: Haake, J.M., Ochoa, S.F., Cechich, A. (eds.) CRIWG 2007. LNCS, vol. 4715, pp. 255–270. Springer, Heidelberg (2007)
8. Morán, E.B., Tentori, M., Gonzalez, V.M., Favela, J.: A.I. Mobility in Hospital Work: Towards a Pervasive Computing Hospital Environment. International J. of Electronic Healthcare (IJEH) 3(1), 72–89 (2007)
9. Neyem, A., Ochoa, S.F., Pino, J.A.: Integrating Service-Oriented Mobile Units to Support Collaboration in Ad-hoc Scenarios. J. Univers. Comput. Sci. 14(1), 88–122 (2008)
10. Sánchez, D., Tentori, M., Favela, J.: Activity Recognition for the Smart Hospital. IEEE Intelligent Systems 23(2), 50–57 (2008)
11. Yusuke, M.: Collaborative Environments for Disaster Relief. Master's Thesis, Massachusetts Institute of Technology (2001)

A Social Matching Approach to Support Team Configuration

Flavia Ernesto de Oliveira da Silva[1], Claudia L.R. Motta[1], Flávia Maria Santoro[2], and Carlo Emmanoel Tolla de Oliveira[1]

[1] PPGI – Federal University of Rio de Janeiro – UFRJ
[2] NP2Tec/PPGI – Federal University of the State of Rio de Janeiro – UNIRIO
flavia_ernesto@yahoo.com.br, claudiam@nce.ufrj.br,
flavia.santoro@uniriotec.br, carlo@nce.ufrj.br

Abstract. Organizations aim to store knowledge about their employees systematically in order to find "who knows what" in a timely fashion and match people by identifying their skills and competences. This paper presents a Social Matching Model, which brings together skills, socio-emotional factors and roles to support the composition of teamwork.

1 Introduction

The idea of teamwork arose when organizations realized that pooling individual knowledge and skills would make it easier to achive their goals. The volume and rate of change of information and the need to acquire new knowledge motivates this type of work [1]. According to Robbins & Finley [2], teams play important roles in our professional and personal life, with high levels of interdependence among their components directed to achieving goals or performing tasks. In this context, the work team generates positive synergy, i.e., the performance level of a team is greater than the sum of individual efforts. Since team members share information and resources, coordinated teams achieve greater productivity, use resources more effectively and solve problems better. The team's composition of is key to its success.

In social relations, we unconsciously combine with others by looking for people with interests, whether professional, personal or social, similar to our own. Organizations aim to store their employees' tacit knowledge systematically so as to discover "who knows what" promptly and match people by identifying their skills and competences. There are difficulties, however: How are specialists' skills to be mapped in order to meet a given demand? How are the roles an expert is able to play in a task to be identified? These issues can be supported partly by Social Matching Systems, which automate recommendation of persons [3]. Their essence is to bring together people who have similar interests and some kind of social compatibility.

This paper presents a Social Matching Model, which brings together skills, socio-emotional factors and roles that a person can play to support the composition of teamwork. This proposal is based on the premise that individuals who have social affinities work together more easily and do not need rules for trading cooperation, because their modes of interaction can be easily understood.

L. Carriço, N. Baloian, and B. Fonseca (Eds.): CRIWG 2009, LNCS 5784, pp. 49–64, 2009.

The paper is organized as follows: Section 2 discusses the concepts guiding the proposal, Section 3 presents the model and proposed Social Matching System, Section 4 describes an experimental application, and Section 5 concludes the paper and points to future work.

2 Team Composition

According to Katzenbach & Smith [4], a team is a group of people with complementary skills, who are committed to a common goal, perform interdependent work, and are collectively responsible for the results. However, an examination of the literature indicates that differences between individual and collective interests can throw up obstacles to forming teams.

It is thus relevant to argue what should be considered in forming a team with a view to implementing a given task. Usually there is a manager responsible for identifying the knowledge or skills necessary among the staff. In addition, other issues must be considered in forming teams, such as the temperament of the individuals involved and the roles to be played. This section will discuss the work relating to these three aspects: skills, temperaments and roles.

2.1 Skills

Fleury & Fleury [5] define competence as "a knowing-how-to-act responsibly that entails mobilizing, integrating and transferring knowledge, resources and skills that add economic value to the organization, and social value to the individual." The authors state that individual competence can be understood from three fundamental pillars: individual, educational and professional experience.

From the educational point of view, Perrenoud [6] defines competence as the mobilization of cognitive resources (knowledge, information, values, intelligence, patterns of perception and reasoning) to solve a problem. Whenever action is instigated, the individual must properly implement his or her cognitive resources, and the experience gathered in past actions influences future decisions [7].

However, organizations that use knowledge requirements as the basis for identifying skills tend to work with a bias: they assume that managers know how to gauge the demand for knowledge without using formal tools or methods. There is a lack of methods to measure skills and allocate people correctly to group tasks.

2.2 Temperaments

One of the fundamental factors in team composition is the individual's behavior in pursuit of an activity, or his temperament. Individuals who have similar temperaments, it is believed, tend to work more collaboratively [8]. For various reasons, individuals are different in their ways, and are endowed with attributes both variable, such as emotion, and invariable and innate to humans, such as temperament. According to Pasquali [9], the literature reflects a concern to identify and typify temperament from observable results. Tests such as MBTI, DISC and Roger Verdier's Educational Evaluation endeavor to map a person's temperament using questionnaires. Identifying correlations among temperaments can help match compatible individuals to work together.

2.2.1 MBTI – Myers-Briggs Type Indicator

The MBTI [10] is a test based on Jung [11], a swiss psychiatrist who created the theory of psychological types. Regarding psychological functions, Jung [12] proposed four types – thinking, feeling, sensation and intuition – that serve to adapt the individual to outside and inside situations of life. These are divided into rational mental processes of judgement and irrational mental processes of perception. All individuals have the four functions, but what differentiates one from another is how the functions develop according to personal or professional experiences. Jung thus defined eight personality types, which combine into 16 functional types to demonstrate individuals' characteristics (Extroversion/Introversion; Feeling/Intuition, Thinking/Feeling, Judging/Perception). The MBTI consists of a questionnaire that highlights the differences among people.

2.2.2 DISC (Dominance, Influence, Steadiness and Conscientiousness)

Marston [13] created the DISC theory, which examines four dimensions of imdividual human behavior – dominance: how we deal with problems and challenges; influence: how we deal with people and influence them; stability: how we deal with changes and set pace; and compliance: how we deal with rules and established procedures – resulting in over 19,000 matchings. A 20-item questionnaire enables individual outlines to be drawn up.

2.2.3 Roger Verdier's Educational Evaluation

Verdier's evaluation infers aspects of personality and leadership. Personality is the quality of a person's character. Some personality traits indicate possible greater or lesser suitability for certain activities and sometimes conditions that are decisive for the process of collaboration. For example, a very rigid person will have great difficulty in an activity that requires flexibility and a communicative person cannot work in isolation for long. According to Justo [14], the key factors of character are: Emotion – individuals are considered emotional (E) if they experience pleasure and pain more easily than the average, otherwise they are not emotional (nE); Activity – if action is a necessity and a pleasure for them, they are considered active (A), otherwise not (nA); Effect – how long events influence consciousness classifies them as Primary (P), where the effect is over almost immediately after the phenomenon, and Secondary (S), if the memory remains deep in consciousness. Verdier shows eight types of temperaments: melancholy, unstable, amorphous, apathetic, social, phlegmatic, active and leader. He examined affinities among them (Table 1).

A case study was conducted in 2008 to examine the feasibility of adopting a test to measure temperament and thereby identify social affinity among individuals in a team. Twenty-seven masters program students taking the discipline of Neuropedagogy and Informatics at Rio de Janeiro Federal University participated in the study. The study was divided into two stages. In the first stage the students answered a test that led to the identification of temperaments. After calculating the temperament of each student, in the second stage we compared the temperaments of individuals within their teams to determine whether they have similar or antithetical behavior in practice. The results were analyzed and incompatibilities were identified. In the final analysis, the results were presented to professors who teach the discipline and they noticed positive signs both on the identification of the temperaments and the correlation made between the allied and opposing temperaments. On the stength of this result, the Roger Verdier test was chosen.

Table 1. Related and antithetical temperaments [14]

Dominant temperament	has affinity with	is antithetical with
Melancholy	Leader, Unstable, Apathetic	Social
Unstable	Melancholy, Amorphous, Active	Phlegmatic
Amorphous	Unstable, Apathetic, Social	Leader
Apathetic	Melancholy, Apathetic, Phlegmatic	Active
Social	Amorphous, Active, Phlegmatic	Melancholy
Phlegmatic	Social, Leader, Apathetic	Unstable
Active	Unstable, Social, Leader	Apathetic
Leader	Phlegmatic, Melancholy, Active	Amorphous

2.3 Roles of Individuals in a Team

Another issue to be considered is the characteristic patterns of an individual's behavior in relation to others as regards the facilitating process within the team, or the roles that individual can play. Individuals play various roles, either simultaneously or sequentially, with greater or lesser intensity at different times. For Belbin [15] getting the right people to form a team does not happen by accident – the biggest challenge

Table 2. Roles defined by Belbin [15]

Role name	Positive points	Negative points
The Plant	Tends to be introverted, serious and idealistic. Independent and visionary, prefers to work alone with concentration.	May need someone to keep his feet on the ground. Does not always manage to convey his ideas.
The Resource Investigator	Creative, outgoing linker strongly attracted by challenges. Goes beyond the bounds of the team in search of resources. Loves to negotiate and argue.	May be discouraged when nothing new is happening. Always wants to be doing something different from the last job completed.
The Coordinator	Team controller, tends to introversion, but able to motivate others and direct them to a common purpose. Highly skilled in resolving conflicts.	The team can put too much trust in him and overburden him. He can spend the whole time trying to satisfy people, which is rather complicated.
The Shaper	Team dictator. Task-oriented and prepared to move mountains to achieve his goals. Outgoing, challenging, frank.	He may offend people. He can forget to give the necessary explanations.
The Team Worker	Concerned with harmony and good relationships within the team. Outgoing, ready to cooperate and seek new tactics. Respects others.	May not be decisive and focused enough to control task execution. Avoids conflict and tries to hide problems under the rug.
The Implementer	Outgoing, likes to have an organized structure to work in. Transforms ideas and strategies into work plans.	Can be inflexible. Sudden changes or unexpected opportunities can leave him dissatisfied.
The Monitor Evaluator	Reserved personality with air of calm and stability. The thinker is quiet and analytical.	May not have a clear focus and delay team progress in order to avoid certain risks.
The Completer Finisher	Considers it total failure not to deliver work on time. Introverted, and super hard-working. Likes the traditional approach and hierarchical order.	When a thinker presents a bright idea, the Finisher may ignore or reject it.
The Specialist	Is the right person to consult on matters specific to the work in hand. Essential support for the team.	May not show any interest in the overall task being performed.

for whoever is assembling a team is precisely whether the members are the best for the team. Belbin [15] empirically identified nine different "team roles" (Table 2), types of behavior that influence team success and are considered "A tendency to behave, contribute and interrelate with others in a particular way".

A second case study culminated in confirmatory findings as regards the correlation between temperament and role. This study was conducted in 2008 with eleven students at Rio de Janeiro Federal University, taking the discipline Agile Modeling (Extreme Programming methodology) for software developers. The study was divided into two stages: the first entailed filling out two questionnaires to measure temperament and the roles an individual can play, and the second was a simulated team game to assess the team formed on the basis of the correlated results from the first stage. We noticed that it is important that teams are structured by role attributes. These attributes, correlated with temperament, allow individuals with personal and professional affinities to come together. Results show that correlating temperaments and roles is feasible and can generate satisfactory matchings to inform recommendations for team formation.

2.4 Social Matching Systems

After performing an analysis of people recommendation systems, Terveen & McDonald [3] coined the term Social Matching Systems, the essence of which is to bring together people with similar interests and some kind of social compatibility. A social matching model should include the following: user profile – containing information on individuals that the system needs in order to identify their capabilities; combination (matching) – a mechanism (algorithm) implemented by the system to express a model of consistency between profiles; presentation – the system promotes meetings between people with compatible profiles, preserving users' privacy until they are available for contact; and interaction – communication tools to motivate virtual or in-person interaction. Berscheid & Reis [16] present an overview of three factors that predict interpersonal attraction, as shown in Table 3.

Table 3. Factors that predict interpersonal attraction [3]

Personal characteristics	Individuals with similar personal characteristics are more likely to be attracted to each other. Personal characteristics such as tastes or preferences should be taken into account in recommending.
Demographic data	Characteristics, such as ethnicity, gender, marital status, occupation, and income, are often used as visible signs of individuals' values and attitudes.
Familiarity	People who are physically close meet and interact more frequently.

3 Team Composer: Social Matching-Based Model

Our proposal is a matching approach based on an organization's social needs when wanting to bring people together to form a team to accomplish a specific task. It thus has to identify and map the skills required in a specific context to ensure that people fit that context, define the roles that each member can play within the team, and investigate the temperament of the people to estimate compatibilities among team members. The purpose of this approach is to provide suggestions for arriving at combinations of people from these requirements. The approach of the Team Composer is described in three steps:

- Step 1 – Recording Profiles: individual profiles are built up by recording skills and completing questionnaires to identify roles and temperament. This information must be provided by the individual.
- Step 2 – Matching Context with Individuals: this automated process generates combinations of organization members from the description of the teamwork context to be achieved. This step is supported by a Genetic Algorithm.
- Step 3 – Presenting Results: the results of the matching list the possible teams and give the manager options from which to choose the team that best suits the needs.

The proposed model is generic and can be applied in various organizations with different scenarios. However, for a prototype test to be constructed, it was necessary to set a domain. The option was for software development teams, since [17] and [18] reported the problem of forming teams in software organizations. The solution is detailed next.

3.1 Identifying Competences

One of the processes used by software development teams is the Rational Unified Process (RUP), which aims to increase the chances that a software is produced with predictable quality, time and cost [19]. The RUP defines thirty-two functional roles, which do not represent people, but the description of the skills necessary for them to assume a functional role.

Table 4. RUP Roles and Skills

RUP Functional Role	Responsibilities	Necessary skills
System Analyst	Requirements identification and case use modeling	Ease of expression and communication; Facility to build relationships; Easy to adapt to change; Initiative in solving problems and developing creative alternatives; Tolerance to pressure; Promptness and initiative; Organization; Proactivity and objectivity
Software architect	Responsible for the general structure of the architecture. Works allied with the project manager	Knowledge and maturity; Vision and reviews sensible and judicious in the absence of complete information; Leadership to lead the technical effort among several teams; Makes important decisions under pressure; Makes sure such decisions are complied with at risk; Communication to gain trust; Power of persuasion and motivation; Guidance for goals, pro-activity and focus on results.
Project manager	Allocate resources, set priorities, coordinate interactions among users and clients and keep the team focused.	Analysis of decisions; Skills of presentation, communication and negotiation; Leadership and development of team spirit; Management of time; Decision-making capacity in situations of stress; Good interpersonal relations; Objectivity in defining and evaluating work, ensuring the participation of the whole team; Honest in assessing results
Developer	Build and test components	Attention to detail and good memory; Ability to concentrate; Ability to solve practical problems; Discipline; Facility with mathematics; Patience; Perseverance; Logical

Rajendran [20] compared the skills proposed by Belbin with those of the most important roles in a software development team: Project Manager, System Analyst, Software Architect and Developer. Output from a questionnaire of thirty-two items shows individuals' skills and associates them to one of the RUP functional roles (Table 4).

3.2 Identifying Roles

The roles used in this proposal were established according to the classification proposed by Belbin [21] and measured through a questionnaire containing 20 objective questions. The questionnaire reports a ranking score, where individuals are associated with roles according to which column they score most points in.

3.3 Identifying Temperaments

We presented various types of tests designed to yield data analyses pointing to a definition of a person's temperament. Among them Roger Verdier's Educational Evaluation was selected. Data from responses to a questionnaire were treated to determine the individual's personality/temperament. The treatment entailed combining the factors resulting from quantification of the results obtained in step 1, which can result in various combinations, such as nEnAS, nEnAP, NEAP, NEAS, EnAP, EAP, or EnAS EAS.

3.4 Implementation

Automation of the model was based on a Genetic Algorithm (GA). The GA is a direct analogy to Darwinian ideas, where individuals with greater capacity to adapt to their environment will have more opportunities to survive and reproduce than those less adapted. After many generations, individuals of the population acquire characteristics that give them greater adaptability to the environment than people from earlier generations [22]. The optimal solution involves a team of individuals appropriate to the criteria set by the manager.

In an GA, the chromosome (genotype) is defined by a string of bits representing a possible solution to the problem. Thus, each chromosome (called the individual in the GA) is a point in the space of solutions to the optimization problem. The solution adopted in the GA process is to generate a large number of individuals, or populations, according to specific rules, in order to promote a scan as large as the space needed for solutions. The following logic is used to support the formation of teams: through an interface, the manager sets the context with the desired criteria (skills, roles, temperament and number of individuals to compose the team); initially, the algorithm considers each criterion in the following order: skills, roles, and temperaments, while it searches the whole organization for individuals to make up the team.

3.4.1 Specification of the GA Items

To optimize team formation, the attributes to be used as criteria for team formation must be defined. In GA, fitness is a calculation that will generate the best individual in the current population. Thus, two attributes were defined in the prototype, the first based on individual profiles, where the manager is responsible for defining which groups are acceptable, and the second based on the cohesion predicted by sociometric rules, which take account of the social affinity among team members.

This can be expressed formally as: let C be a class of k x n (where n is the number of teams and k is the size of teams) and F_1, F_2, F_m ... is a list of factors that will influence team formation. In our case, the manager needs a "good formation" of class C into n teams of k individuals subject to the factors F_1, F_2, F_m ... For that purpose, a group-k is defined in C as a subset (l_1, l_2, ..., l_k) of C with k elements. A division-k x n of C is defined as a subset (g_1, g_2, ..., g_n) with n elements, where each element is a different team-k in C. The next step is to define fitness by a division-k x n D of C taking into account the factors F_1, F_2, F_m ... as a function defined in terms of F_1, F_2, F_m ..., which maps D in a number that measures the fitness, Φ. Φ = FITNESS (F_1, F_2, F_m ...) (D). Thus, the problem of teams is defined as the matching of a division-k x n D of C that maximizes the function FITNESS (F_1, F_2, F_m ...) (D).

3.4.2 Profiles

In this context the individual attribute profile based on skills, roles and temperament is measured. The individual's profile will be represented by a p-tuple <at $^m{}_1$, at $^m{}_2$, ..., at $^m{}_p$>, where each m is at the value of an attribute of the individual. The individual's profile will be based on three attributes: (1) identifying the individual's skills (at $^1{}_1$), (2) identifying the roles that individual can play (at $^1{}_2$) and (3) the individual's temperament (at $^1{}_3$). To determine acceptance by a team, Function 1 is defined in terms of the profiles described previously:

$$P|g|_A^i = \begin{cases} 3 \text{ if } \{P(a_1), P(a_2),...,P(a_3)\} \in I - A \\ 2 \text{ if } \{P(a_1), P(a_2),...,P(a_3)\} \in I \cap A \\ 1 \text{ if } \{P(a_1), P(a_2),...,P(a_3)\} \in A - I \\ 0 \text{ caso contrário} \end{cases}$$

Function 1. Parameters used in the individuals' profiles

Labidi [23] introduced the sociometric test as a technique to determine the degree to which individuals are accepted or rejected in a team, and to discover the relationships among individuals and reveal their structure. It consists of applying a questionnaire to individuals and then analyzing the responses. In this study, the following rules were followed as part of sociometric analysis of the team composition problem: Rule 1 (R1) – Never build teams where there is no relationship between candidate members; Rule 2 (R2) – Teams must meet at least one choice in all attributes; Rule 3 (R3) – Form teams to preserve mutual choices where possible; Rule 4 (R4) – Always allocate an isolated individual to a team where his/her first choice is satisfied. Function 2 maps teams-k in degrees of cohesion, where grade 0 indicates no cohesion and grade 3 indicates strong cohesion.

$$s|g|_R = \begin{cases} 0 \text{ if } \neg R_1(g) \vee \neg R_2(g) \\ 1 \text{ if } R_1(g) \wedge R_2(g) \wedge \neg R_3(g) \\ 2 \text{ if } R_1(g) \wedge R_2(g) \wedge R_3(g) \neg R_4(g) \\ 3 \text{ caso contrário} \end{cases}$$

Function 2. Computation of team mapping through degree of cohesion

3.4.3 Specification of Parameters Used in GA

a) Problem Solution

The problem is how to form teams of software developers. For this problem, each chromosome represents a team of individuals, where each chromosome gene characterizes an individual within a team, as shown in Figure 1.

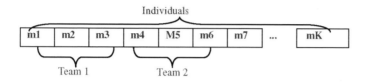

Fig. 1. Chromosome codification of a team for n=3

Thus, the different genetic codes of the chromosomes will be given by reordering of the genes in the teams. The distribution of genes in teams is a very particular feature in the case of optimized team formation. Every n genes characterizes a team; this means that individuals are considered twins only when they have genetic variation within a given team. Thus, the number of possible solutions for C is given by Equation 1, where m is the number of individuals in the organization and n is the number of individuals in the team.

$$C(m,n) = n!/[(n-m)! \, m!]$$

Equation 1. Number of possible solutions

b) Population Size

There are no conclusive studies on optimal population size [24]. In optimized formation of groups, chromosome size (number of genes) depends on the number of individuals in the organization; also the number of individuals in the population influences the performance of the GA. Accordingly, from tests with different chromosome and population sizes, Labidi [25] defines a relationship of chromosome size to population size (Table 5) in order to derive the size of the population.

Table 5. Population size

Number of Individuals in chromossome – n	Number of Individuals in population
n < 5	64
n < 6	77
n < 7	89
n < 8	102
9 < n < 10	128

c) Assessment Function

The assessment function of each population individual will be given by calculating the p and s functions (Function 3).

$$fitness = \sum_{g \in D} F(p[g]^I_A, s[g]_R) = \sum_{g \in D} (x.p[g] + y.s[g])$$

Function 3. Computes each individual's aptitude, p relates to the individuals' profiles; s to sociometric team cohesion; and x and y are weights assigned by the manager.

Combining all the factors above defines how good a division D of the class is, taking into account the definition of profiles acceptable to the manager $p[g]^I_A$, sociometric cohesion and s [g] R, as in the Function 4:

$$FITNESS (p[.]^I_A, s[.]_R)(D) = \sum F(p[.]^I_A, s[.]_R)$$

Function 4. Computes Fitness, where: F(p,s) maps the degree of acceptance and sociometric cohesion, reflecting team quality

d) Selection and Crossover

The selection operator is responsible for finding the best individuals within a population to perform reproduction, prioritizing those most adapted to the environment [24]. Because of its probabilistic features, the GAs may find the best individual at any time in the run, and this information can be lost when genetic operators are applied during the evolutionary process. To solve this problem two operators are merged: the method of roulette-wheel selection and elitism, which preserves the best individuals found during evolution. The crossover operator performs well when applied to problems of matching, since it preserves the position of information during recombination. The process randomly selects one cut-off point and the children receive a portion of the genetic code from each parent at that cut-off. The higher the probability of crossover, the more new individuals will be inserted into the population. For this study we used a crossover probability of 60% to 65%.

e) Mutation Operator

This study used the shift mutation type, which randomly switches the positions of two genes in the chromosomes. The mutation selects two genes at random, removes the first and then shifts all the genes between the empty position of the gene removed up to and including the second selected gene. The gene first removed is then inserted into the former position of the second gene, left empty by the shift. Probability of mutation, ranging from 0% to 100%, indicates the likelihood of the mutation operator's acting. A small probability of mutation makes it possible to reach any point in the search space, while a high value can make the search essentially random. The probability used in this study was between 0.1% and 5%.

f) Stop Criteria

The strategy adopted in this study was: the execution of the algorithm is terminated when a number G of generations have the same solution, i.e., if the GA finds a solution in the i-th generation and it remains the best solution for the G subsequent generations, the algorithm is then considered to have converged and is restarted until it has the three teams with best fitness. The result the manager receives is an indication of teams with highest fitness in terms of the criteria of selection, crossover and mutation. From these results, the managers decide which team to choose according to their personal considerations.

3.5 Prototype

The prototype comprised two layers: the graphical user interface and the controller which communicates between the engine supported by the genetic algorithm and the graphical interface. The graphical user interface layer consists of a web page where the user fills in the questionnaires. The control layer receives the command information from the user and sends the device's application, the genetic algorithm. The control layer contains the heuristic described below.

1. Set the number of members in each team, according to the number of people in the organization, allowing teams of up to 8 people.
2. Generate the initial population, searching the registered users in the database of the organization. The data refer to the criteria set by the manager when he describes the team context (skills, roles, temperament, and number of team members).
3. Carry out the assessment of chromosomes.
4. For teams formed, where one of the members has a temperament opposite to the other members, apply the depreciation function.
5. Sort the chromosomes in ascending order, according to the assessments and reductions allocated.
6. Apply the crossover function and evaluate the new population generated.
7. Present the team formed from the requirements requested in order to assist the manager in supporting the formation of teams.
8. Extract the team from the population of chromosomes formed.
9. Repeat steps 3 to 9, until teams are formed.

The results with the highest fitness generated by the GA are presented.

4 Simulated Experiment

A simulated experiment was conducted for the purpose of evaluating the proposed solution: the results of the teams generated by implementing a GA that supports work team formation. The simulated experiment was conducted in three stages based on Design of Experiments (DOE) by Cox & Reid [26] and culminated in findings on the feasibility of implementing this proposal to conduct and manage matching recommendations to support the process of forming a team. Experimental simulation is suggested when it is considered impractical or even impossible to perform experiments. From analysis of the simulation methodology proposed by Chwif [27], a flowchart was built up to explain the logic of the simulation (Figure 2).

From the real world data, a conceptual model is created, which for this study will be represented by the Team Composer approach. This step entails a clear understanding of the prototype to be simulated. The scope of the approach, its assumptions and level of detail are checked. In the second stage, the conceptual model is converted into a computational model by using a computer language or a commercial simulator. In our case, the prototype (GA) we built was used. What thus occurred was what is called implementation of the conceptual approach. Finally, in the analysis of results stage, the computational model is ready for the experiments to be performed, giving rise to the experimental model design.

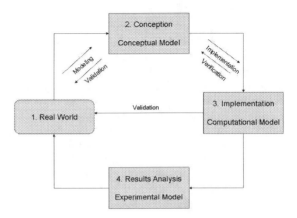

Fig. 2. Methodology used in the simulated experiment

The real world is considered the starting point and is always taken as the basis for validation and verification. In this context, we considered twenty-six individual, hypothetical users with the widest variety of attribute profiles and variance from the lowest to the highest value in the quantification of these attributes, thereby forming a population with variable characteristics. This stage is performed to verify the computational model where it is implemented, noting whether the effects of application are in accordance with the conceptual approach.

In implementing the simulation the following decision was taken: the GA is run 10 times. This was done to improve the reliability of the results, because in a simulation the GA can arrive at a local minimum and therefore not converge. Each simulation supposedly yields a properly developed population and, at the end of 10 simulations, the best individuals are ranked and the 3 teams with the highest fitness values are recommended to the manager.

Analysis of the results sought to provide answers on the correlation between the number of simulations required and the generation in which the GA found the optimal solution, showing the quality of convergence found between the teams. The graphs in Figures 3 and 4 show the results, where the x axis gives the number of simulations and the y axis gives the generation in which the GA found the optimal solution, in other words, the generation where the GA found the teams with best skills.

For purposes of comparative analysis, two approaches were implemented under the GA to generate the initial population. The results shown by the graph in Figure 3 were removed in the preliminary analysis where the concern was not to define the initial population, but to assess the results. Accordingly, in this first analysis the genetic algorithm did not check whether all the teams were in the initial population generation, leaving it as diversified as possible. However, the algorithm can be seen to find the optimal solution after about twenty generations ten times, which indicates slowdown and low convergence between teams.

The graph in Figure 4 shows that, in most cases, the algorithm found the optimal solution in less than ten generations, indicating that the problem should be solved through the way delimitation of the initial population is generated, so as to increase convergence between the teams.

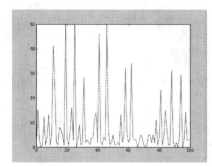

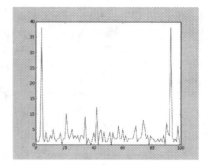

Fig. 3. 1st Results from team generation   **Fig. 4.** 2nd Results from team generation

4.1 Analysis of Results: Benefits and Limitations of the Proposal

Some observations on the case studies and simulated experiment should be discussed. The use of sociometric tests can provide descriptive results on social relations with considerable accuracy. Moreover, it is possible to gain a lot of information with only a few questions, as seen in implementation of the sociometric test. However, such results do not give the grounds and reasons for social relations. Since they are essentially quantitative, qualitative aspects of major importance were not considered.

Applying a sociometric test seems really quite simple, because the number of questions is usually very small. On the other hand, there are a number of difficulties from a technical point of view, because analysis of results becomes very arduous. Furthermore, care must be taken with certain pre-conditions for applying the test, such as the need to have a minimum mutual understanding between team members.

It is essential to ensure the absolute confidentiality of results; they will be known only to the manager. Problems may arise if affinity issues are revealed and this can cause conflict between team members.

In determining the scope of the treatment to test the proposed solution – team of software developers – the third attribute – mapping of individual skills – was delimited most precisely, thus enabling more highly refined assembly of teams that meet the manager's criteria. As the skills mapping has a specific scope it is not possible to perform tests supporting the team formation in other contexts.

5 Conclusions and Future Work

Teamwork is increasingly valued in organizations as a means of promoting knowledge sharing. In this context, organizations can use Social Matching productively to support the process of team formation.

In addition to proposing a social matching model, a prototype was developed to facilitate implementation. The proposed approach is based on the social matching mechanism, to leverage support for team formation. This approach endeavored to state explicitly features designed to assist the manager when deciding the composition of a new team, so as to add value to the organization and contribute to recognition for specialists who possess the desired attributes.

To assess the feasibility of the proposed solution, we first analyzed a case study with data from 27 Masters students taking the discipline of Neuropedagogy and Informatics at Rio de Janeiro Federal University (UFRJ/PPGI). This first study aimed to evaluate the use of questionnaires to measure the temperament of individuals in a team. The results were important to show a first attempt to validate the definition of the test that promoted the attribute temperament.

A second case study was conducted with eleven students taking the discipline of Agile Modeling (UFRJ/PPGI). This case study aimed to assess the correlation between the questionnaires on temperaments and roles. After the participants answered the questionnaires, data were analyzed and two teams were formed: one with similar temperaments and another with identical temperaments. The results showed evidence of the effectiveness of the social matching measure generated by agreement between the questionnaires, which fosters the second attribute addressed in the proposal – the roles – in addition to correlating these with the attribute temperament.

Finally, an experiment was simulated using the last attribute of the proposal in the social mix: skills. This experiment was carried out to validate the purpose of the proposed focus: formation of teams of software developers. This was done using a prototype based on the genetic algorithm technique.

The main contribution of these studies was to confirm that it is possible to apply the two theories tested: temperaments and roles to support the formation of teams. Through the correlation between the results of a questionnaire that guides these two theories, we found input to provide managers with an identification of subjective factors in the profile of each individual. Furthermore, the simulated experiment showed the importance of adding the skills attribute addressed in the third approach, which allows the choice to build on a mapping of individual skills, further refining the matching process.

Another consideration are the results from an application using genetic algorithms and theoretical foundations that have parameters shown to support the formation of teams as requested by the team manager. The result from implementation of Team Composer is the formation of three teams that have most skills based on the attributes set by the manager. However, it is important to consider that the results are not intended to measure the success of the process of support for team formation, because that measurement depends on a number of other factors, such as the manager's assessment of the process of implementation the specific task by the team so selected. This should be addressed in a longer experiment.

Future work should include other attributes relevant to the process of supporting team formation, such as checking the availability of an individual by creating an agenda, allowing the manager to choose only those who have not engaged in other projects, in addition to displaying a ranking of team scores, showing the process of team composition in greater detail, allowing the manager to make a more refined choice.

Analysis of the teams generated can be further refined by studying fitness isotopic curves, which will allow a larger set of parameters and operators in the formula and increase the team equity. In addition, further studies should be conducted to test, monitor and evaluate the quality of results in the formation of teams generated by the prototype.

Finally, we believe that the approach described in this paper has potential for application not only in organizations, but also in other areas involving group formation in virtual environments, such as Education.

References

[1] Malinowski, J., Keim, T., Wendt, O., Weitzel, T.: Matching People and Jobs: A Bilateral Recommendation Approach. In: Proceedings of 39th Hawaii International Conference on System Sciences, vol. 6 (04-07), p. 137c (2006)

[2] Robbins, H., Finley, M.: The New Why Teams Don't Work: What Goes Wrong and How to Make It Right, 2nd edn. Berrett-Koehler Publishers, San Francisco (2000)

[3] Terveen, L.G., McDonald, D.W.: Social Matching: A Framework and Research Agenda. ACM Transactions on Computer-Human Interaction (ToCHI) 12(3), 401–434 (2005)

[4] Katzenbach, J.R., Smith, D.K.: The Wisdom of Teams: Creating the High-Performance Organization. Collins Business, New York (2003)

[5] Fleury, M.T., Fleury, A.: Estratégias Empresariais e Formação de Competências: um quebra-cabeça caleidoscópio da indústria brasileira (in Portuguese). Atlas, São Paulo (2000)

[6] Perrenoud, P.: Dez novas competências para ensinar (in Portuguese). Artmed Editora, Porto Alegre (2000)

[7] Fuks, H., Mitchell, L.H.R.G., Gerosa, M.A., Lucena, C.J.P.: Competency Management for Group Formation in the AulaNet Learning Environment. In: Favela, J., Decouchant, D. (eds.) CRIWG 2003. LNCS, vol. 2806, pp. 183–190. Springer, Heidelberg (2003)

[8] Jones, A., Issroff, K.: Learning technologies: Affective and social issues in computer-supported collaborative learning. University College, London, UK (2005)

[9] Pasqualli, L.: Human Types: Personality Theory (2000), http://CopyMarket.com

[10] Myers, I.B.: Gifts differing. Consulting Psychologists Press, Palo Alto (1980)

[11] Jung, C.G.: The Development of Personality (1967). Routledge, London (1991)

[12] Jung, C.G., Hull, R.F.C.: Psychological Types, a revised edn. Routlege, London (1991)

[13] Straw, J.: The 4 Dimensional Manager: DiSC Strategies for Managing Different People in the Best Ways, 1st edn. Berrett-Koehler Publishers (2002)

[14] Justo, F.S.C.: Teste de Caráter ao Alcance de Todos (in Portuguese). Editora Escola Profissional La Salle, Canoas (1996)

[15] Belsin, R.M.: Team Roles at Work. Butterworth Heinemann, Oxford (1993)

[16] Berscheid, E., Reis, H.T.: Attraction and close relationships. In: Gilbert, D.T., Fiske, S.T., Lindzey, G. (eds.) The Handbook of Social Psychology, pp. 193–254. Oxford University Press, Oxford (1998)

[17] Gorla, N., Lam, Y.W.: Who Should Work With Whom? Building Effective Software Project Teams. Communications of the ACM 47(6), 79–82 (2004)

[18] Karn, J., Cowling, T.: A Follow up Study of the Effect of Personality on the Performance of Software Engineering Teams. In: Proceedings of the 2006 ACM/IEEE International Symposium on Empirical Software Engineering (ISESE 2006), Rio de Janeiro, Brazil, pp. 232–241 (2006)

[19] Kruchten, P.: The Rational Unified Process: An Introduction, 2nd edn. Addison-Wesley Professional, Reading (2000)

[20] Rajendran, M.: Analysis of team effectiveness in software development teams working on hardware and software environments using Belbin Self-perception Inventory. Journal of Management Development 24(8), 738–753 (2005)

[21] Belbin, R.M.: Beyond the Team. Butterworth Heinemann, Oxford (2000)
[22] Koza, J.R.: Genetic programming: on the programming of computers by means of natural selection. MIT, Cambridge (1992)
[23] Labidi, S., Quarto, C.C., Jaques, P.: Inferring Socio-Affective Factors and Cooperation Capacity in Computer Assisted Collaborative Teaching/Learning Environments. In: The 6th IEEE International Conference on Advanced Learning Technologies, Danvers, MA, USA, vol. 60, pp. 608–612 (2006)
[24] Mitchell, M.: An Introduction to Genetic Algorithms. MIT Press, Cambridge (1996)
[25] Labidi, S., Lima, L.C.M., Souza, C.M.: Modeling Agents and their Interaction within SHIECC: A Computer Supported Cooperative Learning Framework. Revue D'information Scientifique et Technique 10(1), 41–54 (2000)
[26] Cox, D.R., Reid, N.: The Theory of the Design of Experiments. Chapman & Hall/CRC, Boca Raton (2000)
[27] Chwif, L., Medina, A.C.: Modelagem e Simulação de Eventos Discretos: teoria e aplicações (in Portuguese). Authors' ed., São Paulo, Brazil (2006)

Understanding Open Source Developers' Evolution Using TransFlow

Jean M.R. Costa, Francisco W. Santana, and Cleidson R.B. de Souza

Faculdade de Computação, Instituto de Ciências Exatas e Naturais,
Universidade Federal do Pará, Belém, Pará, Brasil, 66075-110
wertherjr@gmail.com, {marcel,cdesouza}@ufpa.br

Abstract. Due to the success of many Open Source Software projects, both the industry and the academic community are interested in understanding how such software is produced. Particularly, there is interest in understanding how these communities are organized, maintained, and also how the contributors join and evolve their roles in these projects. However, few studies have been conducted around the evolution of the developers in the communities, i.e., how they reach roles of greater importance, and how the software changes over time through this evolution. This paper describes TransFlow, a tool aimed to support the integrated study of the evolution of both: the software itself and the developers' participation in open source projects. This integrated study is a requirement since the software architecture may support or hinder developers' participation in the project. We describe the rationale for building TransFlow and illustrate how its features can be used to study open source projects.

Keywords: Open Source, Software Evolution, Developers Evolution, Role Migration.

1 Introduction

Over the years, there has been a growing interest on open source software (OSS) due to their quality and reliability [1]. Apache, for example, is the most used web server in the world, currently being used by over 50% of web domains [2]. As several other open source projects become more and more relevant, both the industry and the academic community are interested in understanding how such software is produced. For instance, some research has been conducted focusing on the evolution of open source software [3][4]. Results indicate that open source software do not evolve according to Lehman's laws of software evolution [3].

In this paper we argue that, in addition to understanding the (open source) software evolution, it is necessary to understand *how developers evolve in the project*, i.e., how developers change their contributions and become more or less integrated into the community. This understanding will potentially enable the identification of the required changes in the development process or in parts of the software [5] to create mechanisms that facilitate the engagement of new developers in the community. This will eventually lead to higher quality software systems and a sustainable open source

L. Carriço, N. Baloian, and B. Fonseca (Eds.): CRIWG 2009, LNCS 5784, pp. 65–78, 2009.

community. Moreover, as OSS projects are a typical example of distributed collaborative work, to understand the processes of OSS development can potentially facilitate the adaption of similar approaches in organizations where developers are geographically distributed.

Nevertheless, few studies explore the evolution of developers in these communities. The main studies include [6] [7] and [8]. In [6] open source communities were analyzed to understand how software developers change their roles in the communities by analyzing their contributions in forums and code repositories. [7] presents Augur, a tool that enable the analysis of both the artifacts produced and the activities related to that artifacts. They present examples of shifts from the periphery to the core of the project taking into account the structure of the software. Finally, [8] discuss role advancement and migration processes of three large open source projects: Mozilla, Apache and NetBeans, illustrating how developers' participation in these projects changes over the time.

So far, most of the approaches, however, do not take into account the architecture of the software being created, and this is an important aspect that facilitates or hinders the engagement of new members in both open source and proprietary software projects. For instance, McCormack and colleagues [5] compared the software architecture of an open source project (Linux) and the architecture of another software whose development was, at the time, proprietary (Mozilla). They reported significant differences between the two software architectures – Linux was more modular than Mozilla – and discussed how Linux's modularity facilitated the joining of new contributors in the project. In general, open source software (OSS) tends to have more modular architectures than proprietary software, because this is necessary to enable the development of distributed software [9]. Baldwin and Clark [10] investigated the causes and effects of this increased modularity in open source projects and suggested that modular architectures create more opportunities for developers to contribute and that encourages the entry and permanence of contributors. In sum, previous work suggest that modular architectures allow developers to contribute to the project by focusing in a particular part of it without worrying about the impact of their changes in the rest of the software architecture.

Based on the observations above about the importance of software architecture, we present in this paper TransFlow, a tool aimed to enable a more complete analysis on the evolution of open source projects because it supports the analysis of software developers' evolution by taking into account the architecture of the software being created. In fact, this is the main difference between TransFlow and previous tools: it collects data from configuration management repositories, analyzes this data (for instance, by recreating the software architecture at the time of the commit), and presents different visualizations that allow a researcher to investigate how a software project evolved and how the developers evolved alongside the project. TransFlow allows a researcher to understand the effects of the software architecture on developers' evolution.

The rest of the paper is organized as follows. We begin by presenting a brief literature review including previous work that motivated our approach. Then, TransFlow is described: we present its main features and describe our approach in details. In the following section we present an initial evaluation that we performed showing some usage scenarios using real large OSS projects: JEdit, MegaMek and JBoss. Finally, our conclusions and final remarks are presented.

2   Related Work

The success of many open source projects instigated both the software industry and the academic community to study these projects to understand how such software is produced. One aspect that has been studied is the open source evolution, i.e., the investigation of how these projects evolve, including their creation, the joining of new members and so on. Software evolution is a much-studied subject by the scientific community. The most prominent studies of software evolution have been conducted by Lehman [11] who identified a number of properties that have been described in eight laws, known as "Laws of Software Evolution". Later, additional research [4] [12] [13] reported similar results to those achieved by Lehman et al in different software projects. Meanwhile, recent studies found that open source projects have different characteristics regarding its software evolution. For example, [14] [15] [16] conducted case studies to analyze the evolution patterns of some OSS projects. Their results show super linear trends of growth, which contradicts earlier studies that argue that the software growth rate should be linear or decrease over time.

Most studies that address the software evolution focus on the evolution of the software itself, considering, for instance, the growth in size (LNOC) and the decay of the software architecture. Work by [14] [15] [16] are examples of studies that have adopted this approach. However, to have a full understanding of the open source projects is also necessary to understand the evolution of the community that develops the software and how they drive the system evolution [17]. In this case, only a few studies have been carried out. Among them, it is important to mention Gonzalez-Barahona et al. [16] who show how the information available in CVS repositories can be used to study the structure of the project and the connections between the developers. They conducted a case-study in the Apache project and their results indicate how the growth of the community of developers was proportional to the growing number of modules in the project. Note that in this paper it was investigated the correlation between the evolution of the OSS project with the evolution of the number of developers who develop it. This does not take into account the evolution of the developers in the community. In [18] it is argued that it is not enough to examine the evolution of a system at the highest level, but it is also necessary to consider the evolution of its subsystems because the development rate of the subsystems may be different from the system as a whole. This argument, used to point out the need to study the software structure, can also be used for the developers' community: it is not enough to consider only the community evolution, it is also necessary to investigate the evolution of the individual developers who are part of the community. Examples of studies focusing on developers' evolution include [6], [7], [19] and [20]. Ducheneaut [6] describes a tool, OSS Browser, and approach to study the evolution of open source developers, i.e., how developers moved from the periphery to the core of the project, gaining respect, admiration, and decision power in the Perl project. de Souza et al. [7] present Augur, a tool that allows one to analyze the complexity of software development artifacts and the complexity of the development process of these artifacts. They demonstrate how artifacts can reveal the relationship between the technical and social infrastructure of software development projects. In addition, they describe how Augur can be used to identify movements from the periphery to the core of the project, and vice-versa, through the analysis of developers' contributions to the project. Sarma et al. [19] presents Tesseract, a socio-technical dependency browser that utilizes cross-linked displays to enable exploration of relationships between artifacts, developers, and communications, instead of considering them independently. The tool was evaluated with GNOME project data. Although it supports the connection between artifacts and developers, it does not fully explore the software architecture of the project, for instance, by calculating coupling and cohesion metrics of the architecture. Finally, Gilbert and Karahalios [20] present CodeSaw, a tool that provides social visualization of distributed software development. The tool combines two different perspectives: the information contained in source-code repositories and the communication that takes place during the project. CodeSaw allows the user to compare the participation of the main developers considering the number of code lines added by each developer and the number of words written in e-mails exchanged in the mailing lists of the project.

All tools described above – OSS Browser, Augur, CodeSaw, and Tesseract – allow, to some extent, the correlation between the *software* evolution aspects with *developers'* evolution. However, these tools were not designed with this focus, so, it is not possible to have a full understanding of this correlation. In addition, these tools do not take into full account the software architecture of the project being analyzed[1]. And, as discussed by [9] [10] and [5], this architecture is fundamentally important to allow new developers to join the project: modular architectures allow developers to contribute to the project by focusing in one particular aspect of the software without worrying about the impact of their changes in the rest of the software architecture.

Aiming to enable a more complete analysis of software developers' evolution in OSS projects, we developed TransFlow. Transflow is a tool that collects source-code repositories data, analyze the results to reconstruct the software architecture considering all the historical data and present visualizations that enable the analysis of the *software* evolution *and* the *developers'* evolution during the project.

3   TransFlow

3.1   TransFlow's Visualizations

TransFlow is a tool that processes data collected from source code repositories and presents interactive visualizations of this data. It allows the investigation of how a software system, and its developers evolved over time. The tool obtains data for each commit performed by developers and, based on a historical analysis of the repository, allows one to reconstitute how the entire software was developed. Since the information is obtained from the project, four types of visualizations are presented, allowing the user to analyze the project based on the comparison of a view with the other ones. The four TransFlow visualizations are the following:

Scatter Plot (Figure 1). This visualization uses a vertical and a horizontal axes to show the correlation between variables by plotting data points. By default, each point corresponds to a single commit in the repository, but this can be modified to represent as many commits as the user wants in a single data point. Each item has a color that

---

1 As discussed, Augur and Tesseract provides some support to this aspect, but in a very limited way.

corresponds to a specific developer, allowing the easy identification of a developer's contribution. For example, the user can identify the developers who contributed the most to the project by just looking at the predominant colors in the visualization;

Graph (Figure 2). This shows the dependencies between the software developers with a graph that represents a social network. In a social network, each node represents a software developer, and each edge indicates the coordination requirements between these developers [21]. These requirements, and the social network, are created by analyzing files that have been changed by two or more developers and also the dependencies between the files. This technique is described in details in [21]. In the visualization, edge thickness indicates the degree of required coordination.

Line View (Figure 3). This allows the comparison of two variables using a horizontal and a vertical axes to plot the view. It's the simplest visualization implemented, but can be used to identify trends of growth and decrease of variables in a simple and effective way; and

TreeMap (Figure 4). This visualization presents the software's hierarchical structure using nested rectangles. This view is generated based on the information of which file paths were modified by each developer. For example, if a developer modified a file whose path was /Project/src/io/FileCreator.java, this string would be processed and the hierarchical structure would be added to the visualization. Thus, the user can see how the software structure evolves over time. In addition, the visualization:

- Shows all files modified by one or more developers;
- Identifies the files that were changed more often; and
- Identifies the files that were modified by more developers.

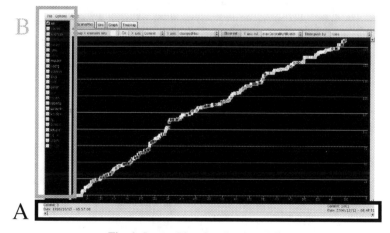

Fig. 1. Scatter Plot visualizations

Each one of these visualizations is a graphical representation of a certain time period of the project. However, the user may not be interested in viewing the information about the entire project, but instead to focus on a given period of time (e.g.,

the period between two software releases). To support that, TransFlow has a slider
(Figure 1 (A)), which enables the user to select the time period of interest. As the user
drags the slider, all views are dynamically modified to reflect the change. Based on
how the visualizations are modified when the slider changes, the user can make inferences about events that occurred in the project over time.

Another way to filter the information presented in the visualizations is by selecting
which developers must be displayed. This can be done from a panel (Figure 1 (B))
that has a listing of all the project developers that worked during the period of analysis. In this panel, there is a label and a checkbox for each developer, and when the
user (un)selects the checkbox, the visualizations are modified to (un)show the corresponding information of the developers. The color of the label is used to identify the
developers in the visualizations. For example, if the developer label is blue, the data
points in the ScatterPlot view that represents the commits made by this developer will
also be blue. Hence, the user can easily identify in the visualizations which developers
were responsible for what changes, allowing the comparison between the contributions of a developer with other developers' contributions.

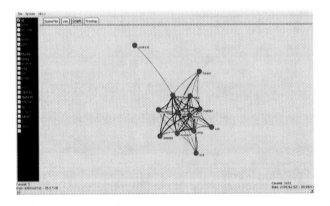

Fig. 2. Graph visualization

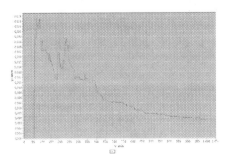

Fig. 3. Line view

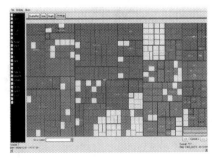

Fig. 4. TreeMap visualization

3.2 Transflow's Software Architecture

TransFlow is designed in a 4-layer architecture: Collection, Analysis, Visualization and Persistence. We designed TransFlow to separate the functions corresponding to each of these layers, so that the changes are restricted to the functions of each layer.

Collection. As configuration management systems repositories maintain records of all commits made by developers, it is possible to obtain data from each commit performed in a project. TransFlow collects data from each commit in a given period of time defined by the user. Currently, the tool supports access to Subversion (SVN) repositories. Based on the data collected, TransFlow generates a graph to represent the software and another graph to represent the dependencies between the developers. The use of graphs has been gaining a huge popularity, due to its intrinsic power to reduce a particular system to its components and relationships [16]. In the graph representing the software, nodes represent the source code files and edges indicate the dependencies between these files. In the graph that represents dependencies between developers nodes correspond to the developers and edges indicate the need for each developer coordinate its activities with the other ones. These graphs are generated for each commit found in the repository and were built using the technique described by Cataldo et al. [21]. The technical graph, whose nodes represent artifacts, is generated taking into account the frequency in which the artifacts have been modified together. Thus, a dependency exists between two artifacts when both were changed together several times. To infer the social dependencies matrix operations are performed.

Analysis. Once the data is collected and the graphs are created, metrics are used to extract relevant information about the software itself and its developers. Some metrics are applied to each file, while others are applied to a set of files, grouped according the software revision associated. The metrics that are displayed depending on the granularity level desired by the user. In the case of metrics applied to a set of files because of a commit, it is possible to choose the average or the maximum value found in the set. For instance, the degree centrality metric [22] when applied to a software revision considers the centrality of each modified file and, finally, considers either the average or the maximum centrality obtained for all files. Some of the metrics implemented in TransFlow obtain information directly from the data in the repository, such as the number of added or modified source code files. Other metrics are implemented through the analysis of the generated graphs. Examples of these metrics include: degree centrality, betweenness centrality, density and clustering coefficient [22].

Persistence. When all data are collected and all information is obtained, everything is persisted in a relational database. To persist this information is important because it makes easier to retrieve and handle this information without accessing the configuration management repository again.

Visualization. Finally, the last layer is responsible for the displaying the visualizations: data is obtained from the database and presented in 4 different types of view (see previous section). For each of these visualizations there are filters that the user can use to view only the information that interests him.

3.3 TransFlow's Evaluation

By using TransFlow, the user is able to identify evolution patterns in the software itself and, more importantly, he can also identify how developers are evolving in the project. In this section we present some examples that illustrate how TransFlow can be used to analyze software development projects. All examples are based on the analysis of real OSS projects.

Developers' Evolution
We initially focus on TransFlow's main contribution: the possibility to investigate not only the software evolution, but also the evolution of the developers involved in the software. Therefore, we present some ways to investigate the evolution of developers' participation.

- **Source files added/modified (by developer).** The measure of the number of source code files added /modified can be used to identify how the developers begin to modify *new files*, i.e., how they expand their contributions to other parts of the software. Figure 5 shows an example of this metric being used to investigate the JBoss project. It is possible to note the evolution of 2 developers. One, who started to contribute earlier, clearly evolved faster than the other one: he changed a larger number of files (this information is represented in the y axis). In addition, the visualization allows the user to identify the frequency in which new files have been modified by the developers based on the distance between a square and another one in the view (the x-axis which represents time). For instance, in Figure 5 both developers stopped submitting contributions for the project during a time period: this can be inferred by the gaps between squares (commits) in the figure.

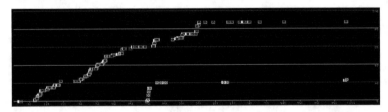

Fig. 5. Developers' evolution in JBoss Project considering the source code files added

- **Maximum centrality achieved.** All software projects have a set of most important files, which several other depend on directly or indirectly [23]. These files usually form the core of the project and require extra attention when changed because they can impact several other files. To identify how the developers start to modify these central files is crucial to understanding how they evolved in the project. Based on the analysis of the maximum degree centrality or the maximum betweenness centrality [22] achieved by each developer, it is possible to identify how each developer "reached" these important artifacts. Figure 6 presents an example of this metric, considering the maximum degree centrality. In the figure, the gray points correspond to the maximum values reached by a developer, while the white points indicate the maximum centrality in the project when the commit was performed. This indicates that, during the period of analysis, this particular developer never changed the most important files. In contrast, Figure 7 presents a different developer who started changing less important files and, during the period of analysis, was able to change the central files. This can be observed because his commit (the gray one) is located alongside the white point.

– **Developers' specialization/generalization.** Developer's specialization refers to which software module a developer modifies [24]. High specialization indicates that a developer modified the same modules over time, while generalization indicates that another developer modified multiple modules. While specialization is considered the best way to innovate, generalization is seen as a way for the developer to acquire a better knowledge about the architecture, allowing a better management of the interfaces and coupling between the artifacts [24]. Figure 8 shows an example of TransFlow being used to identify specializations/generalizations using the metric "number of source files added / changed". In the figure, it is possible to distinguish two periods of contribution from a developer. In the first period, the ascending sequence of squares shows that the developer changed at least a new file for each commit, thus characterizing a generalization. In the second period, the horizontal sequence of squares shows that the developer was changing files that he had already modified, thus characterizing a specialization: the same modules were being changed during the period.

– **Shift from core to periphery.** Since TransFlow allows the identification of the dependencies between the developers for any time period of the project, it is possible to investigate how the developers shift from the periphery to the core of the project, i.e., how the code they produce start to assume roles of greater importance in the project. In this paper, the core and periphery groups are defined according to the pattern of interactions between the developers. So, a group whose the interactions network is dense and cohesive is defined the core group, while the one that is more sparse and unconnected is called the periphery group [25]. Figure 9 shows the graphs corresponding to two release periods of the MegaMek project. Looking at the pictures is possible to see that the developer *hawkprime*, which in figure 9(A) was in the periphery, shifted to the core, as shown in Figure 9(B).

Entry Points
An important aspect to be considered in the analysis of OSS development is where in the software the first contributions of the developers are performed. By analyzing how and where the developers start to contribute to the project, it is possible to identify patterns of joining that can be used as a reference for new developers who want to know which part of the software they can start modifying. We illustrate this aspect using TreeMap visualizations of two large open-source projects: JEdit and MegaMek. In those views, we used a feature of TransFlow to identify which files have been modified in the developers' first contributions. Each rectangle is a file, and the ones that are inside thicker lines are files that are changed in the first commits of each developer. For these rectangles, outlined by thicker lines, the color indicates how many

developers changed the file: the darker the color, the larger number of developers who modified that file. Thus, a white rectangle indicates that a single developer changed the corresponding file, while rectangles with darker colors indicate that many developers began their project participation modifying these files. In MegaMek, whose TreeMap is shown in Figure 10, it is possible to see that there are several files that tend to be modified by the developers in their early contributions. In other words, MegaMek has several "entry points" for new developers.

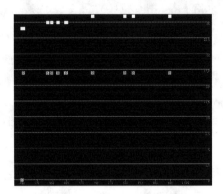

Fig. 6. Maximum centrality reached

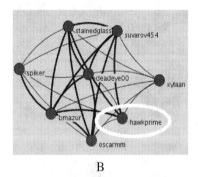

Fig. 7. Developer evolving and reaching the central files

Fig. 8. Specialization and generalization in the project JBoss

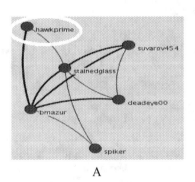

A
B

Fig. 9. Graphs showing a developer shifting from periphery to core

Figure 11 presents the TreeMap of the JEdit project. In the figure it is possible to see that few files were changed by more than one developer in their first contributions. In other words, if a developer wants to join JEdit there is no clear "entry point" where most developers start. He can start in any particular package. Of course, based on the Figure, it is also possible to infer that some particular packages are usually developers' first option.

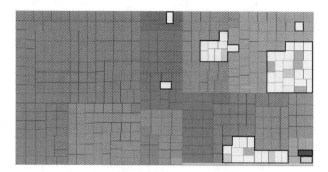

Fig. 10. MegaMek TreeMap

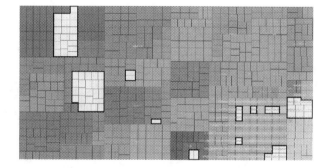

Fig. 11. JEdit TreeMap

Developers' Knowledge

Another aspect that TransFlow allows one to analyze is developers' knowledge about the software architecture. As a software system evolves, it becomes increasingly difficult to understand it completely, so it is common for developers to specialize in specific modules. Uncovering developers' degree of knowledge about the software modules is interesting not only for the OSS community members, that could know who is the best person to make specific modifications or answer questions about a particular set of modules, but also for researchers, who could study knowledge management issues related to specific modules.

The following aspect – specialization in the software architecture – is illustrated in Figures 12 and 13 below. Figure 12 displays the files modified by two developers of the JEdit project, while Figure 13 displays the files changed by two developers of the MegaMek project. Comparing the JEdit figures with the MegaMek figures is possible to see that JEdit developers tend to specialize more than those of MegaMek. In MegaMek, their changes often occur spread in the entire software.

Software Evolution

One of the main features of TransFlow is to present visualizations that allow the analysis of how the software evolved over time. As metrics are applied to each commit in the configuration management repository, is possible to evaluate how the metrics values changed as the software was being developed. Next, we present some of the metrics implemented in TransFlow that can be used to analyze the software evolution.

- **Source files added/modified.** One way to analyze the software evolution is from the files counting of source code that were added/modified over time [26]. Figure 1 shows an example of using this metric, considering the files that have been modified.

- **Density.** This metric indicates the level of interdependence between the software artifacts. Using this visualization it can be identified whether the software artifacts are becoming more or less uncoupled over time. For this reason, this metric can be used as a way to identify the need for refactoring in a system. Figure 3 shows a line view in which is possible to see the evolution of the JBoss project density.

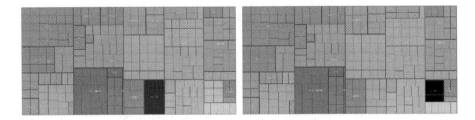

Fig. 12. JEdit TreeMaps indicating specialization

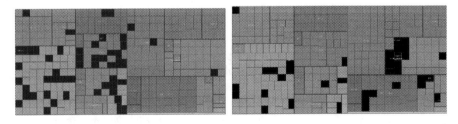

Fig. 13. MegaMek TreeMaps indicating generalization

4   Conclusions and Future Work

This paper described TransFlow, a tool that collects data from source code repositories, analyzes the software (by calculating different metrics) and present interactive visualizations of a project evolution. TransFlow's main contribution is the ability to support the analysis of how the software evolved and, more importantly, *how the developers evolved in the project*. We described TransFlow's main features and its software architecture. In addition, we presented TransFlow's initial evaluation by presenting examples from three open source software projects, namely: JEdit, MegaMek and JBoss. These examples illustrate how TransFlow can be used to identify certain aspects of the development of OSS projects.

Through this initial evaluation it is possible to observe that TransFlow can be used to analyze different aspects of the software evolution and developers' participation. For instance, Figures 6 and 7 present two different developers' participation and clearly illustrate how only one of them was able to reach the "core" of the project, i.e., only one of them was able to change the most central files in the open source project.

Nevertheless, during the evaluation we identified the need for further improvements in our visualizations. In fact, currently there are four different types of visualization in TransFlow: line view, scatter plot, graph and TreeMap. Each of these views focuses on the analysis of one or more aspects of the software development. Thus, other types of visualization could be used to examine additional aspects. For instance, a pie chart would be useful, because could be used to compare the proportions between the database values. For example, whereas the graph entire circumference represent the total number of commits made in the system, the pie "slices" indicate the number of contributions made by each developer. Then, it would be possible to easily identify which developers were more active in the development. We are extending TransFlow to support new visualizations and configuration management repositories, but we are also analyzing additional open source projects.

Acknowledgments

This research was supported by the Brazilian Government under grants CNPq 479206/2006-6, CNPq 473220/2008-3 and by the Fundação de Amparo à Pesquisa do Estado do Pará (FAPESPA) through "Edital Universal N.° 003/2008".

References

1. Hoepman, J., Jacobs, B.: Increased Security Through Open Source. Communications of the ACM, 79–83 (2007)
2. Netcraft, http://www.netcraft.com
3. Scacchi, W.: Understanding Open Source Software Evolution: Applying, Breaking, and Rethinking the Laws of Software Evolution (2003)
4. Bauer, A., Pizka, M.: The Contribution of Free Software to Software Evolution. In: Proceedings of the International Workshop on Principles of Software Evolution (IWPSE), Helsinki (2003)
5. MacCormack, A., Rusnak, J., Baldwin, C.Y.: Exploring the Structure of Complex Software Designs: An empirical Study of Open Source and Proprietary Code. Management Science 52(7), 1015–1030 (2006)
6. Ducheneaut, N.: The reproduction of open source software programming communities, Ph.D. dissertation, Berkeley (2003)
7. de Souza, C., Froehlich, J.E., Dourish, P.: Seeking the Source: Software Source Code as a Social and Technical Artifact. In: Conference on Supporting Group Work (2005)

8. Jensen, C., Scacchi, W.: Role migration and advancement processes in ossd projects: A comparative case study. In: ICSE '07: Proceedings of the 29th International Conference on Software Engineering, pp. 364–374. IEEE Computer Society, Washington (2007)
9. Raymond, E.S.: The Cathedral & the Bazaar. O'Reilly, Sebastopol (1999)
10. Baldwin, C.Y., Clark, K.B.: The architecture of participation: Does code architecture mitigate free riding in the open source development model? Management science Working paper; Harvard Business School 52(7), 1116–1127 (2006)
11. Lehman, M., Ramil, J., Wernick, P., Perry, D., Turski, W.: Metrics and laws of software evolution-the nineties view, pp. 20–32 (1997)
12. Bennett, K.H., Rajlich, V.T.: Software Maintenance and Evolution: a Roadmap. In: Finkelstein, A. (ed.) The Future of Software Engineering, ICSE 2000, Limerick, Ireland, June 4-11, pp. 75–87. ACM Press, New York (2000)
13. Pfleeger, S.L.: The Nature of System Change. IEEE Softw. 15(3), 87–90 (1998)
14. Godfrey, M.W., Tu, Q.: Evolution in open source software: a case study. In: Proceedings of the International Conference on Software Maintenance, pp. 131–142 (2000)
15. Robles, G., Gonzalez-Barahona, J.M., Centeno-Gonzalez, J., Matellan-Olivera, V., Rodero-Merino, L.: Studying the evolution of libre software projects using publicly available data. In: Proceedings of the 3rd Workshop on Open Source Software Engineering (2003)
16. Gonzalez-Barahona, J.M., Lopez, L., Robles, G.: Community structure of modules in the Apache project. In: Proceedings of 4th Workshop on Open Source Software Engineering (2004)
17. Nakakoji, K., Yamamoto, Y., Nishinaka, Y., Kishida, K., Ye, Y.: Evolution patterns of open-source software systems and communities. In: Proceedings of the International Workshop on Principles of Software Evolution, pp. 76–85. ACM Press, New York (2002)
18. Fernandez-Ramil, J.C., Perry, D.E. (eds.): Software Evolution and Feeback: Theory and Practice, ch. 9. John Wiley & Sons, Chichester
19. Sarma, A., Maccherone, L., Wagstrom, P., Herbsleb, J.: Tesseract: Interactive Visual Exploration of Socio-Technical Relationships in Software Development. In: ICSE 2009 (to appear, 2009)
20. Gilbert, E., Karahalios, K.: CodeSaw: A Social Visualization of Distributed Software Development. In: Baranauskas, C., Palanque, P., Abascal, J., Barbosa, S.D.J. (eds.) INTERACT 2007. LNCS, vol. 4663, pp. 303–316. Springer, Heidelberg (2007)
21. Cataldo, M., Wagstrom, P., Herbsleb, J.D., Carley, K.: Identification of coordination requirements: Implications for the design of collaboration and awareness tools. In: ACM Conference on Computer-Supported Cooperative Work, pp. 353–362 (2006)
22. Wasserman, S., Faust, K.: Social Network Analysis: Methods and Applications (Structural Analysis in the Social Sciences). Cambridge University Press, Cambridge (1995)
23. De Souza, C., Redmiles, D.F.: An Empirical Study of Software Developers Management of Dependencies and Changes. In: International Conference on Software Engineering, pp. 241–250 (2008)
24. Von Krogh, G., Spaeth, S., Lakhani, K.: Community, joining, and specialization in open source software innovation: a case study. Research Policy 32(7), 1217–1241 (2003)
25. Borgatti, S., Everett, M.: Models of core/periphery structures. Social Networks 21, 375–395 (1999)
26. Fernandez-Ramil, J.C., Perry, D.E. (eds.): Software Evolution and Feeback: Theory and Practice, ch. 9. John Wiley & Sons, Chichester

Exploring the Effects of a Convergence Intervention on the Artifacts of an Ideation Activity during Sensemaking

Victoria Badura[1,2], Aaron S. Read[1], Robert O. Briggs[1], and Gert-Jan de Vreede[1]

[1] Center for Collaboration Science, University of Nebraska Omaha
[2] Chadron State College
vbadura@csc.edu, {aread,rbriggs,gdevreede}@mail.unomaha.edu

Abstract. Organizations must enlist the efforts of groups to solve important problems. Six patterns of collaboration describe group behavior as they work towards solutions. The convergence patterns of collaboration-- reduce and clarify are key in helping a group focus effort on issues that are worthy of further attention. These group behaviors have not been extensively studied in the literature. In the current study, we further this research effort by exploring and characterizing the effects of a fast focus intervention on an ideation artifact. Researchers conducted an observational case study of executives addressing a real task within a large organization. Analysis of the problem statements generated during a problem identification and clarification session revealed several implications about convergence activities. The FastFocus thinkLet was found to reduce the number of concepts from 246 down to 30, a reduction of 76%. Ambiguity was reduced from 45% in the ideation artifact to 3% in the converged artifact. A serendipitous event in the field allowed researchers a window into comprehensiveness, showing that the FastFocus thinkLet may not contribute to comprehensiveness as much as was previously thought. Finally implications for brainstorming instructions were identified that may contribute to reduced ambiguity in ideation artifacts.

Keywords: Collaboration, Convergence, Group Support Systems, Information Overload.

1 Introduction

IS/IT professionals and the decision makers they serve face a continuing need to solve problems. Problem solving has, therefore, been the focus of much research in the IS/IT and Management literature. Authors in a number of disciplines have proposed systematic frameworks that characterize decision-making and problem-solving processes (e.g. Ackoff 1970; Brightman 1980; Bross 1953; Dewey 1933; Dunker 1945; Simon 1977). All of these models begin with one or more phases during which actors attempt to make sense of their problem(s). Ackoff (1970) for example, includes a phase for formulating objectives, while Brightman (1980) include a diagnosis phase. Weick (1993) describes sensemaking as an effort to create order and make retrospective sense of events. In this work, we use the term, sensemaking, as a label for activities a group executes in order to identify unacceptable conditions or symptoms, to understand what

L. Carriço, N. Baloian, and B. Fonseca (Eds.): CRIWG 2009, LNCS 5784, pp. 79–93, 2009.
© Springer-Verlag Berlin Heidelberg 2009

may have caused those conditions or symptoms to exist, and to understand which of those causes are under the control of or can be influenced by the group.

Collaboration Engineering researchers have identified six patterns of collaboration that manifest as groups work through a problem solving process, and characterize how groups move through their activities (Kolfschoten et al. 2006): Generate, Reduce, Clarify, Organize, Evaluate, and Build Consensus. In the context of sensemaking, for example, groups generate symptoms of and possible causes for their problems. They reduce that set to the symptoms and causes they deem worthy of further attention. They clarify their problem statements to establish shared meaning for the concepts with which they work. They organize concepts in a variety of ways (e.g. identifying cause and effect relationships, classifying symptoms by stakeholders affected). They evaluate which of the symptoms and causes are most plausible, and build consensus around which of the problem statements are most plausible, and so should be carried forward into the rest of the problem solving activities. Some authors combine the reduce and clarify patterns under the more general heading, convergence (Davis et al. 2007).

Much work has also been undertaken to understand consensus building, e.g. (Dunlop 1984; Innes and Booher 1999; Rosenau 1962). A great deal of research has also been done on the idea generation pattern by groups (e.g. Diehl and Strobe 1987; Diehl and Stroebe 1991; Fjermestad and Hiltz 1998-1999; Fjermestad and Hiltz 2001; Graham 1977; Kolfschoten and Santanen 2007; Lindgren 1967; Osborn 1963). Ideation activities, however, often generate more ideas than a group will find useful. Indeed, some ideation techniques encourage group members to contribute poor ideas in addition to good ones (Osborn 1963). In the knowledge economy, however, attention may be an organization's scarcest resource (Grover and Davenport 2001). To reduce cognitive load, therefore, idea generation activities are typically followed by convergence activities. In an effort to improve the team productivity, therefore, researchers have recently begun to focus on convergence activities, - reduction activities intended to move a group from having many ideas to a focus on fewer that they deem worthy of more attention, and clarification activities intended to move the group from less to more shared understanding of the concepts in their shared set (e.g. Davis et al. 2008; Davis et al. 2007).

Researchers have noted that group members sometimes find convergence activities to be painful and time consuming process (Chen et al. 1994; Easton et al. 1990). Facilitators report that guiding a group through a convergence process requires more facilitation skills than any of the other patterns of collaboration (Vreede and Briggs 2005). A better understanding of convergence and the techniques used to invoke it might provide a conceptual foundation for creating new convergence methods that require less time and effort to execute, and that produce superior results, and so improve the performance of groups working together toward high-value goals. Better understandings of convergence might also lead to better understandings of the other patterns that often precede and follow convergence activities. It may be, for example, that observed variations in the artifacts of ideation activities would be discovered to correlate with the ease and effectiveness of subsequent convergence activities. Such insights could lead to better understandings of ideation. Observed variations in converged artifacts may also be discovered to correlate with variations in the organizing, evaluating, and consensus building activities that follow a convergence, and so lead to better understandings of those patterns.

In this paper, we report an exploratory study to address the research question, "What is the effect of a convergence intervention on the artifacts of an ideation activity?" In the next section we summarize recent literature on convergence activities. We then report a field study where we observed senior executives for a multi-national financial institution as they generated a set of possible barriers to achieving the organization's strategic goals, and then converged on a set of problem statements that they would use as the basis for action planning. We next analyze and compare the artifacts of the ideation activity and the artifacts of the convergence activity to characterize the effects the convergence activity had on the artifact of the ideation session. We discuss the implications of our exploration for research and practice, and conclude by outlining directions for future research.

This paper makes two key contributions to the convergence literature: a) it provides reliable, repeatable approaches to for measuring and comparing the effects of convergence interventions on the artifacts of ideation sessions; and b) it characterizes those effects in detail for a particular ideation technique and a particular convergence technique as they were used by senior executives for a real sensemaking task in a real workplace. Such exploratory studies may help clarify the phenomena of interest for theory building and experimentation in future convergence research.

2   Background

2.1   Convergence

The convergence pattern of collaboration is composed of two sub-patterns: reduce, which is defined as moving a group from having many ideas to a focus on fewer the group deems worthy of more attention, and Clarify, which is defined as moving a group from less to more shared understanding of the concepts they are using (Davison et al. 2005). The reduce sub has two sub-patterns:

- *Select-* choosing among a subset from a pool of concepts that will receive further attention;
- *Abstract* - reducing the number of concepts under consideration through generalization, abstraction, or synthesis and then eliminating the lower-level concepts in favor of the more general concept

2.2   Importance of Reduction

Davis and colleagues (2007) proposes that one of the main purposes of convergence is to reduce the cognitive load of follow-on activities. Cognitive load theory proposes an architecture for human memory to explain why humans have difficulty wielding many concepts simultaneously (Sweller et al. 1998). This architecture specifies a working memory with limited capacity for holding information. Approximately seven items of information can be held in working memory and processed at a time (Miller 1956). This set of information is what we can process at any given moment. The pieces of information in this set can be combined and stored in long-term memory as schema.

When these pieces of information are successfully stored in Long-Term memory, learning has taken place. Groups, therefore, who can wield fewer concepts at one time, may have more cognitive power to devote to other reasoning.

2.3 Importance of Clarification

Another important goal of convergence is to create shared understanding amongst group members (Valacich and Junh 2006). Shared understanding allows individuals with differing expertise and information to create new ideas and solution (Arias et al. 2001). Pendergast and Haynes (1995) conceptualize shared understanding as a shared context model which puts bounds around what a group focuses on in a meeting. Groups that do not have shared understanding of concepts may devolve into unproductive conflict. Briggs, Reinig, and Nunamaker (2003), for example, cited a case where a group negotiating requirements for a new online bookstore reached an impasse over system rights that should be granted to "affiliates." It turned out that there were five orthogonal meanings for the word, "affiliate" in that group. Until they reached clarity on those five concepts, they could not move forward with decisions about access rights.

3 Methods

3.1 The Research Venue

This study took place in the headquarters of a large Financial Services organization headquartered in the Midwestern United States. The company is more than 100 years old, and has more than 2,000 employees.

3.2 Participants

Nine senior executives from across the organization participated in the work practice we observed. Participants came from both the headquarters (Office) and from the sales force (Field). They represented several different departments from customer service and underwriting to information technology. Participants ranged in age from 37 to 50 years of age, with an average age of 42.9 years. On average, participants had 23 years of experience in this industry. Eight of the executives were female; one was male. All were born in the U.S.A., and all were U.S. citizens who spoke English as their first language. All but one had lived in the U.S. for their entire lives. One had lived 24 of her 44 years overseas.

3.3 The Task

The executives met at the world headquarters to identify barriers that prevented the organization from achieving its strategic objectives. Over time, the financial services industry in general, and the insurance segment in particular have changed. The company hired an outside paid facilitator to design and conduct the workshop. The workshop design took participants through several activities to create a list of high-priority problem statements organized under key themes.

In the first workshop activity, the group generated ideas for 15 minutes in response to the question, "What are the key problems that block us from obtaining our strategic objectives?" This activity was based on an idea generation technique called Free Brainstorm (Briggs and Vreede 2001). For this technique, each participant starts with a different page. Participants may add a single idea to a page. After contributing their first idea, participants must exchange pages with another participant. They may add one more idea, and then they must exchange pages again. For this interaction, participants used group support system software that managed the page-swapping automatically. Participants received the following instructions, in accordance with the rules of the FreeBrainstorming technique:

"...After you submit each idea, the computer will take your page away, and bring you back a different page. You may other people's ideas on the new page. You may respond to other people' ideas in one of three ways:

- *You may argue with the idea. All you contributions are anonymous, so nobody's feelings were get hurt if you challenge a problem statement*
- *You may agree with a problem statement by building on it or adding detail*
- *You may be inspired to a completely new problem statement.*

Your goal is to generate as many possible problem statements as you can in the short time we have available."

The brainstorming technique allowed all participants to contribute simultaneously and anonymously. It yielded an artifact containing 124 comments. Some of the comments contained a single problem. Some contained multiple problems. Some contained no problems. This artifact we refer to as the Raw Data Set.

The second workshop activity was based on a convergence technique called Fast-Focus (Briggs and Vreede 2001). During this activity, each participant held a different page from the brainstorming activity. At the beginning of the activity, the facilitator displayed an empty list on a public projection screen where all participants could read it. The participants received the following instructions in accordance with the rules of the FastFocus technique:

"...Each of you is now looking at a different page. You each hold a different part of our brainstorming conversation in your hands. In a moment, I will call on each of you in turn. I will ask you, 'What is the most important problem on the page in front of you that blocks us from achieving our strategic objectives?'

The moderator then began a round robin, calling on each participant in turn. Each participant contributed a single problem statement to the public list. In accordance with the rules of FastFocus, the facilitator screened their contributions for five things:

> *Redundancy*: Participants could only add ideas that were not already on the list
> *Relevance*: Participants could only add problem statements to the list
> *Clarity*: Ideas had to be expressed concisely and unambiguously
> *Levels of abstraction*: Contributions could not be so vague as to be inactionable, nor could they be so specific that they obscured the root causes of the problems.

Criticism: Participants were not allowed to contribute criticisms of the ideas added by other participants. They were told they would have an opportunity to evaluate the ideas later.

When the facilitator was satisfied that a problem statement was non-redundant, relevant, clear, and at a useful level of abstraction, he typed it onto the public list. When every participant had had a chance to contribute one idea to the public list, the facilitator asked participants to switch pages. The facilitator then said, "What is the most important problem on the page in front of you that has not already been added to the public list?" Participants added new items to the list as they discovered them.

The facilitator repeated the cycle of contributions and page exchanges until no participant contributed any new problem statements for three successive exchanges of pages. The FastFocus activity lasted 45 minutes and yielded an artifact containing 29 problem statements. This artifact we refer to as the Reduced Data Set.

The participants continued working with the facilitator on a number of subsequent activities. These, however, are outside the scope of this study and so are not reported.

3.4 Metrics

(Davis et al. 2007) proposed five measures for comparing convergence techniques:

1. *Speed* – The time required to execute the convergence activity for a given data set.
2. *Comprehensiveness* – The degree to which all useful ideas from the idea generation artifact are included in the converged artifact, and the degree to which less-useful ideas are excluded from the converged artifact.
3. *Shared Understanding* – The degree to which group members ascribe similar meanings to the concepts in the converged artifact.
4. *Reduction* – The degree to which the converged artifact contains fewer concepts than the idea generation artifact.
5. *Refinement* – the degree to which concepts are expressed parsimoniously and un-ambiguously.

Speed is an attribute of an activity, and shared understanding is an attribute of a group. Comprehensiveness, reduction, and refinement are all attributes of artifacts. It is on these measures that this exploratory study focuses.

In order to measure the effects of the convergence intervention on the ideation artifact, we conducted a multi-step step analysis three phases: a) characterizing the ideation artifact; b) characterizing converged artifact; and c) comparing the converged artifact to the converged artifact.

3.5 Characterizing the Ideation Artifact

Several steps were required to characterize the Ideation artifact.

First, two coders counted the number of comments contributed to the brainstorming activity.

Next, in order to allow researchers to count the number of unique problem statements in the brainstorming artifact, the coders disaggregate the Raw Data Set into single unique problem statements. For this activity, two coders independently evaluated each contribution contained in the Raw Data Set Coders were required to disaggregate the contributions according to the 15 rules presented in Appendix A. To summarize these rules, coders were required to extract from each contribution each unique noun/verb/object combination that identified a desired state or outcome that had not yet been achieved. Coders were directed to mark a contribution as ambiguous if its wording allowed for two or more grammatically correct interpretations, which would lead to differences in the disaggregation. When coders deemed a comment ambiguous, they were required to demonstrate its ambiguity by writing down at least two different grammatically correct interpretations allowed by the wording of the comment.

Coders trained for 4 hours on a sample data set, until they were able to achieve concurrence that exceeded 90%. They then each independently disaggregated R with an inter-coder reliability of .93. Disaggregation required about 4 hours for each coder.

The coders met to resolve their differences. In all cases where coders disagreed on the disaggregation of unambiguous comments, they were able to demonstrate that one or the other of the coders had broken a disaggregation rule. In those cases, they agreed to the disaggregation that met the constraints of the rules.

In two cases, the coders discovered that the statement upon which they disagreed was, in fact ambiguous, each coder having derived a different grammatically sound interpretation meaning for the contribution, leading to different interpretations. The coders marked those items as ambiguous. Coders then reviewed the ambiguous statements together to determine the most plausible interpretation, given the context of the brainstorming problem. Then they disaggregated the statement. For those where no single interpretation seemed more plausible than the others, coders did not disaggregate the statement. If the two coders disagreed on the interpretation of an ambiguous contribution, a third coder resolved the issue. Comments were also deemed to be ambiguous if they contained a rhetorical question rather than a problem statement (e.g. Do we really know what our customers want?). Statements were also deemed to be ambiguous if they offered a solution disguised as a problem or (e.g. "We should get a faster computer," implying that current computers are too slow to handle our current workload), or a solution disguised as a problem (e.g. "We have not yet switched over to a service oriented architecture," offering a solution, but not implying any particular problem the solution would help).

Resolution of disagreements and ambiguities required about 5 person hours. Fifty-six of the 124 comments were discovered to be ambiguous by the coders. Of those, 41 were discovered by both coders, 13 by a single coder, and 2 were discovered to be ambiguous during the resolution phase.

Finally, two coders worked together to identify redundant problem statements in the disaggregated set. They followed a constraint that if either coder found a contribution to be non-redundant, it was to be given the benefit of the doubt and retained in the set. This activity yielded the Disaggregated Raw Data Set (R-D). Researchers then counted the number of unique problem statements included in the Ideation Artifact.

3.6 Characterizing the Converged Artifact

Researchers followed the same protocol to characterize artifact of the Convergence Activity, referred to as the Reduced Data Set, as they used to characterize the artifact of the Ideation Activity, referred to as the Raw Data Set. This data set contained 30 items, of which only 1 was discovered to be ambiguous by both reviewers.

3.7 Comparing and Contrasting the Ideation and Converged Artifacts

Three coders worked together to determine whether each of the unique disaggregated problem statements from the Raw Data Set were reflected in the text of the problem statements in the Reduced data Set, and which were not. This activity required about 5 person hours. Researchers then counted how many unique problem statements from the Ideation Artifact were included in the Reduced artifact.

Table 1. Characteristics of the Ideation Artifact

Description of Variable	Variable Name	ID	Value
Number of Raw Comments Contributed	Raw Comments	RC	124
Number of Raw Off Topic Comments (no problem statement disaggregated from the comment)	Raw Off-Topic	RC-Off	1
Number of Raw On-Topic Comments	Raw On-Topic	RC-On	123
Number of Raw Comments deemed unambiguous	Raw Unambiguous	RCU	70
Number of Raw Comments deemed to be ambiguous	Raw Ambiguous	RCA	56
Ratio of Ambiguous Comments to Raw comments (RCA / RC)	Raw Ambiguity Ratio	A/RC	0.45
Number of Raw Disaggregated Problem Statements	Raw Disaggregated Problems	RCD-Problems	275
Number of Unique Raw Disaggregated Problem Statements	Raw Disaggregated Unique	RCD-Unique	246
Number of Redundant Raw Disaggregated Problem Statemnts	Raw Disaggregated Redundant	RCD-Redundant	29
Number of Raw Problem Statements disaggregated from unambiguous Raw Comments	Raw Disaggregated Problems - From Unambiguous Comments	RCD-Unambig	157
Number of Raw Problem Statements disaggregated from Raw Ambiguous Comments	Raw Disaggregated Problems - From Ambiguous Comments	RCD-Ambig	151
Average Number of Problem Statements Disaggregated from each Raw Comment (RCD-Unique / RC)	RC Disaggregation Ratio	U/RC	1.98
Average Number of Problem Statements Disaggregated from Each Unambiguous Raw Comment (RCD-Unambig / RCU)	RC Disaggregation Ratio - unambiguous	U/RC-Unambig	2.2

4 Analysis

Table 1 presents a collection of variables that characterize the Ideation Artifact. Participants contributed 124 comments, of which only one was deemed to be off-topic. Fifty six of the comments (45%) were discovered to be ambiguous. Seventy of the comments (55%) were demonstrated to be unambiguous when independent coders produced identical disaggregations of their content.

The 124 comments disaggregated into 275 simple problem statements. Twenty-nine of those were deemed to be redundant, yielding 246 unique disaggregated problem statements. Approximately half the unique problem statements were disaggregated from unambiguous comments; the other half from ambiguous comments. On average, coders identified 2.2 problem statements per unambiguous contribution, and 2.6 unique problem statements per ambiguous contributions.

Table 2. Characteristics of the Ideation Artifact

Description of Variable	Variable Name	ID	Value
Number of Reduced Problem Statements	*Reduced Problems*	*RD*	30
Number of Reduced Problem Statements deemed unambiguous by coders	*Reduced Problems - Unambiguous*	*RDU*	29
Number of Reduced Problem Statements deemed to be ambiguous by any coder	*Reduced Problems- Ambiguous*	*RDA*	1
Ratio of Ambiguous Reduced Problem Statements to Number of Reduced Problem Statements	*Reduced Ambiguity Ratio*	*A/RD*	0.03
Number of Reduced Disaggregated Problem Statements	*Reduced Disaggregated Problems*	*RDD-Problems*	59
Number of Unique Reduced Disaggregated Problem Statements	*Unique Reduced Disaggregated Problems*	*RDD-Unique*	58
Number of Redundant Reduced Disaggregated Problem Statements	*Redundant Reduced Disaggregated Problem Statements*	*RDD-Redundant*	1
Number of Reduced Problem Statements disaggregated from unambiguous Reduced Problem Statements	*Reduced Disaggregated Problems - Unambiguous*	*RDP-Unambig*	55
Number of Reduced Disaggregated Problem Statements from ambiguous Reduced Problem Statements	*Reduced Disaggregated Problems - Ambiguous*	*RDD-Ambig*	\ 4
Average Number of Problem Statements Disaggregated from each Reduced Problem Statement (RDD-Unique / RD)	*RD Disaggregation Ratio*	*U/RD*	1.96
Average Number of Problem Statements Disaggregated from Each Unambiguous Raw Comment (RDD-Unambig / RDU)	*RD Disaggregation Ratio - unambiguous*	*U/RD-Unambig*	1.90
Average Number of Problem Statements Disaggregated from each Ambiguous Raw comment (RDD-Ambig / RDA)	*RD Disaggregation Ratio – ambiguous*	*U/RD-Ambig*	4.00

Table 2 presents a collection of variables that characterize the Converged Artifact. This set contains 30 Reduced Problem Statements, of which only one was discovered to be ambiguous by the coders. Those 30 statements disaggregated to 59 simple problem statements. One statement was found to be redundant. Four of these items were disaggregated from ambiguous statements.

5 Discussion

The Ideation artifact contained 246 unique problem statements, while the Convergence Artifact contained 59 unique problem statements, a 76% diminution. Considering the six patterns of collaboration The FastFocus technique therefore produced a demonstrable reduction effect.

Forty-five percent of the contributions in the Ideation Artifact were discovered to be ambiguous. Only three percent of the problem statements in the Convergence Artifact were discovered to be ambiguous. The FastFocus technique therefore produced a demonstrable decrease in ambiguity. Given that ambiguity in problem statements would increase the likelihood that different people in a group would ascribe different meanings to the same problem statements, a reduction in ambiguity is likely to correspond to an increase in shared meaning. Therefore, the FastFocus technique may also have produced a clarification effect.

Davis and colleagues 2007 proposed five possible measures for comparing convergence techniques. Among those were Reduction, Comprehensiveness, and Refinement. In this paper contributes an approach to measuring reduction effects. The measures we advanced for measuring ambiguity contribute an approach to measuring an aspect of shared understanding.

Table 3. Examples of Ideas Eliminated from the Reduced Data Set and Their Votes Two Months Later

There are insufficient human resources and training resources for the workforce	10
The delivery of our customer service is sub-standard	8
The bi-weekly cutoff detracts from customer service (also a work process consideration)	7

Given the constraints of working with real groups on real tasks in the field, we did not anticipate that we would gather any measures of comprehensiveness. However, a serendipitous event in the field, however, provided a modest indicator. Another group of executives in the organization conducted a workshop on the same topic. Two months after the initial session, both groups combined their Converged Artifacts and voted on the importance of the problem statements from both lists. Some of the items from the second group did not appear in the Converged List of the first group, but did appear in their Ideation Artifact. When the members of the first group rated the items they had excluded from their reduced artifact, they rated several of their excluded ideas higher than most of the ideas in their Converged set (See Table 3). This result has two possible explanations. First, it may be that the FastFocus technique is not fully comprehensive in the extent to which it extracts all useful ideas from the ideation set. The other possibility is that something transpired during the intervening two months that caused participants' priorities to change. Given that the FastFocus technique is often selected for its comprehensiveness, this aspect of the technique should be explored more fully in future research.

Ambiguity was determined to be a significant factor in problem statement generation. By identifying a method for recognizing grammatical ambiguity, the researchers identified implications for brainstorming techniques. Ambiguity is a barrier to shared understanding. It is an indicator that an artifact is not as refined as needed. Creating thinkLets that reduce ambiguity will increase refinement and at a minimum remove barriers to ambiguity.

5.1 Implications for Brainstorming Techniques

The study provides several interesting implications for Brainstorming instructions. One implication is that results should be less ambiguous if participants are instructed not to use the word "Need" in their responses. Several of the coded responses became ambiguous because they included the word "need." Including "need" in a problem statement, leads directly to ambiguity. If for example, a participant types, "We need faster computers." There is automatic ambiguity. Is this a solution to a problem, perhaps of too much work or is this an implied problem "the computers are too slow?" Instructing participants to structure problem statements without the word need will avoid this source of ambiguity. A second implication for Brainstorming Instructions is that participants should be instructed to keep problem statements as short as possible. Causal statements should also be kept short. If a participant identifies a causal statement of one or two layers deep, they should be instructed to break those problems up into simple related causes. For example "There is too much work, so people are working overtime, which results in reduced moral, leading to lower productivity" could be broken up by participants into several single layer causal statements. Analysis of the disaggregated artifact showed that there were significantly higher numbers of ideas disaggregated from complex problem statements than from simple problem statements.

5.2 Limitations and Future Research

We note that our dataset was limited to one group. The reliability of our findings would be strengthened by analysis of multiple groups using the same thinkLet for reduction in the same context. We also note that it would be useful to compare the performance of thinkLets on the same dataset. We also did not control for how the group worked. We only observed them. This may limit our ability to generalize our findings to other settings.

We were not in a position to rate every idea, including the (unnecessary) ideas to see whether or not all the necessary ideas made it into the final converged set of ideas. To be able to better understand how well a convergence thinkLet captured statements that the group deemed worthy of further attention, future research should have group members rate every statement for importance, not just those that made it to the fast focus. This would provide solid evidence that important ideas were included and not excluded.

Future research should investigate the effects of prompting a group for proper problem statement phrasing (for example if, beginning each problem statement by "a lack of" would reduce ambiguities in the artifact). If ambiguity was significantly reduced, this would be an important finding for practitioners. Future research should also seek to understand the causal mechanisms that link convergence activities to an effective reduction of ideas and ambiguity that is measured here.

5.3 Conclusion

By exploring a reduced artifact generated by a group in a natural organizational setting, we have demonstrated several aspects of the Fast Focus thinkLet's convergence performance in a field setting. While some of these relate to previous research, we have found a way to describe the ambiguity of a converged document that will be useful in future research. We have discovered implications for brainstorming techniques that should improve the output for future groups. We have also provided several directions for future research.

References

Ackoff, R.L.: A concept of corporate planning. Wiley-Interscience, New York (1970)

Arias, E., Eden, H., Fischer, G., Gorman, A., Scharff, E.: Transcending the individual human mind: Creating shared understanding through collaborative design. ACM Transactions on Computer-Human Interaction 7(1), 84–113 (2001)

Briggs, R.O., Vreede, G.J., Nunamaker, J.F.: Collaboration Engineering with thinkLets to Persue Sustained Success with Group Support Systems. Journal of Management Information Systems 19(4), 31–63 (2003)

Briggs, R.O., Vreede, G.J.d.: ThinkLets: A Pattern Language for Collaboration GroupSystems Corporation (2001)

Brightman, H.J.: Problem solving: A logical and creative approach Business Publishing Division, College of Business Administration, Georgia State University, Atlanta (1980)

Bross, I.D.F.: Design for Decision. The Macmillan Company, New York (1953)

Chen, H., Hsu, P., Orwig, R., Hoopes, L., Nunamaker, J.F.: Automatic concept classification of text from electronic meetings. Communications of the ACM 37(10), 56–73 (1994)

Davis, A.J., Badura, V., Vreede, G.J.d.: Understanding Methodological Differences to Study Convergence in Group Support Systems Sessions. In: Proceedings of the 14th Collaboration Researchers' International Workshop on Groupware, Omaha, Nebraska (2008)

Davis, A.J., Vreede, G.J.d., Briggs, R.O.: Designing thinkLets for convergence. In: Proceedings of the 13th Americas Conference on Information Systems (AMCIS-13), Keystone, Colorado (2007)

Davison, R.M., Vreede, G.J.d., Briggs, R.O.: On peer review standards for the information systems literature. Communications of the AIS 16(4), 967–980 (2005)

Dewey, J.: How We Think Health, New York (1933)

Diehl, M., Strobe, W.: Productivity Loss in Brainstorming Groups: Towards the Solution of a Riddle. Journal of Personality and Social Psychology 53(3), 497–509 (1987)

Diehl, M., Stroebe, W.: Productivity Loss in Idea-Generation Groups: Tracking Down the Blocking Effect. Journal of Personality and Social Psychology (61), 392–403 (1991)

Dunker, K.: On Problem Solving. Psychological Monographs 58(5), 47–90 (1945)

Dunlop, J.T.: Dispute resolution: Negotiation and consensus building. Auburn House, Westport (1984)

Easton, G.K., George, J.F., Nunamaker Jr., J.F., Pendergast, M.O.: Using two different electronic meeting system tools for the same task: An experimental comparison. Journal of Management Information Systems 7(1), 85–101 (1990)

Fjermestad, J., Hiltz, S.R.: An Assessment of Group Support Systems Experimental Research: Methodology and Results. Journal of Management Information Systems 15(3), 7–149 (1999)

Fjermestad, J., Hiltz, S.R.: Group Support Systems: A Descriptive Evaluation of Case and Field Studies. Journal of Management Information Systems 17(3), 115–159 (2001)

Graham, W.K.: Acceptance of ideas generated through individual and group brainstorming. Journal of Social Psychology 101(2), 231–234 (1977)

Grover, V., Davenport, T.H.: General Perspecitives on Knowledge Management: Fostering a Research Agenda. Journal of Management Information Systems 18(1), 5–21 (2001)

Innes, J.E., Booher, D.E.: Consensus Building and Complex Adaptive Systems. Journal of the American Planning Association 65(4), 412–423 (1999)

Kolfschoten, G.L., Briggs, R.O., Vreede, G.J.d., Appleman, J.H.: Conceptual foundation of the ThinkLet Concept for Collaboration Engineering. International Journal of Human-Computer Interaction 64(7), 611–627 (2006)

Kolfschoten, G.L., Santanen, E.L.: Reconceptualizing generate thinkLets: The role of the modifier. In: Proceedings of the Hawaii International Conference on System Sciences, pp. 6–16 (2007)

Lindgren, H.C.: Brainstorming and the facilitation of creativity expressed in drawing. Perceptual and Motor Skills (24) (1967)

Miller, G.A.: The magical number seven plus or minus two: Some limits on our capacity for processing information. Psychological Review 63(2), 81–97 (1956)

Osborn, A.: Applied Imagination: Principles and Procedures of Creative Problem Solving Scribner, New york (1963)

Pendergast, M.O., Hayne, S.C.: Alleviating convergence problems in group support systems: The shared context approach. Computer Supported Cooperative Work 3(1), 1–28 (1995)

Rosenau, J.N.: Consensus-building in the American National Community: Some hypotheses and some supporting Data. The Journal of Politics 24(4), 639–661 (1962)

Simon, H.A.: The New Science of Management Decision. Prentice-Hall, NJ (1977)

Sweller, J., van Merrienboer, J.J.G., Paas, F.G.W.C.: Cognitive architecture and instructional design. Educational Psychology Review 10(3), 251–296 (1998)

Valacich, J.S., Junh, J.H.: The effects of individual cognitive ability and idea stimulation on idea-generation. Group Dynamics: Theory, Research, and Practice 10(1), 1–15 (2006)

Vreede, G.J.d., Briggs, R.O.: Collaboration Engineering: Designing repeatable processes for high-value collaboration tasks. In: Proceedings of the 38th Hawaii International Conference on System Sciences, pp. 1–10. IEEE, Los Alamitos (2005) (On CD)

Weick, K.E.: The Collapse of Sensemaking in Organizations: The Mann Gulch Disaster. Administrative Science Quarterly 38(4), 628–652 (1993)

Appendix A

RULES of DISAGGREGATION

DEFINITIONS:

PROBLEM: A problem is a desired state or outcome that has not yet been attained (e.g. Our customers do not feel satisfied, although we want them to).

SYMPTOM: A symptom is some unacceptable condition that implies some desired state or outcome that has not yet been attained (e.g. Customers are returning products)

GUIDLINE: Any time you must infer missing words to make a complete noun-verb-object problem statement, add them explicitly to the problem statement in parenthesis.

RULES:

1. Identify Verbs and Nouns

- UNIQUE NVO: Each unique noun/verb/object combination that identifies a state or outcome that has not yet been attained will be disaggregated into simple problem statements.
- ACCEPT SYMPTOMS: identifying a symptom is an important aspect of framing a problem, so they are acceptable as problem statements and disaggregated using the same rules.
- MEANINGLESS VERB: Objects will not be disaggregated when doing so renders the verb meaningless. E.G. "We feel torn between our duties to home and work." We cannot disaggregate the problem to "we feel torn between our duties to home" and "we feel torn between our duties to work," because doing so renders the concept, "feeling torn between" meaningless.
- ACCEPT REDUNDANCY: If people say the same thing in multiple ways in the same comment, disaggregte both wordings into simple problem statements. Redundancy will be removed in a later activity. Tag it as redundancy for later review.

2. Break Phrases

- BREAK OUT FIRST CAUSES from CAUSE-AND-EFFECT: When presented with a causal chain, disaggregate first causes into stand-alone problem statements. For example, in the comment, "Under-staffing leads to overwork which leads to low morale," "(we have) understaffing" would be disaggregated as a separate problem statement.
- DISTRIBUTE CAUSES: Distribute first causes across their consequent problem statements to make stand-alone problem statements. Add the first cause to the problem statement in parentheses so that the ideas can be understood in subsequent analysis steps. e.g (rushed work) causes low satisfaction. For example, in the comment, "Under-staffing leads to overwork which leads to low morale," The understaffing cause would be paired with the overwork effect, and overwork as a cause would be paired with low morale, as follows: "(Understaffing) leads to overwork, and "(Overwork leads to) low morale." Thus, this comment would be broken out into three problem statements.

- MULTIPLE CAUSES: All first causes must be combined when distributing across consequent problems, because don't know whether either of the causes would invoke the effect on its own. For example, in the statement, "Understaffing and overwork cause low morale," it is not possible to know whether understaffing or overwork each cause low morale, or whether both together cause low morale. Therefore, in this case, both Understaffing and Overwork would be broken out as first causes (see above) but low morale would be broken out, "(Understaffing and overwork) cause low morale. Thus, this comment would be disaggregated into three problem statements, but they would differ from the three statements illustrated in the previous rule.
- NO THREE DEEP: Distributed causes will not span more than one cause and one effect - there will be no causal chains of three or more clauses in the disaggregated problem statements.
- RETAIN DEPENDENT CLAUSES - don't break out dependent clauses as separate problem statements unless they are part of a causal chain (Dependent clauses explain what, how or when) (the dependent clause may contain a problem statement when it begins with "to")

3. Determine ambiguity

- BRACKET AMBIGUITY: If you find the language of the contribution allows multiple grammatically-sound interpretations that could lead to different disaggregation structures, depending on which interpretation you adopt, make your best interpretation, disaggregate accordingly, but put your responses in brackets. Explain the ambiguity by stating at least two possible grammatically sound interpretations allowed by the wording of the original comment.
- BRACKET SOLUTIONS DISGUISED AS PROBLEMS: If a problem statement is rhetorically stated as a solution, break it out as a problem statement and bracket it. It's it is not possibly reliably distinguish solutions from problems during disaggregation. They can be sorted out in a subsequent activity. Explain your brackets.
- IGNORE THE POSITIVE: Do not include positive clauses or phrases in the set of disaggregated problem statements.
- RHETORICAL QUESTIONS: If a statement contains a rhetorical question that can be re-framed as a problem statement, mark the comment as ambiguous and re-frame it as a problem statement, and disaggregate it.

4. Ways to Resolution conflicts and ambiguities.

- NEGOTIATE. Discuss the possible interpretations. If one is clearly more plausible ACCEPT that interpretation and disaggregate accordingly.
- SYNTHESIZE a better interpretation together.
- DISALLOW the comment as too ambiguous if it is not possible to determine which interpretation is more plausible, DO NOT DISAGGREGATE the comment.
- ARITRATE. If coders discover that they have an irreconcilable disagreement about the meaning of the comment, submit it to a third coder for resolution.

Social Knowledge Management in Practice: A Case Study

Ricardo A. Costa[1,2], Edeilson M. Silva[2], Mario G. Neto[2], Diego B. Delgado[1,2], Rafael A. Ribeiro[2], and Silvio R.L. Meira[1,2]

[1] C.E.S.A.R – Recife Center for Advanced Studies and Systems
[2] UFPE – Federal University of Pernambuco
{rac,ems5,mgn,dbd,rar,srlm}@cin.ufpe.br

Abstract. This case study describes the effects of using a Web Based Social Network (WBSN) approach to Knowledge Management in a Brazilian software development organization. During this work it was verified how the previous approach to Knowledge Management, a wiki based approach, did not work to this organization and how users voluntarily migrated from this environment to the WBSN. In addition to the WBSN environment, a Knowledge Management process has been proposed, with some metrics related to its phases. These metrics have been monitored since October 2006 and analyzed in order to verify the efficiency of this approach. In order to give a better understanding of the concepts related to Social Networks and Knowledge Management, it is presented a brief introduction to each one of them, including an evaluation of existing Knowledge Management approaches.

1 Introduction

Organizations are aiming to increase industrial competitiveness. Wherefore they have been searching for new ways to improve their productivity, the quality of their products and cost reduction. One of the many ways to achieve those goals is to manage and share efficiently the knowledge built by their members. In this context, social networks have shown signs of being an efficient tool to proliferate individual and explicit knowledge, even improving the tacit knowledge dissemination, helping to capture organizational knowledge based on the knowledge of each of its employees.

Staab [12] has affirmed that social networks environments are a very good mechanism to promote more interactivity between individuals. The capturing of tacit knowledge trough interaction tools inside the Web Based Social Networks also allows extending this tacit knowledge. According to Davenport [6], in order to create new knowledge is necessary to expose humans to new information, so they can process it and generate new knowledge inside their minds.

This work presents the a.m.i.g.o.s, a Web Based Social Network (WBSN) environment, and the experience of its use as the main tool to foster the communication, collaboration and knowledge management inside C.E.S.A.R, an Innovation Institute located in Brazil.

Besides this introduction, the reminder of this paper is organized as follows: a short explanation about knowledge management concepts, including its proposed process and metrics; definitions and benefits regard the use of a Social Network environment; the case study developed during this work, with a overviews of C.E.S.A.R Organization,

L. Carriço, N. Baloian, and B. Fonseca (Eds.): CRIWG 2009, LNCS 5784, pp. 94–109, 2009.

a.m.i.g.o.s web based social network, a mapping between a Knowledge Management (KM) process and the a.m.i.g.o.s functionalities and effects of its use in C.E.S.A.R during the two phases of this case study; Finally it presents the concluding remarks and directions for future works.

2 Knowledge Management

According to Choi and Lee [5], Knowledge Management (KM) in a software corporation is an opportunity to create a default perception language among software developers so they can interact, deal and share knowledge and experiences. The reduction in loss of Intellectual Capital from employees who leave the company; the cost reduction on the development of new products; and the increased productivity by making knowledge easily accessible to all employees are some of several benefits in using a Knowledge Management strategy.

Polanyi [11] categorized knowledge in tacit and explicit. The latter is basically what can be easily documented and distributed, while the tacit knowledge resides in the human mind, behavior and perception, and thus is difficult to be formalized and distributed. As expected, the traditional knowledge managements approaches usually focuses on explicit knowledge, but some authors indicate that it is necessary to capture, process and transfer tacit knowledge in order to fully understand an organization process [4].

There are several definitions of stages, or phases, composing a KM process. In 2004, Bose [3] presented his definition of a cyclic knowledge management process, including the following stages: Create knowledge; Capture knowledge; Refine knowledge; Store knowledge; Manage knowledge; Disseminate knowledge.

Independent of its process, each KM program needs balance on the kind of knowledge it is focusing. Based on this focus it can be categorized in one of the four KM styles defined by Choi and Lee [5]:

- Passive: It is not managed in a systematic manner - Little interest in KM;
- System-oriented: Put emphasis on codifying and reusing knowledge, in consequence, increases codifiability through IT;
- Human-oriented: Emphasizes on acquiring and sharing tacit knowledge and interpersonal experience;
- Dynamic: Exploits both tacit and explicit knowledge, and does it in a dynamic fashion, similar to a communication-intensive organization.

Choi and Lee [5] performed an empirical study in order to validate the four KM styles. The work analyzed the effects of each method on corporate performance. As expected, the dynamic style was the most effective while the passive one resulted in significantly lower performance. In addition, no significant difference was noted between the system and human-oriented styles.

2.1 Knowledge Management Metrics

If in one hand research shows that there is already a good knowledge base on performance indicators for knowledge management [3], on the other hand Liebowitz [8] reviewed some of the performance indicators present in the literature and stated that "many of the cited metrics lack "creativity" in terms of determining the size and growth of the organization's knowledge base".

Nonaka [10] states that some managers have difficulties in fitting into the model of a knowledge driven company since they believe that the only useful knowledge is the quantifiable one and that the company is a sort of information processing machine. This difficulty, as pointed by Nonaka, is probably the reason behind the lack of higher level performance indicators.

Ahmed [1] also points the same difficulties for having higher level performance indicators:

"Traditional performance measures have focused on outputs, whereas there is a need to look towards the enablers that lead to the production of results. Enablers can be multifaceted and include elements such as leadership, people, systems, strategy, communication, etc. It is from these enablers that improvement benefits flow. Unfortunately whilst managers understand the concepts and the need for measurement many organizations fail to develop performance measurement systems to support their development[…]".

This only reinforces the lack of sensitiveness a financial indicator suffers, since such categories of indicators are only able to provide the whole picture of the Knowledge Management area, being unable to measure every component area. In order to have a sound management of the whole Knowledge Management process, corporations need to be able to measure every stage of the whole process.

Concerned with these limitations, apart from reviewing current indicators present on literature, Liebowitz [8] proposed the following indicators:

- The number of new colleague to colleague relationships spawned;
- The reuse rate of "frequently accessed/reused" content;
- The capture of key expertise in an online way;
- The dissemination of knowledge sharing to appropriate individuals;
- The number of knowledge sharing proficiencies gained;
- The number of new ideas generating innovative products or services;
- The number of lessons learned and best practices applied to create value-added;
- The number of (patents + trademarks + articles + books + talks at conferences) / employees;
- The number of "apprentices" that one mentors, and the success of these apprentices as they mature in the organization;
- Interactions with academicians, consultants and advisors.

If Liebowitz indicators are analyzed in light of what Ahmed has pointed, it is easy to notice that these indicators are fully aligned with what Ahmed argues. It is also worth mentioning that Liebowitz category of indicators focus only on Knowledge Management while some of the indicators present on literature focus on the whole Intellectual Capital area.

3 Social Network

Since the middle of the 90's, Web Based Social Networks (WBSN) is proliferating in an impressive fast pace, including both, the number of WSBN and its scope. There is a really large set of WSBN, each one with its defined scope, ranging from business and entertainment until pet relationships.

One of the reasons justifying the interest of organizations in social networks is how these networks are quite efficient to share the knowledge of each individual [12]. Once that knowledge, that is relevant to the members of the Social Network, is documented, it can be reused, avoiding wasting of effort from employees.

According to Domingos [7], the way that users publish information in social networks is impressive and without any precedent. Therefore, these networks can be seen as a huge data repository, which contains information from each user. Besides that, it also permits to find some unusual correlations, which, according to Staab et. al. [12], could allow finding new information about quantitative and qualitative aspects of social connections.

Because using social networks is an efficient way to share and distribute individual knowledge, its use has become an interesting approach to support a knowledge management initiative. Moreover, using social networks can bring other benefits, like providing an interactive and informal environment where users can express themselves, easily enriching the organization memory.

4   Case Study

This section details the case study, and its two distinct phases, done during this research. The first phase focused on the comparison between the previous Knowledge Management initiative, a wiki based with support of mailing lists, and the approach proposed by this work, a web based social network approach to Knowledge Management. The second phase focused on defining some KM metrics to the a.m.i.g.o.s KM process and analyzing its evolution during a six month span.

In the next sections it is presented a brief introduction of C.E.S.A.R and its knowledge management context, an introduction to a.m.i.g.o.s social network, a mapping between a.m.i.g.o.s functionalities and the cyclic KM process proposed by Bose [3], the results found during the first phase, the proposed KM metrics to the a.m.i.g.o.s KM process and the values of these metrics from July to December of 2008.

4.1   C.E.S.A.R

C.E.S.A.R is a world-class private institution that develops products and processes, provides services, and cradles innovative new companies in their early stages using Information and Communication Technologies (ICT). It is a 600-people organization associated with a Computing centre and R&D departments from the private sector.

C.E.S.A.R. is comprised of five buildings, four in Recife and one in São Paulo. It has more than 50 projects in different areas, like Digital TV, Embedded Systems, Mobility, software reuse, and others. Each project varies from 3 to 50 collaborators geographically distributed throughout its buildings. In environments like C.E.S.A.R, a KM tool should help collaborators to communicate more efficiently and manage the knowledge generated by them.

4.2 Wiki and Mailing Lists Approach

When first defining its Knowledge Management approach, the C.E.S.A.R organization decided to apply a more system-oriented approach for knowledge management. After analyzing the work presented by Choi [5], and because the organization did not have any previous experience in knowledge management initiatives, the organization judge that a system-oriented strategy could bring greater benefits by taking a cheaper and less risky approach to a Software Development organization.

This first initiative was a wiki-based environment where employees could publish papers, documents, videos and any other interesting resources so they could be available to any other employee. The employees also could create forums and initiate discussions to create and share new knowledge. Initially it was defined that these forums would be maintained by members of a group called consultants – people with a high level influence throughout the organization. Each consultant was in charge of keeping his forum active.

After one year, the consultants group verified that this wiki solution was not working as expected. In a brainstorm session, the consultants concluded that neither employees nor consultants felt compelled to write or post information and experiences on it. They have also concluded that the usage of the wiki by C.E.S.A.R employees transformed it in a very impersonal environment.

Electronic mailing lists have been used to minimize wiki communication deficiencies. This mailing list approach did not permit storing the knowledge created and shared by its members, which was a major disadvantage when compared to other KM systems.

C.E.S.A.R used to have four mailing lists to broadcast information through the entire organization, one for institutional messages composed by all employees, one for all project managers and one for all system engineers and other for general purpose. Moreover, each project in C.E.S.A.R also has its own mailing list.

4.3 A.M.I.G.O.S

In order to achieve a better result in its knowledge management strategy, C.E.S.A.R. has developed and incorporated the use of a WBSN tool named a.m.i.g.o.s, which was first deployed in October 2006. This new initiative is trying to add some human-oriented aspects to a system-oriented approach, reducing the gap to a dynamic-oriented approach.

Portuguese acronym for Multimedia Environment for Integration of Groups and Social Organizations, the main goal of a.m.i.g.o.s is to provide a software infrastructure to support the creation of WBSN. The a.m.i.g.o.s also intends to stimulate the knowledge creation and sharing by its members, providing many features in order to be used as a knowledge sharing tool. Some of these features are presented on next sections.

4.3.1 Profiles

Each member has a personal profile. This profile is composed of a set of static information provided by user on registration, like physical address, languages, e-mail address, and a brief personal description, focusing on interest areas.

However, the most important part of the user's profile is not filled explicit by him/her, but inferred by the system. This information includes: (1) the user's activity index, which is calculated through the amount of activities that produce or consume knowledge; (2) a set of subjects the user usually writes about, which are identified by the most relevant terms posted by the user; (3) a set of subjects the user usually likes to read about, which are identification by the most relevant terms from the items rated greater or equal three stars (in a range from zero to five).

4.3.2 Stories and Objects

Stories are intended to register, compile and disseminate emergent knowledge throughout the WBSN. Any user can add his success or failures stories spontaneously. Each story can be enriched with several kinds of objects, as text files, slide presentations, audio or video files. In a similar way, a story can be related to other stories to allow users build bigger ones composed by many small stories.

Every member can also act as a reviewer of the content added by his peers, evaluating contributions. This evaluation can be done in two ways: (1) by adding comments to stories, improving them with new knowledge and creating a dialogue around the added knowledge; (2) rating its relevance from 1 to 5 stars, allowing stories to be shown according to its relevance for the WBSN.

The system also allows adding knowledge through various kinds of objects. In the a.m.i.g.o.s environment, every file that can store knowledge or be used to enrich existing knowledge is seen as an object, like text documents, papers, spreadsheets, audio and video files, and URLs to external resources.

4.3.3 Virtual Communities

In the a.m.i.g.o.s context, virtual communities can be seen as groups of users who have some interests in common. The system supports the creation of such communities by any WBSN member.

A community has three main mechanisms to support knowledge creation and sharing between its members. The main mechanism is the forum, where the members can start new topics about any interesting subject. A second mechanism is to create stories related to the community. A third one is to add objects to the community.

4.3.4 Recommendation and Folksonomy

As a mechanism to turn the knowledge dissemination easier, the system supports recommendations of stories, objects, communities, discussion topics and even users to members. There are two kinds of recommendations, the first one is made explicitly by the user who suggests content to his contacts, and the second is made automatically by the system based on the likelihood the user find the recommended content relevant.

For the automatic recommendation, the system gets the list of terms, which defines the user's interests. Using the Vector Space Model [2], the system tries to identify contents or users who have a similarity measure greater than 50%.

In order to provide classification capabilities, the a.m.i.g.o.s uses the concept of folksonomy [9], where content can be socially and collaboratively classified by users, who add tags to the content based on their understanding. Every user can add tags to communities, stories, objects, comments and discussion topics. All tags are available through the system tagcloud.

4.4 Mapping a Cyclic KM Process on A.M.I.G.O.S

As one of the main goals of A.M.I.G.O.S. environment is to provide a tool to foster the knowledge acquisition, combination and sharing through the organization, it was made a comparison between the KM process proposed by Bose [3] and the existing functionalities, generating the a.m.i.g.o.s KM process (Fig. 1), presented as follows.

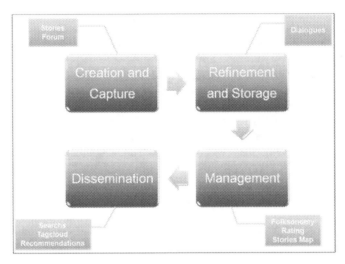

Fig. 1. a.m.i.g.o.s KM process

This proposed process is composed by 4 phases, each focusing in a set of functionalities and its usage. On the following sections the phases of Creation and Capture, Refinement and Storage, Management, and Dissemination are presented in more details.

4.4.1 A.M.I.G.O.S. Knowledge Creation and Capture

Bose proposed that the knowledge comes from experiences and skills of the employees and must be stored in raw form to be useful to others members of the organization.

In the a.m.i.g.o.s environment there are basically two ways to users create and capture knowledge. The first source of knowledge is the reported user's experiences. Every user is encouraged to describe his experiences, lessons learned and any other information as stories, with its related stories, images, videos, audios or any other electronic document. These stories can, additionally, be related to specific communities, increasing the contextual information of the described experience.

Another source of knowledge is the chat module. When taking to others using the a.m.i.g.o.s. chat server, the user has the possibility to store the entire chat content (or a subset of it) in his personal profile, and even publish it so others can have access to it.

4.4.2 A.M.I.G.O.S. Knowledge Refinement and Storage

The refinement and storage of knowledge are made through a diversity of activities that take place inside the a.m.i.g.o.s. environment.

Users are encouraged to develop new knowledge through dialogues about any subject related to a community. The entire dialogue is immediately stored and made accessible by any other member of the community, and, depending on permissions policy, to all a.m.i.g.o.s. users.

The refinement can also occur inside stories created by users, where others can add more information, or even refute what is written. This also is immediately stored in the repository.

4.4.3 A.M.I.G.O.S. Knowledge Management

Because it is basically a social network, there is not a tool in the a.m.i.g.o.s. environment that can be used to allow knowledge management made by a group of specific users, usually the KM team. Instead, several mechanisms were developed to allow the users to manage knowledge implicitly.

The first mechanism is the rating tool. Every time a user reads a piece of knowledge (stories, topics, documents, objects, etc.), he can rate it so the users can be directed to the most relevant pieces of knowledge. The second mechanism is the stories sorting mechanism. Through these filters it is easy for users to identify which are the newest stories, the most recent commented, the most active (a mixture of most commented and most recent), and so on.

Another efficient mechanism used to manage the knowledge is the folksonomy, which allows any user to classify any existing piece of knowledge with some special keywords, called tags. These keywords are used by users to filter and find knowledge related to a specific subject.

4.4.4 A.M.I.G.O.S. Knowledge Distribution

According to Bose, an effective knowledge management needs that every piece of knowledge is accessible to every member in an organization, at anytime and in an organized way.

In a.m.i.g.o.s., the distribution process is focused in searches for knowledge (stories, forum, topics, messages, communities and experts) according to the permissions set for each one of them. But what makes a.m.i.g.o.s. a different social network environment is the recommendation engine that runs in background according to actions taken by users.

The base to this recommendation process is the profile inference module. Every content written or added (as electronic documents) to the a.m.i.g.o.s social network is analyzed and the relevant keywords, with its weight, are added to the user profile. Additionally, every read story which was positively rated is also added to the user profile. So every user in the environment has two groups of relevant keywords, the keywords related to subjects which the user likes to write about, and another set of keywords related to subjects which the user is currently interested in reading.

The mechanism of automation of the recommendations passed by the following developments: (1) recommendations for similar content based on the information that users usually post / write in communities, stories and objects (documents or sites), (2) recommendations for taking similar stories as a basis the stories read by the user, and (3) recommendations of relevant users (experts).

These mechanisms aim to attend the Bose objectives, which are: foster the organization knowledge through the creation of new tacit knowledge (through exposing the employees to new explicit knowledge) and increase the employee qualification.

4.5 First Phase Results

The main goal of this first phase of the case study was to verify if the social network approach to KM would be more successful than the previous approach, based on a wiki and electronic mailing lists. During this phase some metrics related to knowledge creation, capture and dissemination were monitored. The following metrics were used:

- Number of new Knowledge Published: Which on the wiki approach is basically the number of file uploads and page creations, and on the a.m.i.g.o.s approach is the number o stories related and uploaded objects.
- Number of Discussions: Which for the first initiative is the number of messages sent to the general mailing lists plus the number of messages sent to the engineering mailing lists, and for the a.m.i.g.o.s approach is the number of topics and messages initiated and posted on public communities.
- Number of Recommended Knowledge: Which is basically the number of recommendations on a.m.i.g.o.s but do not have an equivalent on the wiki and mailing list approach.

This first phase was initiated on October 2006 and finished on June 2008. After this period the metrics related to the a.m.i.g.o.s were still monitored until December 2008, wich was part of the second phase.

When the a.m.i.g.o.s was first installed on C.E.S.A.R, in October 2006, there were a total of 300 users registered users. But, as we can conclude comparing the metrics presented on Fig. 2, Fig. 3 and Fig. 4, the use of a.m.i.g.o.s by the organization was not so better than the wiki and mailing lists initiatives, and it was necessary to do a major makeover focused mainly on usability and user interface aspects. Almost one year after the first release, the system was released again and after a month there were a total of 508 registered users, almost 80% of the total organization employees.

It is possible to notice a change of direction regarding the knowledge creation and sharing throughout the organization. That is an indication that, except for the data has an intrinsically collaborative nature, like courses pages and articles created through collaborative building, almost all the knowledge created, captured and disseminated by the organization shifted from wiki and public mailing lists to a.m.i.g.o.s.

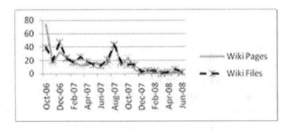

Fig. 2. Wiki monthly metrics

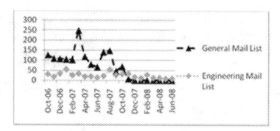

Fig. 3. Public mailing lists monthly metrics

This analysis indicates that, in this particular scenario, a WBSN approach to knowledge management was far more effective than a wiki based approach, and the fact that the migration from the wiki based approach to the WBSN approach occurred spontaneously and without external influences is a strong indication that this kind of solution is more appealing to users than the previous one.

4.6 Second Phase Results

In July 2008 the second phase of the a.m.i.g.o.s case study was initiated. Because the previous phase had already indicated that the WBSN approach to KM was the better approach to C.E.S.A.R, the second phase main goal was to have a better understanding about the behavior of a.m.i.g.o.s' users on the Knowledge Management perspective.

To achieve this better understanding it was necessary to define some new metrics based on the work made by Liebowitz, in 2000. So, during this phase the following metrics were defined and monitored:

- Number of added contacts: This indicator is based on "the number of new colleague to colleague relationships spawned";
- Number of times a topic or history is read: This indicator is based on "the reuse rate of 'frequently accessed/reused' content". For this to be calculated it was defined that a story or topic is read if it stays open on the browser for at least five seconds;
- Number of stories and topics positively qualified: This indicator is based on "the capture of key expertise in an online way". For this to be calculated, it is considered any story or topic with an average rate greater than or equal to three stars (which can vary from zero to five stars);
- Number of knowledge classifications: This indicator tries to measure how frequently the created knowledge is classified using tags. For this to be calculated, it was counted the number of tags applied to any item (stories, topics, objects and comments) inside the environment;
- Number of qualifications: This indicator tries to measure how often a knowledge item (stories and topics) is rated by users, independent of the rate;
- Number of access to elements through tags: This indicator tries to measure how often users look for items related to a specific subject, represented by a tag;
- Tagcloud visualizations: This indicator tries to measure how often users look for knowledge through the list of subjects available, which is represented by the environment tagcloud;

- Number of unique users: This indicator tries to measure what is the real impact of the system on the organization. For this it calculates how many different users log in and uses the system in a month.

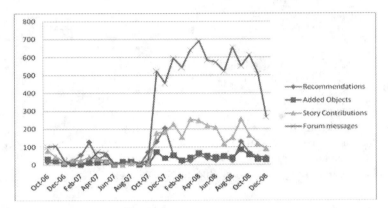

Fig. 4. a.m.i.g.o.s monthly metrics

During this second phase the knowledge creation metrics defined for the first phase were also collected, so it would be possible to monitor the behavior related to activities of knowledge creation, capture and sharing inside the social network environment.

When analyzing the knowledge creation metrics it is possible to verify that there is a pattern on knowledge creation. For every month after November 2007, except December 2008, there were at least 500 messages inside communities, and 200 stories added to the environment.

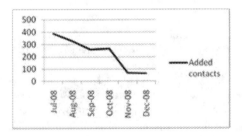

Fig. 6. Number of new Added Contacts

Because the Brazilian summer vacations are usually in the last two weeks of November and all weeks of December, it is expected a decrease on all collected metrics related to these months. But it is necessary to monitor these indicators for at least six more months in order to find a more accurate pattern.

Looking to the number of added contacts, which is presented on the Figure 6, it was possible to verify that, in a span of six months, the number of new relationships varied from 385 in July 2008 to 65 new relationships in December 2008. Except for the last two months, there were more than two hundred new relationships created by the social network users.

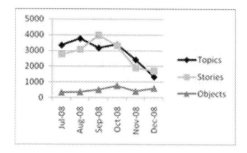

Fig. 7. Number of new read items

As it was verified, the numbers of new relationships for the last two months of the experiment were far lesser than the overall for the four months before. This behavior probably is also related to the Brazilian summer vacations, and also to some kind of stability on this number. This kind of stability is expected as the social network mature and the number of new members starts to decrease.

Observing the variation of the number of read items (Figure 7), it is possible to verify a really active community. The number of read topics and stories varied from 3000 to 4000 monthly, which is more than 15 times the number of created stories and topics during the same span of time. It was also noted a decreasing on the number of read items for the last two months of the year, as expected.

It is important to notice that the number of downloaded objects do not suffered on the last two months. A more deep analysis would be necessary to understand why this happened. These data also suggests that the objects do not take part on the day-to-day activities as does the topics and stories.

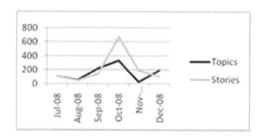

Fig. 8. Number of qualifications

But even with a high number of read items, the number of rated items, that is, the number of topics and stories evaluated by users with a rate from one to five stars, (Figure 8) was not so encouraging. It was noted that an average of only 10% of users who read an item usually rate it. After some interviewing, it was identified that only 20% of interviewed users did know about the existence of the rating functionality or how they could benefit from it, which indicates that there is a problem related to the functionality usability.

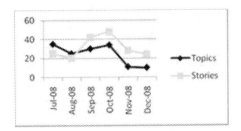

Fig. 9. Number of positively rated items

This behavior of not rating items also can be verified on the number of items with a positive average rate (Figure 9). Only a few topics and stories created did receive rates greater than or equal to three stars. This number is usually around 10% of all items created monthly.

But far more encouraging than the number of qualifications or the number of positively rated items is the number of classified items (Figure 10). The created stories have an average of approximately 2.1 tags for each story. This number is even higher when applied to objects, which presents an average of 2.9 tags for each object. Even topics have an average of 1.3 tags for each created topic.

These numbers also indicates a certain level of awareness of the importance of knowledge classification, so a piece of knowledge could be more easily found by another user who would need this knowledge.

Besides a really high number of knowledge classifications, users do not take full advantage of this classification. Even with almost 500 different tags, the number of accesses to a tag, listing items tagged with a keyword (Figure 11) is really low.

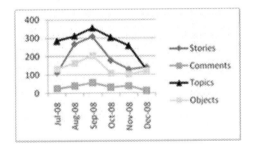

Fig. 10. Number of classified items

When looking to the numbers of items classifieds and confronting them to the number of access to the classification lists (tags and tagcloud), is possible to verify that users do care about classification, but apparently do not know how to access this classified item through the classifications mechanisms, which indicates that there is a problem on the classifications mechanisms, more precisely related to the interface for accessing the classified content.

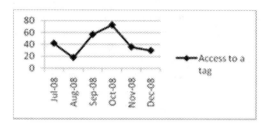

Fig. 11. Number of accesses to a tag

Another interesting indicator is the number of unique visitors to a.m.i.g.o.s. Because the a.m.i.g.o.s participation is still voluntary, that is, users are not obligated to use it, the number of unique visitors is not near the number of employees. As mentioned before, C.E.S.A.R has approximately 600 employees, but a little more than 50% of them are regular users of the WBSN, even that more than 90% of them has logged in at least once.

These numbers suggests that it still is necessary to bring and keep more users inside the WBSN environment. When interviewed, some users affirmed that sometimes they have trouble accessing the a.m.i.g.o.s because this tool is not present yet on their day-to-day activities.

That kind of behavior is really an eye-opener, suggesting that probable the most appealing characteristic of an environment like that is to be part of the members' day-to-day activities inside their companies. This probably can be achieved if all projects and departments inside the organization have their activities, discussions, documents, agendas, for example, inside the WBSN environment.

After this second phase it was possible to identify that the WBSN approach to KM is still working well, but metrics indicates the need for enhancements in certain aspects of the WBSN, like the classification mechanism, the rating mechanism and new functionalities which would bring users' day-to-day activities inside the environment.

5   Conclusion and Future Works

A large number of companies are choosing a WBSN approach on their knowledge management strategies. This work presented the use of a WBSN called a.m.i.g.o.s as the main tool for communication and knowledge management in a Brazilian software development company: C.E.S.A.R. The main goal of this work is to present and evaluate an alternative to align individual interests in an organizational knowledge management initiative, turning it a natural process.

The a.m.i.g.o.s WBSN has been used by C.E.S.A.R since October 2006 and the results indicated that this approach has been well-succeed to give cooperation, collaboration and knowledge sharing support in C.E.S.A.R projects. These can be concluded after the review of data regarding the user of the a.m.i.g.o.s platform compared to the data regarding the existing approaches of wiki and electronic mailing lists, and also through the monitoring of some knowledge management metrics proposed by this work.

During this work it could be seen an indication that WBSN is really an alternative strategy to do knowledge management inside organizations. Using a WBSN the organization adds some human-oriented perspectives on a traditionally system-oriented approach, which makes the use of WBSN as a KM tool quite promising, decreasing the gap to a dynamic approach to KM.

Also according to the data collected during the second phase of this case study, four main improvements shall be made to the system in order to better attend users' needs. The first one is the development of a new feature related to day-to-day activities. Because C.E.S.A.R is a project-oriented company, one of the best ways to support these day-to-day activities is related to project support inside the WBSN. With that kind of support, every project would have a special environment which would support its activities, and where every project member should access in order to keep up to date with the project status, discussions, documents, stories, etc.

The second improvement is related to better exploit the classification mechanisms provide by the folksonomy module. According to the collected metrics, users usually classify knowledge with tags, but they do not use them when looking for information. Based on this behavior it is clear that the folksonomy module shall be modified in order to facilitate the access to the classified content.

A third improvement is related to document management capabilities. The knowledge creation metrics shows little improvement and little activity regarding the uploaded files and bookmarked sites, which together constitutes the user objects. The development of a version control module allowing the uploaded files to have its versions managed would probably lead to a larger knowledge base, bringing files that without this feature would be placed outside the WBSN. Some improvements to the objects user interface probably would also lead to increasing document accesses.

The fourth major improvement shall be made regarding recommendations. The data presented in this work indicates a modest number of both, manual and automatic recommendations. However, with a larger knowledge base, the automatic recommendation probably would have more relevant data to work on.

It also should be interesting to add "yellow pages" or "knowledge maps" based on inferred ontology so the process to allocate people to projects can be more efficient, as the time to solve questions. These ontologies would be extracted from the existing folksonomy, as proposed by Mika [9], and based on user profile the system would be able to associate each user with a term, or set of terms, on the ontology. There are also some ongoing experiments in this area.

In general this work can be used as a reference to other initiatives focusing on the use of WBSN as the main tool in a KM approach. The metrics defined are real and have been monitored for six months, setting some parameters regarding the KM efficiency of a WBSN approach in an environment like that on C.E.S.A.R.

Acknowledgements

This work was supported by the National Institute of Science and Technology for Software Engineering (INES[1]), funded by CNPq and FACEPE, grants 573964/2008-4 and APQ-1037-1.03/08.

---

[1] http://www.ines.org.br

References

1. Ahmed, P., Kwang, L., Mohamed, Z.: Measurement practice for knowledge management. Journal of Workplace Learning: Employee Counselling Today 11(8), 304–311 (1999)
2. Barros, F.A., et al.: Similar Documents Retrieval to Help Browsing and Editing in Digital Repositories. In: Communications, Internet and Information Technology - 2002. St. Thomas, US Virgin Islands (2002)
3. Bose, R.: Knowledge management metrics. Industrial Management & Data Systems 104(6), 457–468 (2004)
4. Caro, R., et al.: Knowledge Management in Action. In: Proceedings of The 4th International Symposium on Energy, Informatics and Cybernetics: EIC 2008, Florida, USA (2008)
5. Choi, B., Lee, H.: An Empirical Investigation of KM Styles and their Effect on Corporate Performance. Information & Management 40, 403–417 (2003)
6. Davenport, T., Prusak, L.: Conhecimento Empresarial: Como as Organizações Gerenciam seu Capital Intelectual, 256 p. Campus, Rio de Janeiro (1998)
7. Domingos, P., Richardson, M.: Mining the network value of customers. In: Proceedings of the seventh ACM SIGKDD international conference on Knowledge discovery and data mining, pp. 57–66 (2001)
8. Liebowitz, J., Suen, C.: Developing knowledge management metrics for measuring intellectual capital. Journal of Intellectual Capital 1(1), 54–67 (2000)
9. Mika, P.: Ontologies are us: A unified model of social networks and semantics. Journal of Web Semantics 5, 5–15 (2007)
10. Nonaka, I., Takeuchi, H.: The Knowledge Creating Company. Oxford University Press, New York (1995)
11. Polanyi, M.: The tacit dimension. In: Knowledge in Organizations, pp. 135–146. Butterworths, London (1967)
12. Staab, S., Domingos, P., Mike, P., Golbeck, J., Ding, L., Finin, T., Joshi, A., Nowak, A., Vallacher, R.R.: Social Networks Applied. IEEE Intelligent Systems 20, 80–93 (2005)

Tailoring Collaboration According Privacy Needs in Real-Identity Collaborative Systems

Mohamed Bourimi[1], Falk Kühnel[2], Jörg M. Haake[1],
Dhiah el Diehn I. Abou-Tair[3], and Dogan Kesdogan[3]

[1] FernUniversität in Hagen, Cooperative Systems, 58084 Hagen, Germany
{mohamed.bourimi,joerg.haake}@fernuni-hagen.de
[2] justread GmbH&Co. KG, 53127 Bonn, Germany
fk@justread.de
[3] University of Siegen, Chair for IT Security, 57076 Siegen, Germany
{aboutair,kesdogan}@fb5.uni-siegen.de

Abstract. Nowadays, collaboration and social interaction among people become everyday activities in our evolving information age. In many learning platforms, collaborative platforms in the educational and industrial field or social networks like LinkedIn or Xing, users have to disclose private information and reveal their identities. Working with those systems allows them to create user profiles which could reveal more information about the user, than he wants to give. Furthermore, such environments may construct profiles about users' interaction, which may be used for attacks; thus preserving privacy is an essential component of such environments. In this paper, a decentralized group-centric approach for tailoring collaboration according privacy needs is introduced. The main idea of our approach lays in its construction. In contrast to traditional collaboration environments with central hosting, our approach gives each group the whole responsibility of hosting the collaboration environment by using their own technical means. The feasibility of our approach is demonstrated through a lightweight ubiquitous collaboration platform. The experiences gathered are discussed.

Keywords: Shared workspaces, adaptation, group context, privacy, tailoring.

1 Introduction

Due to the rapid evolution and sinking costs of computing systems and networking technologies, the use of the Internet as a global information sharing and communication infrastructure cannot be contested. The Internet is now accepted as the de facto information support system in many important sectors of our life activities such as business, health care, education, etc. Furthermore, we can observe a shift in its usage paradigm. Indeed, the Internet is currently performing a shift from single-user-centered usage to support multi-user needs and hence covering many collaboration forms and social aspects. Those collaboration forms and social aspects are not only important for the success of companies, institutions and organizations in the era of globalization, but also for individuals. Users are interested in the collaborative construction of information and knowledge sharing as well as in social interaction with

L. Carriço, N. Baloian, and B. Fonseca (Eds.): CRIWG 2009, LNCS 5784, pp. 110–125, 2009.
© Springer-Verlag Berlin Heidelberg 2009

each other; and this in both of their professional and leisure activities. Nearly 700 studies indicate that collaboration in the educational sector results in higher achievement, greater productivity and social competence, more caring and committed relationships, and self-esteem for students [35].

Information exchange, collaboration and social interaction are supported through collaborative systems and, more recently, social software (e.g. wikis, blogs, etc.). Both provide support for multi-user collaboration and interaction. Compared to collaborative systems, social software focuses on communities as well as networks of users, instead of teams [25]. Furthermore, social software is mostly web based and presumes users' willingness (1) to generate content and (2) to give up their anonymity partially or completely by revealing their real identity in the network [33]. Collaborative systems are in principle used to support real world activities in respective domains, e.g. in CSCL or CSCW scenarios. However, groups of users are first-class objects in such systems as teams, (sub)communities and user's networks. Today's collaborative systems include features of social software and vice versa.

In some collaborative systems, the member's identity must be disclosed; for example, in case of universities using learning platforms or industrial companies using various collaborative systems. In the case of social networks supported by collaborative platforms such as LinkedIn or Xing, users have not only to reveal their real identity but also reveal some private and sensitive information on their profiles (e.g. private and work telephone numbers, addresses and professional references etc.) in order to receive job offers or get some recommendation letters [19]. In this way, users loose effective control over their private spheres, consciously leave private information, and may also leave personalized interaction traces they are unaware of [9]. Risks, abuses and threats which might arise from accidental or intentional disclosure of such information are well-known and have not to be cited here. User-centric approaches for privacy control (such as information retention and anonymization) do not meet the requirements and purposes of collaborative systems. In our opinion, existing server-centric approaches for collaborative systems dealing with real identities do not offer flexible solutions in order to sufficiently respect users' privacy. Furthermore, they do not fully exclude all potential risks and threats, which can arise, since they presume full trust in the platform used. Thus, privacy policies primarily imposed through the platform manager intend to overcome these issues and enforce the platform's business models and interests based on private information and disclosure of social interaction. Considering that anonymization possibilities (i.e., through removing sensitive data like names, addresses, etc) in highly popular social networks like Facebook, Flickr, MySpace or Twitter can be broken, and that users can be identified across distinct social networks, collaborative systems dealing with real identities are even more vulnerable in this respect. E.g., users with accounts on Flickr and Twitter could be identified with an error rate of just 12% as shown in [28].

Based on our experiences in the development and usage of a collaborative system at our university and in the industrial field for many years, and collected end-users' feedback, we detect a bigger need for privacy control. In order to increase the trust in our platform and encourage collaboration inside it, we propose a decentralized group-centric approach to our end-users, which considers their needs as well as the platform developers' needs. The approach consists mainly of empowering the end-users to host their collaborative environments by using their own technical means. Furthermore, end-users can decide which collaboration and interaction results are shared (or not).

We first present the identified needs for privacy control form the end-user and developer perspective. Next, we present related work and then we describe our approach to design and implement a group-centric privacy control in collaborative systems. Finally, we present our conclusions.

2   Problem and Requirements Analysis

The privacy needs we address in this paper were identified by using the CURE platform [21] for typical CSCL and CSCW scenarios. Thus, in this section we describe briefly the main concepts and features of CURE, followed by analyzing the needs from the end-users' and developers' perspective. Thereby, we consider similar situations cited in numerous literatures, our experience as well as feedback reported to us from everyday situations in the industrial field.

2.1   CURE in a Nutshell

A typical collaborative system supports users by constructing collaborative environments. Those environments provide shared workspaces, where collaborative activities are carried out in form of different collaborative work use cases like document sharing, group formation and communication, etc. In order to support different learning scenarios at the German Distance Learning University we developed such a collaborative system. The result of our research is the CURE (collaborative universal remote education) platform [21]. The analysis of intended scenarios within the University showed a great variance. To accommodate this, CURE supports run-time tailoring at different levels (i.e. tailoring at the structural, content, and functional level) [22]. In CURE, end-users manage and control access rights to their workspaces and flexibly organize their work themselves. For dynamic group formation without prior planning by the system administrator, end-users are able to form groups (1) by assignment, (2) by invitation, (3) with free enrolment, and (4) with enrollment confirmed by the members of the respective groups [23].

The conceptual design of the CURE system is shown in Figure 1. CURE uses the room metaphor to model shared workspaces for groups. The virtual key metaphor is used to determine access rights and interactions allowed within a given room. Users who have keys to a certain room can form a group, cooperate and work in the room together. Structuring collaborative environments is carried out by connecting rooms. A special room, called Hall, represents the central access point to the learning portal with its learning environments. From the Hall, adjacent rooms can be recursively accessed. A CURE room provides materials for group members in form of pages, which can be regular content pages containing text or those including binary data such as Word or PDF documents. A wiki editor is used for the creation and asynchronous editing of content pages. Changes created by this editor are immediately visible to other users in the room. The built-in versioning system manages previous versions of pages and supports conflict resolution. The permanent storage of content within the rooms allows users to work asynchronously.

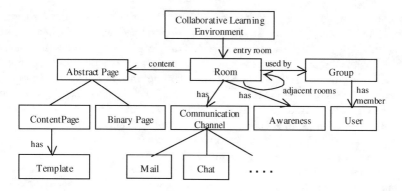

Fig. 1. Conceptual design of CURE

Since CURE is a web based collaborative system, rooms are displayed as web pages. A retrofitted version of CURE enhanced the adaptability and usability of the original CURE platform by supporting direct manipulation techniques for navigation and tailoring [5, 27]. The new user interface is shown in Figure 2.

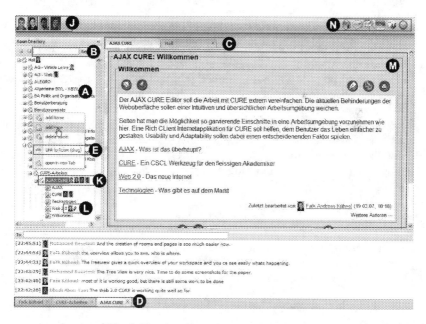

Fig. 2. User interface of the retrofitted CURE

Well-known user interface concepts were used to support, e.g., faster navigation (e.g. tree view, search field, tabbed pages and chat discussions in Figure 2A, B, C, and D, respectively) and visual tailoring by direct manipulation (e.g. Figure 2E). New forms of visualizing synchronous awareness and synchronous communication were also introduced. For instance, users can see who is actually in a given room (Figure 2K) and what s/he is doing (e.g. editing a page in Figure 2L). All active users in the platform are visualized for everyone (Figure 2J). In case of choosing a room, a particular page (the welcome page of this room) will be displayed (Figure 2M) while other functionalities remain accessible through the fish eye menu (Figure 2N).

Interaction rights are determined, as mentioned before, by the virtual keys of the user. Hence for example the navigation tree view (Figure 2A) is generated and shown for every user depending on his actual keys (e.g. users can only see rooms to which they have access to). Due to access restrictions, some users may not be able to use the communication means in the room, or could only view pages but not edit them. Users with sufficient rights, i.e. for creating adjacent rooms, passing on or copying their virtual keys, and editing the content, can therefore at any time adapt the collaborative environment according to their needs (also during the collaboration process).

Since fall 2004, CURE is an integral part of the University's virtual learning space and has currently more than 2000 registered users. Because of its high tailorability and flexible group formation mechanisms, CURE can be used to support a very wide range of CSCL and CSCW settings. Thus, we use the retrofitted version now also in the industrial field, primarily for evaluation and gathering new requirements. The retrofitted version uses Web 2.0 technologies and can in addition to the built-in wiki easily integrate other kinds of social software [27]. This makes it a perfect representative system for our purposes in this paper.

2.2 Privacy-Related Needs from the End-User's Perspective

CURE was developed following an agile software development process. From the beginning this process considered end-users' feedback of the participating departments at our university. For instance, users from various disciplines like mathematics, electrical engineering, computer sciences and psychology were participating in the usage and evaluation of the first prototypes. The developers, who designed and implemented the functionality since 2003, were at the same time instructors, using CURE for their own teaching purposes as well as for their development activities. During this long time period, a lot of log data was collected and we got very valuable users feedback, which helped us to improve the functionality of CURE and extend our research.

The collection of log data as well as observation of user's activities within a system is an established methodology in the field of CSCW and CSCL. We gathered the end-users' interaction information with their consent. The privacy supervisors at our university followed the development and influenced the design of some features. Where possible, we resorted to the usage policy to enable some features (such as user list or activity indicators) since collaborative settings need some degree of user's information disclosure within the platform in order to achieve the intended goals. Indeed, Palen and Dourish [29] mention that some level of disclosure is needed to sustain social engagement. The room metaphor used in combination with virtual keys helped to provide a good degree of privacy. However, the persistent chat and some awareness functions were not appreciated by some users, who expressed this mostly in an indirect way. So, for example, during some organized meetings and retrospectives during or at the end of lab courses, some participants communicated to their instructors in form of jokes, that they did not contributed because they did not want to be monitored. Even though the instructors possessed no keys to the students' private group rooms, students remained uncertain if the system provided back doors to the instructors. In many courses we observed also a stagnation in collaboration, which was sometimes comprehensible (e.g. due to the group's composition and competencies of the members etc.), but we guess that some of the reasons may also relate to privacy and trust concerns of the end-users. Many students restricted their interactions in the collaborative environments to the minimum (i.e. needed interaction for deliverables or for communication triggered by the instructors). Some of them used in parallel their own collaboration tools (e.g. wikis or forums) instead of the provided collaborative environment hosted on our learning platform. However, it must be said that this was mostly observed with computer science students, who seem to be more sensible to privacy and trust concerns. Some end-users did not care at all.

Studies and research results show that users' inhibition regarding privacy and trust concerns may negatively affect their interaction with and trust in the platform. A very representative sentence for similar concerns in the e-learning field can be found in [4]:

"The goal of security in e-learning is to protect authors' e-learning content from copyright infringements, to protect teachers from students who may undermine their evaluation system by cheating, and to protect students from being too closely monitored by their teachers when using the software. Since these intertwined requirements are not met by existing systems, new approaches are needed."

A very deep analysis of privacy-related literature for collaborative environments considering different perspectives is given in [7] and [8]. However, we want to cite here some of the most important findings related to our analysis. Tang et al. [34] state that users are often cautious about how the system handles their privacy and are afraid that their mistakes will affect their reputation. Identity revelation and information sharing are related to the concepts of awareness and privacy and depend on the goal of the interaction [32]. People do not necessarily want to reveal details about their current work tasks to other people [12]. This is still mostly in conflict with the aim of awareness provision, which intends to use knowledge about who is in the collaborative environment, what is s/he working on, and what s/he is doing [20]. In this respect, too, Boyle and Greenberg state in [7] that there are trade-offs between privacy and awareness; and two problems are generally associated with providing awareness: (1) privacy violations and (2) user disruption. Concerning the first point, Bellotti and Sellen [3] argue that rational people respect the privacy of others, but accidental violations are known to happen from time to time. As an example for the second point, the author argues in [33] that more managers are active in Xing than in LinkedIn, because of disruption, since LinkedIn allows for flexible addressing of third-people. In Xing, consent is needed for such communication schemes.

The previously mentioned tendency of students to work in trustworthy groups and using their own means for collaboration and communication is widely spread in the industrial field, too. Based on our experience in industrial projects and reported feedback of many consultants we interviewed, this is due to many factors, but mainly related to confidentiality, privacy and trust. E.g., many companies also use collaborative systems in order to profit from and retain the knowledge of their employees. In some companies, the usage of the collaborative systems is encouraged, for example, through bonus points which are reflected in the incentives or salaries. In some cases the usage of those systems is prescribed for important projects. According to our own observations participating in some of those systems, participants often restricted collaboration to a minimum. Here, the collaborative platform is used as an entry point, where only the official project-related information is kept and further links for communication means or environments could be found. Various companies involve external consultants in their developments due to the current out-sourcing models. Furthermore, projects inside the same company possess different levels of confidentiality, so that different employees of the same company have different access to the projects. In summary, we recognized the need for new trustworthy usage models in order to support collaboration among customers and consultants.

In order to satisfy end-users' needs in some projects, we redesigned respectively added/removed some functionality. For example, the data logs were anonymized to a given degree and in addition to the persistent chats, instant messaging (IM) functionalities were added to the retrofitted CURE. IM allow contacting users while they are situated in other rooms and it is not anymore necessary, for instance, to create private rooms only for private communications or change access rights for such purposes. IM conversations avoid user inhibition while better respecting privacy concerns. Furthermore, user profiles and room properties were extended to support different privacy settings (e.g. notifications settings, disabling the usage of persistent chats, showing the presence to others etc.).

From the previous analysis we summarize high-level requirement 1 (HLR1): there is still the need for providing improved privacy control to end-users in terms of better elimination of their concerns related to sharing information and interaction aspects. New approaches are needed, which allow groups working in a larger collaborative work environment to deal with their privacy issues within the smaller, more trustworthy group. By reaching consent on the extent of privacy data exchange individual users privacy worries are taken care of.

2.3   Privacy-Related Analysis from the Designers' and Developers' Perspective

Distributed systems are hard to design, implement and maintain. Adding new support or functionality requires adding and/or modifying a lot of source code. This often complicates the API and requires a redesign of the underlying domain models [11]. Systems used to support collaborative work and social interactions represent a special category of distributed systems. The conceptual design of CURE depicted in Figure 1 is common to many other collaborative systems. There are main entities (e.g. user, shared workspace, content etc.) which have to be modeled independently from the underlying architecture (e.g. object-oriented, service-oriented, etc.). E.g., if an object-oriented model is used to capture these entities, CURE's conceptual design can be mapped into classes of this model. Thus, functionality added to such systems is reflected in the complexity of respective classes. Since citizens' privacy becomes priority in the digital age [17], related directives and laws are continuously introduced. Therefore, the extension or retrofitting as well as the integration of new components in such systems represent realistic scenarios, which have to be considered in terms of development costs. Privacy is important to human-computer interaction design of collaborative environment [14, 7]. Thus, design for privacy is a key requirement from this standpoint.

A way to solve privacy issues consists in using access control for restricting access to sensitive information. Another way consists of providing a possibility to refer to members without revealing their true identities, i.e. with pseudonyms closely linked to partial identities. True and partial identities are mostly provided through identity management systems. Those systems deal with the transmission of personal information. For this, credential systems are used that have to support cryptographic credentials to assert the correctness and confidentiality of transmitted information. Access control based on credentials is provided by policy systems. Policies are often application specific and have to be implemented or adapted to the application context. So, different privacy protection strategies can be built into the collaborative systems when considering the previous means. Functionality provided with the help of these means has to be reflected in the domain model and user interface, since end-users balance their privacy when using them. Furthermore, especially awareness functions and the involvement of third-parties (e.g. payment services) raise a tension with privacy in collaborative systems. Including mechanisms for better end-users' privacy control is not a trivial task from the software engineering perspective. Poor design can results in privacy violations. Researchers in CSCW generally assume that privacy issues arise due to the way systems are designed, implemented, and deployed [8].

Consequently, we define high-level requirement 2 (HLR2): new approaches to reduce development costs related to the implementation and integration of privacy-respecting strategies in collaborative systems and social networks are needed.

3 Related Work

Privacy Enhancing Technologies (PETs) refer to a variety of technologies that protect personal data by minimizing or eliminating the collection of personal data as well as to those technologies, which protect its confidentiality, integrity and availability [18]. Such technologies are very appreciated especially in collaborative environments and systems. The most available implementations of PETs use either a server-centric architecture or a client-centric/user-centric architecture. Client-centric approaches are not sufficient and suitable for collaborative environments since they always require exchange of information. Server-centric approaches imply that the server is the central point of information exchanging. However, such approaches require a great amount of trust into the server [1]. In order to bypass this problem concepts and solutions have been developed within the EU FP6 integrated project PRIME ("Privacy and Identity Management for Europe") [31]. The PRIME project provides tools on both, the user and services/server side in order to enhance the trust of users in an IT processing environment of a service provider [9]. Additionally, PRIME provides procedures to find out about the trustworthiness of service provider [30]. Our approach differs from an identity management system such as PRIME in that it deals with real identities instead of pseudonyms. In terms of privacy policies languages, P3P [13] enables users to express their privacy expectations in form of policies and thus define the level of privacy a service provider should support in order to access the provider services. The P3P approach's weak point lies in the fact that this mechanism cannot guarantee that personal data, once they are collected, will be only applied as regulated in the specified privacy policies [16, 24, 10]. One more schema to represent privacy policies is EPAL [15], which is a formal language that helps applications and enterprises to exchange privacy policies in a structured format. Analogous to P3P, EPAL does not provide mechanisms to ensure that the correct privacy policy is enforced [2].

In summary, we state that existing approaches provide different access control as well as privacy models and mechanisms to protect shared information and to blur interaction traces. However, the linkability is made easy in systems dealing with real identities (as in our case) since they do not completely eliminate the possibility of linking the users with their real identities. Thus, those systems can be used to infer sensible user information. Gathering a sufficient amount of private and interaction information can result in building full-fledged profiles about users' preferences, habits etc. This provides a perfect starting point for potential man-in-the-middle attacks. Some advanced approaches aim at minimizing the information disclosed while collaborating and interacting in those systems in order to hinder observation, monitoring and surveillance techniques. This is mostly done introducing trusted nodes at the server-side and anonymizing involved parties to each other [9]. Nevertheless, those approaches assume full trust in the platform used and that the organizations hosting it are not interested in such inferences. However, this assumption remains often in conflict with the business models and interests of the organizations hosting the platform since they constantly adapt their service portfolio according to collected user's profile information like by Amazon. In the case of commercial social networks, information about members is made available to other members who pay for it. The nature of disclosed information is defined through the underlying privacy policy of the respective platform. As mentioned previously, such approaches can be seen as expensive in terms of design, implementation, operation as well as retrofitting costs in order to support emerging privacy and trust requirements. Thus, they do not satisfy the identified needs and do not fulfill our high level requirements 1 and 2, especially because our systems deal with real identities (even though an extension to support partial identities into the collaborative system is possible, as done in [4]).

A K.O. criterion, which could be cited every time, is that existing systems and approaches do not fully eliminate accidental or intentional risks and threats, which can arise through the analysis or reconstruction of personal information and interaction traces. The building of a user's full-fledged profile is still possible at least through the judicial authorities, which force, for example, service providers to allow dispute resolution means in order to recognize frauds. The citation of point of views and arguments related to privacy and trust issues (which seem sometimes paranoid but are realistic and possible) go beyond the scope of this paper.

The novelty of our approach is indicated by the usage of a decentralized group-centric approach. That is, the collaborative environment is divided in trustworthy small groups of interest where the exchange of information and interaction is limited to the associated group. Members of such a group of interest can communicate through their real identities within the group and can be sure that all communicated and generated data as well as interaction traces will be kept within the group and the collaborative environments used for this purpose. The connection to the server is only used to coordinate the work inside the main collaborative system (i.e. between groups). To the best of our knowledge there is no such approach that constructs small trusted groups within a collaborative environment.

4 Our Approach

In order to meet the high-level requirements 1 and 2, we first introduce a decentralized group-centric approach supporting several usage models. Secondly, we present our proposed usage model and the concrete requirements resulting from it.

4.1 A Decentralized Group-Centric Approach

A decentralized group-centric approach is the key to satisfy both high-level requirements, HLR1 and HLR2. The structure of traditional collaborative environment has the form of a star network topology. This means that each collaborative environment user is connected to all other users and thus it is easy to access and exchange data. The collaborative environments act as nodes, where sharing data, collaboration and interaction can take place. All collaborative environments are hosted on a collaborative system or platform. Our approach decentralizes the collaborative environments and moves them to nodes, which are fully under group control and hosted on their own hardware (cf. Figure 3). Thus, we give groups of users the whole control over their data and interaction traces (HLR1). Furthermore, we reduce development costs by decentralization, since we do not have to integrate expensive privacy and trust mechanisms into the system (HLR2).

For this, we presuppose a main collaborative system, which provides a remote interface to its services. Services may access and manipulate a (shared) data model. The services are accessible, e.g., through a web services layer or a remote programming API. Calls using such remote interfaces lead to model changes in the respective collaborative system or platform. A ubiquitous platform providing a sub-set of the features of the main platform is used in order to provide collaborative environments to different groups of users. The ubiquitous platform is hosted on the hardware of the

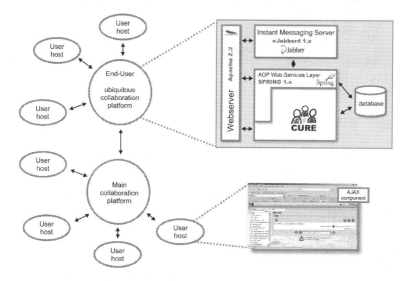

Fig. 3. Proposed decentralized group-centric architecture

end-users and uses interfaces provided for interaction with the main platform (cf. Figure 3). Therefore, groups are able to share their data with the main collaborative platform. The resulting topology can be seen as a dedicated star topology, where the center node represents the main platform and the surrounding nodes represent the end-users' ubiquitous platforms. Depending on the usage model, several topologies can emerge. In the following subsection, we describe our proposed usage model.

The architecture of the ubiquitous platform is similar to the architecture of the retrofitted CURE, which is shown in Figure 3. For further details concerning the architecture and implementation, we refer here to [6, 26, 27].

4.2 The Proposed Usage Model and Requirements Gathering Process

The proposed usage model is based on the analysis of scenarios to be supported. In our case, these are CSCL and CSCW scenarios used in our teaching as well as in research and industrial projects.

In distance learning scenarios, the instructors construct their learning environments with the aim of reducing redundancy, time scalability problems, learning location dependencies as well as communication problems. Shared workspaces are made available to students using access control mechanisms. In such shared workspaces, learning materials are provided and learning activities take place. In our case, CURE supports self-organized learning (cf. Section 2.1). The organization of courses is mostly described informally. E.g., the instructor may define deadlines and the participating students have to deliver their results in time using the capabilities of the collaborative environments (i.e. workspaces). In this scenario, neither workflows have to be followed nor need automatic execution mechanisms to be provided by the platform. When students are involved in distance learning courses, their main goal consists of achieving the targets of the learning process. In our case, at the end of the semester and the course activities the collaborative environments are still available (e.g. for later use in rehearsal). However, most of those environments stagnate after the end of the courses. Group members revisit the environment mostly in order to get some valuable information they miss (e.g. answers of specific problems, implementation tips, descriptions etc.). This phenomenon is well known in the usage of social software like wikis. Because of this, our end-users required an export and import facility in order to continue sharing data across collaborative environments. This usage model is common to many other scenarios for distributed groups support.

In order to gather more specific requirements for intended scenarios following our usage model, we proposed our approach to selected end-users and conducted interviews, walkthroughs and collected first usage experiences with the prototypes implementing our approach. Thereby, we followed an iterative requirements gathering process composed of various phases. The selected users were instructors and students from different departments, designers and developers of our learning platform CURE as well as consultants and partners in the industrial field. For this we implemented a prototype consisting of a main collaborative platform and its ubiquitous pendant. The latter is installed, managed and hosted by the individual groups themselves using their own hardware. As main platform we used retrofitted CURE described briefly in Section 2.1 [6, 27, 26]. The ubiquitous pendant is a slimmed variant of retrofitted CURE. Using this approach, we support many CSCL and CSCW scenarios.

According to our suggested usage model, the end-users have to download the ubiquitous pendant collaborative platform and install it on their own machines. Our usage model suggests also that only one member of the group will do this for all group members on his/hers own hardware. Our teaching experiences show that this model is feasible due to today's low hardware and communication costs. At the German Distance Learning University, e.g., the usage of internet is a pre-requisite. Our users expect that the installation and operation of the platform must be easy. It should be reasonable in size and the installation routines should be manageable by a lay computer user. Configuration must be simple and the operation should ideally only need a single start/stop mechanism. Specific details of the involved software packages should be hidden from the user (R1).

In the case of our industrial partners, a mechanism to check the integrity of the software package in order to prove authenticity and to prevent harm through tampered installation packages was required (R2).

Accessing the ubiquitous platform and using it should be easy for end-users. It should not require any installation or configuration other than a standard Web browser in our case (R3).

After installation the group member who hosts the ubiquitous platform has to start the platform and invite the rest of the group. Since the ubiquitous platform is not hosted in a central manner, notification mechanisms have to be provided in order to inform the group members about the availability of the ubiquitous platform. Flexible group formation mechanisms are needed for ad hoc group forming (R4).

Since data is only kept on the group's node (on the ubiquitous platform), group members need import and export mechanisms to transfer information and artifacts to other collaboration environments via the main collaboration platform. Different levels of privacy should be accommodated (anonymous data, history data, secret data). This way, data can be shared with a different environment, without disclosing unwanted information (R5).

5 Discussion

In this section, we discuss how the proposed approach fulfills the requirements from sections 2 and 4.

The ubiquitous platform does not integrate, e.g., persistent chat, threaded mail functionality, student gallery functionality allowing to see all platform users, and many other unnecessary features for small group interactions. This was done in order to reduce the size and the complexity of the software package. This allowed us to simplify the installation and configuration procedure for the ubiquitous platform. We provide an installer, which handles the installation of several standard software packages that are integrated into the CURE platform and which reduces the configuration options to a necessary minimum, by providing a predefined configuration setup. The installation routine creates a program group under the start menu and an icon on the desktop, to allow starting and stopping the ubiquitous platform with a single double click. Therefore we fulfill (R1).

Expert users and industrial partners are able to check out the source code from the repository and then manually install all necessary software packages. Thereby they

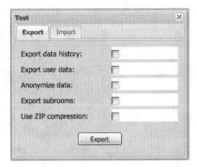

Fig. 4. Invitation window (left) and Export window (right)

can take a look at the source code and check its integrity. They can also use more detailed configuration options, if their installation needs to accommodate existing firewall configurations and VPN access. However, this presumes an advanced knowledge of the CURE platform. Consequently we fulfill (R2).

Group members can work with the ubiquitous platform, using a standard web browser. No additional software is needed on the client-side, therefore fulfilling (R3).

To allow for simple ad hoc group formation, we included a basic email invitation mechanism, which allows a user to send an invitation to the other prospective group members. In the main menu we offer a simple invitation button, which will open up an invitation window (Figure 4 left). There the invitees email address has to be entered as well as an optional message, and the invitee will receive an email with a link to the registration page of the system. This interface supports inviting several members at the same time. An email notification is sent by each start or shutdown of the ubiquitous platform which includes the new generated IP-based URL (since the majority of the providers periodically change assigned IP addresses). In this way the users can simply invite new collaborators to collaboration environments. By inviting people who may or may not respond we support R4.

Since we want to interact with a main collaborative platform as well, we need a way to transfer our work to the main collaborative platform. To accommodate our privacy and trust needs we need a way to transfer only that information that we want to be seen to the main collaboration platform. The retrofitted CURE and its ubiquitous pendant were extended to support the export and import mechanisms implemented in [5]. We refined those mechanisms and added them to the ubiquitous platform. While the import simply reads and includes the given data, the export mechanism allows filtering the information that is exported. Basically, the exported data can be completely anonymized, stripping all user information from the export data set. Additionally, all history information from the versioning system can be removed to only export the current data. If the exported data should include personal user information, this has to be explicitly requested when exporting the data. The current implementation allows only members with full access rights to do this. Export is provided through the user interface shown in Figure 4 (right), fulfilling (R5). The URI of the target platform is configured by the installation.

6 Conclusions

The main question addressed in this paper is how to preserve and enhance privacy in collaborative and social networks, especially when such environments deal with real identities. Our proposed solution to this problem is based on the way in which such environments are constructed. We propose a star topology where the center node represents the main platform of the collaborative system (allowing sharing between group-focused environments) and the surrounding nodes represent the end-users' ubiquitous platforms (hosting the group-focused environments). By doing this, user groups are able to share their data with other groups or even other collaborative environments, having complete control over their data. Our approach encourages users to work together without the pressure that all interaction traces can be monitored and evaluated, which can lead to full-fledged user profiles. Information can be shared with explicit consent. We developed a prototype consisting of a main collaborative platform and its ubiquitous pendant. The latter is installed, managed and hosted by the individual groups themselves using their own hardware. This methodology was applied on some chosen CSCL and CSCW scenarios and shows great promise. In order to port our approach to systems supporting social networks we expect that those systems provide remote interfaces for interaction and exchange of data with the ubiquitous pendants.

References

1. Agrawal, R., Kiernan, J., Srikant, R., Xu, Y.: Implementing P3P using database technology. In: ICDE, pp. 595–606 (2003)
2. Antón, A.I., Bertino, E., Li, N., Yu, T.: A roadmap for comprehensive online privacy policy management. CACM 50(7), 109–116 (2007)
3. Bellotti, V., Sellen, A.J.: Design for privacy in ubiquitous computing environments. In: de Michelis, G., Simone, C., Schmidt, K. (eds.) Proceedings of the Third European Conference on Computer Supported Cooperative Work - ECSCW 1993, pp. 77–92. Kluwer, Dordrecht (1993)
4. Borcea-Pfitzmann, K., Liesebach, K., Pfitzmann, A.: Establishing a privacy-aware collaborative elearning environment. In: Proceedings of the EADTU Annual Conference 2005: Towards Lisbon 2010: Collaboration for Innovative Content in Lifelong Open and Flexible Learning (2005)
5. Bourimi, M.: Collaborative design and tailoring of web based learning environments in CURE. In: Dimitriadis, Y.A., Zigurs, I., Gómez-Sánchez, E. (eds.) CRIWG 2006. LNCS, vol. 4154, pp. 421–436. Springer, Heidelberg (2006)
6. Bourimi, M., Lukosch, S., Kuehnel, F.: Leveraging visual tailoring and synchronous awareness in web-based collaborative systems. In: Haake, J.M., Ochoa, S.F., Cechich, A. (eds.) CRIWG 2007. LNCS, vol. 4715, pp. 40–55. Springer, Heidelberg (2007)
7. Boyle, M., Greenberg, S.: The language of privacy: Learning from video media space analysis and design. ACM Trans. Comput.-Hum. Interact. 12(2), 328–370 (2005)
8. Boyle, M., Neustaedter, C., Greenberg, S.: Privacy factors in video-based media spaces. In: Harrision, S. (ed.) n Media Space: 20+ Years of Mediated Life, pp. 99–124. Springer, Heidelberg (2008)

9. Camenisch, J., Crane, S., Fischer-Hbner, S., Leenes, R., Pearson, S., Pettersson, J.S., Sommer, D., Andersson, C.: Trust in PRIME. In: Proceedings of the Fifth IEEE International Symposium on Signal Processing and Information Technology, December 2005, pp. 552–559 (2005)
10. Cha, S.C., Joung, Y.J.: From P3P to data licenses. In: Dingledine, R. (ed.) PET 2003. LNCS, vol. 2760, pp. 205–221. Springer, Heidelberg (2003)
11. Cheng, L.T., Patterson, J., Rohall, S.L., Hupfer, S., Ross, S.: Weaving a social fabric into existing software. In: AOSD 2005: Proceedings of the 4th international conference on Aspect-oriented software development, pp. 147–158. ACM Press, New York (2005)
12. Clement, A.: Considering privacy in the development of multimedia communications, pp. 907–931 (1996)
13. Cranor, L., Langheinrich, M., Marchioriand, M., Presler-Marshall, M., Reagle, J.: Platform for privacy preferences (p3p). In: W3C Recommendations (2002)
14. Dourish, P.: Culture and control in a media space. In: ECSCW 1993: Proceedings of the third conference on European Conference on Computer-Supported Cooperative Work, Norwell, MA, USA, pp. 125–137. Kluwer Academic Publishers, Dordrecht (1993)
15. EPAL: The enterprise privacy authorization language, epal 1.2 specification. IBM (2004)
16. EPIC, Junkbuster: Pretty Poor Privacy: An Assessment of P3P and Internet Privacy (2002)
17. EU News: Citizens' privacy must become priority in digital age, says EU Commissioner Reding (2009),
    http://ec.europa.eu/informationsociety/newsroom/cf/
    itemdetail.cfm?itemid=4881
18. Fischer-Huebner, S.: Privacy-Enhancing Technologies, Lecture Notes in Computer Science. LNCS. Springer, Heidelberg (2001)
19. Gross, R., Acquisti, A., Heinz III, H.J.: Information revelation and privacy in online social networks. In: WPES 2005: Proceedings of the 2005 ACM workshop on Privacy in the electronic society, pp. 71–80. ACM, New York (2005)
20. Gutwin, C., Stark, G., Greenberg, S.: Support for workspace awareness in educational groupware. In: CSCL 1995: The first international conference on Computer support for collaborative learning, Hillsdale, NJ, USA, pp. 147–156. L. Erlbaum Associates Inc., Mahwah (1995)
21. Haake, J.M., Schümmer, T., Haake, A., Bourimi, M., Landgraf, B.: Supporting flexible collaborative distance learning in the cure platform, vol. 1, p. 10003a. IEEE Computer Society, Los Alamitos (2004)
22. Haake, J.M., Schümmer, T., Haake, A., Bourimi, M., Landgraf, B.: Two-level tailoring support for CSCL. In: Favela, J., Decouchant, D. (eds.) CRIWG 2003. LNCS, vol. 2806, pp. 74–81. Springer, Heidelberg (2003)
23. Haake, J.M., Haake, A., Schümmer, T., Bourimi, M., Landgraf, B.: End-user controlled group formation and access rights management in a shared workspace system. In: CSCW 2004: Proceedings of the 2004 ACM conference on Computer supported cooperative work, Chicago, Illinois, USA, November 6-10, pp. 554–563. ACM Press, New York (2004)
24. Isenberg, D.: GigaLaw Guide to Internet Law. Random House Inc., New York (2002)
25. Koch, M., Richter, A.: Enterprise 2.0: Planung, Einführung und erfolgreicher Einsatz von Social Software in Unternehmen: Planung, Einführung und erfolgreicher Einsatz von Social Software in Unternehmen. Oldenbourg, München, Wien (2007)
26. Kühnel, F.: Visual tailoring and synchronous awareness support in web-based collaborative systems. Master's thesis, FernUniversität in Hagen (January 2008)

27. Lukosch, S., Bourimi, M.: Towards an enhanced adaptability and usability of web-based collaborative systems. International Journal of Cooperative Information Systems, Special Issue on 'Design, Implementation of Groupware, 467–494 (2008)
28. Narayanan, A., Shmatikov, V.: De-anonymizing social networks (2009)
29. Palen, L., Dourish, P.: Unpacking "privacy" for a networked world. In: CHI 2003: Proceedings of the SIGCHI conference on Human factors in computing systems, pp. 129–136. ACM Press, New York (2003)
30. Pearson, S.: Towards automated evaluation of trust constraints. In: Stølen, K., Winsborough, W.H., Martinelli, F., Massacci, F. (eds.) iTrust 2006. LNCS, vol. 3986, pp. 252–266. Springer, Heidelberg (2006)
31. PRIME (June 2009), http://www.prime-psroject.eu.org/
32. Smale, S.K.: Collecting and Sharing Transient Personal Information Online. Master's thesis, University Of Calgary (November 2007)
33. Szugat, M., Gewehr, J.E., Lochmann, C.: Social Software, 1st edn. Entwickler Press (2007)
34. Tang, J.C., Isaacs, E.A., Rua, M.: Supporting distributed groups with a montage of lightweight interactions. In: CSCW 1994: Proceedings of the 1994 ACM conference on Computer supported cooperative work, pp. 23–34. ACM, New York (1994)
35. Woolf, B.P.: Building Intelligent Interactive Tutors. Morgan Kaufman, San Francisco (2008)

Why Should I Trust in a Virtual Community Member?

Juan Pablo Soto, Aurora Vizcaíno, Javier Portillo-Rodríguez, and Mario Piattini

Alarcos Research Group
Escuela Superior de Informática
Information Systems and Technologies Department
Indra-UCLM Research and Development Institute
University of Castilla-La Mancha
Ciudad Real, Spain
juanpablo.soto@inf-cr.uclm.es,
{aurora.vizcaino,javier.portillo,mario.piattini}@uclm.es

Abstract. A huge amount of virtual communities focusing on different topics currently exist. In this paper we centre on those virtual communities in which people share knowledge and experience. However, the level of knowledge shared may decrease when there is no face to face communication and when members do not have the chance to meet each other personally. In order to reduce this problem we propose a trust model with which to help community members decide whether another person is trustworthy or otherwise.

Keywords: Virtual Communities, Trust, Software Agents.

1 Introduction

The development of groupware technologies and the Internet has led to a new kind of community, "virtual communities", in which members may or may not meet one another face to face and may exchange words and ideas through the medium of computer networks [1]. According to the definition of Rothaermel and Sugiyama in [2] a virtual community can be seen as a group in which individuals come together around a shared purpose, interest, or goal.

The knowledge shared in virtual communities is highly important. It is therefore essential to encourage contributions if the community is to be successful and sustainable. Virtual community practitioners have developed various mechanisms in the hope of encouraging member participation and contribution. Nevertheless, since the people in present-day virtual communities are usually geographically dispersed they do not have a face to face communication and this situation could be problematic since the main knowledge sources in virtual communities are the members themselves. We consider that it is highly important to be able to discover how trustworthy a knowledge source (i.e. another member) is. This knowledge will help members to decide whether or not a document is valuable depending on the knowledge source from which it originates.

Despite the importance of virtual communities, large numbers of them fail. Participation is often sub-optimal, with only a small minority contributing. Under-contributing is a problem even in those communities that do survive [3]. For instance,

L. Carriço, N. Baloian, and B. Fonseca (Eds.): CRIWG 2009, LNCS 5784, pp. 126–133, 2009.

in open source development communities, four percent of members account for 50 percent of answers on a user-to-user help site [4], and four percent of developers contribute 88% of new code and 66% of code fixes [5]. Other problems in this kind of environment are related to communication and coordination, and are made more difficult as a result of differences in culture, timetable, language, etc [6].

Furthermore, although virtual communities are a focus of knowledge sharing there is hardly ever any quality control of the knowledge generated in the community. In order to avoid these situations we propose a trust model to discover which knowledge sources are trustworthy. Moreover, we intend to implement this trust model in a multi-agent system in which one software agent represents one member of the community. The software agent will therefore be able to use the trust model to recommend trustworthy members, knowledge, etc., to the user

The remainder of the paper is organized as follows. The following section presents two important concepts related to our work: trust and reputation. Section 3 presents our model of virtual communities. Section 4 then describes the trust model that we propose for use in virtual communities. Later in Section 5 the prototype based on the virtual community model is outlined. Finally in Section 6 conclusions and future work are presented.

2 Trust and Reputation Models

There are many recent proposals for reputation mechanisms and approaches to evaluate trust in P2P systems in general [7, 8], and multi-agent systems in particular [9-11, 8]. However, there is no universal agreement on the definition of trust and reputation. Since the main goal of our work is to rate the credibility of information sources and of knowledge in virtual communities, it is first necessary to define these two important concepts.

Trust is a complex notion whose study is usually of a narrow scope. This has given rise to an evident lack of coherence among researchers in the definition of trust. For instance in [7], Wang and Vassileva define trust as a peer's belief in another peer's capabilities, honesty and reliability based on his/her own direct experiences.

Another important concept related to trust is reputation. Several definitions of reputation can be found in literature, such as that of Barber and Kim who define this concept as the amount of trust that an agent has in an information source, created through interactions with information sources [12], and that of Mui et al [13] which defines reputation as a perception a partner creates through past actions about his intentions and norms. This may be considered as a global or personalized quantity [13].

The concepts of trust and reputation are sometimes used interchangeably. However, recent research has shown that there is a clear difference between them, whilst accepting that there is a certain amount of correlation between the two concepts in some cases[14, 15].

In our work we intend to follow the definition given by Wang and Vassileva which considers that the difference between both concepts depends on who has previous experience, so if a person has direct experiences of, for instance, a knowledge source we can say that this person has a trust value in that knowledge.

The main differences between previous reputation/trust models and our approach are that most of previous models need an initial number of interactions to obtain a good reputation value and it is not possible to use them to discover whether or not a new user can be trusted. A further difference is that our approach is oriented towards collaboration between users in virtual communities. Other approaches are more oriented towards competition, and most of them are tested in auctions. Before describing the trust model proposed, in the following section we shall define the virtual community model to be used in organizations whose employees are organized in communities.

3   Community Virtual Model

This model is based on the Isakovic and Sulcic proposal [16]. In this proposal the authors consider two factors (purpose and people). However, we consider that trust is another important factor that must be considered in this kind of communities.

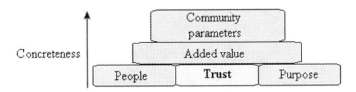

Fig. 1. Virtual community model

Therefore, the bases for our model are three community factors: people, trust and purpose, as is shown in Figure 1.

- The *Purpose* factor defines the purpose of the community in the organization.
- The *People* factor defines the community members' roles and type of participation. For instance, in a virtual community a person can play three types of roles:
    - The person contributes new knowledge to the communities in which s/he is registered. In this case the person plays the role of Provider.
    - The person uses knowledge previously stored in the community. This person will therefore be considered as a Consumer.
    - The person helps other users to achieve their goals, for instance by giving an evaluation of certain knowledge. In this case the role is that of the Partner.
- The *Trust* factor is in charge of generating a trust value for the knowledge sources with which a person interacts in the virtual community. It is of interest to note that members of a community are frequently more likely to use knowledge built by their community team members than those created by members outside their group. This occurs because people trust more in the information offered by a member of their community than in that supplied by a person who does not belong to that community. Of course, the fact of belonging to the same virtual community already implies that these people

have similar interests and perhaps the same level of knowledge about a topic. Consequently, the level of trust within a community is often higher than that which exists outside the community. The aforementioned reasons have led us to consider that the implementation of a mechanism in charge of measuring and controlling the confidence level in a community in which the members share information is of great importance.

Based on community purpose, it is possible to define the community *Added Value*. For instance, in our case, the community purpose is based on providing the users with a friendly environment in order to allow them to share, reuse and learn from their own experience.

After the main community factors have been defined, we define the *Community Parameters* used to specify the community details in more concrete terms, for instance, social norms, profiles, events, rewards, etc. In order to show the feasibility of this model, in the following section we shall describe the trust model proposed for use in virtual communities.

4 Trust Model in Virtual Communities

One of ours aims is to provide a trust model based on real world social properties of trust in virtual communities.

Most previous trust models calculate trust by using only the users' previous experience with other users, but several factors, such as shared social norms, repeated interactions, and shared experiences, have been suggested to facilitate the development of trust [17]. Because of this we propose some social factors such as:

- *Position.* employees often consider information that comes from a boss as being more reliable than that which comes from another employee in the same (or a lower) position as him/her [18]. Such different positions inevitably influence the way in which knowledge is acquired, diffused and eventually transformed within the local area.
- *Expertise.* This is an important factor since people often trust experts more than novice employees. In addition, "individual" level knowledge is embedded in the skills and competencies of the researchers, experts, and professionals working in the organization [19]. The level of expertise that a person has in a company or in a CoP could be calculated from his/her CV or by considering the amount of time that a person has been working on a topic. This is data that most companies are presumed to have.
- *Previous experience.* This is a critical factor in rating a trust value since previous experience is the key value through which to obtain a precise trust value. However, when previous experience is scarce, or it does not exist, humans use other factors to decide whether or not to trust in a person or a knowledge source. One of these factors is intuition.
- *Intuition.* This is a subjective factor which, according to our study of the state-of-the-art, has not been considered in previous trust models. However, this concept is of great importance since when people do not have any previous experience they often use their "intuition" to decide whether or not they are

going to trust something. We have attempted to model intuition according to the similarity between personal profiles: the greater the similarity between one person and another, the greater the level of trust in that person as a result of intuition.

There are three different ways of using these factors, which depend upon the agent's situation:

1. If the agent has no previous experience, for instance because it is a new user in the community, then the agent uses position, expertise and intuition to obtain an initial trust value and this value is used to discover which other agents it can trust.

2. When the agent has previous experience obtained through interactions with other agents but this previous experience is low (low number of interactions), the agent calculates the trust value by considering the intuition value and the experience value. For instance, a person who has to choose between information from two different people will normally choose that which comes from the person who has the same background, same customs etc. as him/her. By following this pattern, the agents compare their own profiles with those of the other agents in order to decide whether a person appears to be trustworthy or not. We could say that an agent 'thinks' "I do not know whether I can trust this agent but it has similar features to me so it seems trustworthy". The agents' profiles may alter according to the community in which they are working.

3. When the agent has sufficient previous experience to consider that the trust value it has obtained is reliable, then the agent only considers this value.

The trust model is translated into a value by using the following formula:

$$T_{ij} = w_e * E_j + w_p * P_j + w_I * I_{ij} + \frac{1}{n}\sum_{i=1}^{n} QC_{ij}$$

where T_{ij} is the trust value of j in the eyes of i, and E_j is the value of expertise which is calculated according to the degree of experience that the person upon whose behalf the agent acts has in a domain. P_j is the value assigned to a person's position. I_{ij} denotes the intuition value that agent i has in agent j, and is calculated by comparing each of the users' profiles.

Previous experience should also be calculated. When an agent i consults information from another agent j, agent i should evaluate how useful that information is. This value is called QC_{ij} (Quality of j's Contribution in the opinion of i). To attain the average value of an agent's contribution, we calculate the sum of all the values assigned to these contributions and we divide it between their total. In the expression n represents the total number of evaluated contributions.

Finally, w_e, w_p and w_I are weights with which the trust value can be adjusted according to the degree of knowledge that one agent has about another. Therefore, if an agent i has had frequent interactions with another agent j, then agent i will give a low weight (or even zero) to w_I since, in this case, previous experience is more important than intuition. The same may occur with w_e, w_p. The weights may therefore have the value of 0 or 1 depending on the previous experience that an agent has.

5 Prototype

A prototype has been constructed to offer virtual community members the possibility of obtaining document recommendations. The prototype also offers the possibility of registering in a community, connecting to a community and sending/evaluating documents.

In order to illustrate how the prototype works, let us look at an example. If a user selects a topic and wishes to search for documents related to that topic, his/her user agent will contact other user agents which have documents concerning said topic, and the user agent will then calculate the trust value for each agent, which means that these agents are considered to be knowledge sources and the user agent needs to calculate which "knowledge source" is more trustworthy. Once these values have been calculated, the user agent only shows its user the documents which have come from the most trustworthy agents (see Figure 2).

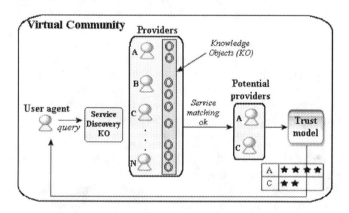

Fig. 2. Trust model integration in a virtual community

This method of rating trust helps to detect an increasing problem in companies or communities in which employees are rewarded if they contribute with knowledge in the community. Thus, if a person introduces non-valuable documents with the sole aim of obtaining rewards, the situation can be detected since these documents will have low values and the person will also be considered to be less trustworthy. The agent will not, therefore, recommend those documents. Moreover, this model implies the reduction of users' overload when they use knowledge management systems, since with this model the user agent only recommends the most adequate and trustworthy knowledge.

6 Conclusion and Future Work

In this paper we have presented a virtual community model and a trust model to create a trustworthy environment for community members. We have also designed a prototype to support virtual communities in which knowledge sources are rated by using the trust model proposed, and which is to be used solely in virtual communities. In this prototype virtual community members can introduce documents and the software agents must decide how trustworthy those documents are for the user that they represent.

One important contribution of the prototype (described in Section 5) is that it detects experts in a community, since those knowledge sources with high trust values are supposed to be people who contribute with valuable knowledge. The trust model used also helps to detect fraud when users contribute with non-valuable knowledge. Another important feature of our trust model, and that which makes it different from previous models, is that even when a user is new to the community and other agents do not have any previous experience of working with him/her, the trust model allows agents to obtain a preliminary trust value by considering other factors such as the new agent's position and level of expertise, along with the intuition that each agent has with regard to the new member. We thus attempt to model human features, since when a person has to evaluate something and s/he has no previous experience that person uses other aspects such as his/her intuition in order to decide whether or not to trust in it.

In future work, we plan to extend our experiments to consider each of the trust model factors (previous experience, intuition, expertise and position) separately, in order to detect the trust value's variability with regard to the factor used. For instance, trust models that use only direct experiences typically require a great deal of time to achieve stable performance. Furthermore, we shall focus on using different trust models in the virtual community model proposed in order to make a comparison and to measure the feasibility of our trust model with regard to other models.

Acknowledgments. This work is partially supported by FABRUM project, Ministerio de Ciencia e Innovación (grant PPT-430000-2008-063), the MELISA (PAC08-0142-3315) and ENGLOBAS (PII2109-0147-8235) projects, Junta de Comunidades de Castilla-La Mancha, Consejería de Educación y Ciencia, in Spain and CONACYT (México) under grant of the scholarship 206147 provided to the first author.

References

1. Geib, M., Braun, C., Kolbe, L., Brenner, W.: Measuring the Utilization of Collaboration Technology for Knowledge Development and Exchange in Virtual Communities. In: Proceedings of the 37th Annual Hawaii International Conference on Systems Sciences (HICSS), vol. 1 (2004)
2. Rothaermel, F., Sugiyama, S.: Virtual Internet communities and commercial success: Individual and community-level theory grounded in the atypical case of timezone.com. Journal of Management 27(3), 297–312 (2001)
3. Beenen, G., Ling, K., Wang, X., Chang, K., Frankowski, D.: Using Social Psychology to Motivate Contributions to Online Communities. In: CSCW, vol. 6(3), pp. 212–221. ACM Press, New York (2004)
4. Lakhani, K., Hippel, E.: How Open Source Software Works: "Free" user to user assistance. Research Policy 32, 923–943 (2003)

5. Mockus, A., Fielding, R., Andersen, H.: Two case studies of open source software development: Apache and Mozilla. ACM Transactions on Software Engineering and Methodology 11(3), 309–346 (2002)
6. Boland, D., Fitzgerald, B.: Transitioning from a Co-Located to a Globally-Distributed Software Development Team: A Case Study at Analog Devices Inc. In: 3rd International Workshop on Global Software Development, Edinburgh, Scotland (2004)
7. Wang, Y., Vassileva, J.: Trust and Reputation Model in Peer-to-Peer Networks. In: Proceedings of the 3rd International Conference on Peer-to-Peer Computing, pp. 150–157 (2003)
8. Yu, B., Singh, M., Sycara, K.: An evidential model of distributed reputation management. In: Proceedings of the first international joint conference on Autonomous agents and multiagents systems (AAMAS), pp. 294–301. ACM Press, New York (2002)
9. Huynh, T., Jennings, N., Shadbolt, N.: FIRE: An Integrated Trust and Reputation Model for Open Multi-agent Systems. In: Proceedings of the 16th European Conference on Artificial Intelligence, ECAI (2004)
10. Sabater, J., Sierra, C.: Reputation and Social Network Analysis in Multi-Agent Systems. In: Proceedings of the first international joint conference on autonomous agents and multiagent systems (AAMAS), pp. 475–482. ACM Press, New York (2002)
11. Taecy, W., Chalkiadakis, G., Rogers, A., Jennings, N.: Sequential Decision Making with Untrustworthy Service Providers. In: Proceedings of 7th International Conference on autonomous Agents and Multiagent Systems (AAMAS), pp. 755–762 (2008)
12. Barber, K., Kim, J.: Belief Revision Process Based on Trust: Simulation Experiments. In: 4th Workshop on Deception, Fraud and Trust in Agent Societies, Montreal, Canada, pp. 1–12 (2004)
13. Mui, L., Mohtashemi, M., Halberstadt, A.: A Computational Model of Trust and Reputation for E-businesses. In: Proceedings of the 35th Hawaii International Conference on Systems Sciences (HICSS), vol. 7, p. 188. IEEE Computer Society Press, Los Alamitos (2002)
14. Jøsang, A., Ismail, R., Boyd, C.: A Survey of Trust and Reputation Systems for Online Services Provision. Decision Support Systems 43(2), 618–644 (2007)
15. Sabater, J., Sierra, C.: Review on Computational Trust and Reputation Models. Artificial Intelligence Review 24, 33–60 (2005)
16. Isakovic, J., Sulcic, A.: A Value-Added Methodology for Defining Virtual Communities for Enterprises. In: Proceedings of the 12th international conference on entertainment and media in ubiquitous era, pp. 159–161 (2008)
17. Meyer, R., Davis, J., Schoorman, F.: An integrative model of organizational trust. Academy of Management Review 20(3), 709–734 (1995)
18. Wasserman, S., Glaskiewics, J.: Advances in Social Networks Analysis. Sage Publications, Thousand Oaks (1994)
19. Nonaka, I., Takeuchi, H.: The Knowledge Creation Company: How Japanese Companies Create the Dynamics of Innovation. Oxford University Press, Oxford (1995)

Antecedents of Awareness in Virtual Teams

Chyng-Yang Jang

University of Texas at Arlington

Abstract. This study examined the antecedents of awareness in geographically distributed teams. Task structure, group propensity, communication frequency, groupware usage, and subjective reliance on various communication tools were proposed to affect the perceived level of awareness. Based on data collected from seven distributed student engineering teams, results showed that perceived task interdependence and communication frequency were positively associated with the level of awareness, but group propensity and groupware usage were not. Further analysis found the reliance on synchronous meetings mediated the relationship between task interdependence and awareness level. Implications for virtual team management are discussed.

Keywords: Virtual team, awareness.

1 Introduction

The separation of time and space among virtual team members creates various communication barriers. Particularly, the lack of physical proximity makes it difficult for virtual team members to have frequent encounters and spontaneous conversations to exchange project-related information [1-3]. As a result, maintaining awareness regarding the current project status and distant teammates' task progress becomes a major challenge to distributed work groups [4, 5]. Without timely updates on each other's progress or problems, team collaboration suffers from greater process loss. Therefore, managing awareness needs is a critical issue for virtual teamwork.

The issue of awareness was first recognized by computer-supported cooperative work (CSCW) and groupware research communities. A number of groupware tools have been designed to address this problem of lack of awareness for distributed teams, including Porthole [6], Awareness Widgets [7], TeamSCOPE [2], Sideshow [8], Classroom BRIDGE [9], and FASTDash [10]. These tools emphasize on the design issues, such as how to represent different types of awareness information and how to minimize cognitive loads for the users [11]. While these innovations attempt to address various aspects of awareness supply in distributed teams, factors related to team members' awareness demands were seldom a part of these investigations. However, it is the demand for timely updates that drives distant collaborators to select and use various awareness tools. A better understanding of what motivates the awareness demand should lead to valuable lessons for virtual team management. This study aimed to inquire about the antecedents of awareness in distant collaboration. Literatures from work group study, group communication, and social psychology helped to orient this investigation.

L. Carriço, N. Baloian, and B. Fonseca (Eds.): CRIWG 2009, LNCS 5784, pp. 134–141, 2009.
© Springer-Verlag Berlin Heidelberg 2009

2 Literature Review and Hypotheses

Awareness is essential for collaborative work. Ethnographic research on workplace found that collaborators constantly keep each other updated on task-related activities [12, 13]. By staying aware of the current state of the project and the activities of co-workers, people are able to adjust and orient their own and their partners' work toward a common goal. In this study, awareness is considered as possessing knowledge about the current status and actions of the various components in a collaborative system, including people, tasks, shared objects, communication media, collaborative software tools, and the work environments [2]. In the context of distributed work, the difficulties in both collecting and delivering awareness information translate into higher costs for collaboration. People working in virtual teams have to be motivated to overcome those barriers. This study looked at two possible motivators of awareness conducts: task interdependence and group propensity. Each will be discussed below. In addition, due to geographical dispersion, virtual team members no longer physically share a workspace. All work exchanges will be carried out using electronic media. The impact of media and groupware use will also be examined.

Task Interdependence. Task interdependence is one of the most important group variables and is considered a dominant factor that dictates how collaborators interact with each other [14, 15]. At the group level, task interdependence is viewed as the extent to which group members are dependent upon one another to perform their individual tasks due to the structural relationship between team members and the nature of the task [16]. Literature suggests that task interdependence has predictive power on group communication and information exchange [15, 17]. Higher mutual dependence requires tighter coordination which, in turn, posts higher demand for awareness information. The more teammates depend on each other, the more they need to stay updated in order to coordinate their works. Task interdependence is, therefore, a motivator that prompts team members to provide and acquire awareness information.

H1: Higher level of perceived task interdependence will lead to higher level of perceived awareness.

In addition, the mutual dependency embedded in the task structure can be seen as a source of uncertainty and, accordingly, a source of information demand [15]. Both lead to more frequent and flexible interaction among collaborators. As reported by Staples and Jarvenpaa [18], the use of electronic media increased when people perceived a higher level of task interdependence with one another. In virtual teams with groupware support, more frequent and flexible interaction should be represented by higher communication frequency and more usage of the groupware.

H2a: Higher level of perceived task interdependence will lead to higher communication frequency.
H2b: Higher level of perceived task interdependence will lead to higher groupware usage.

Group Propensity. Group propensity refers to a general preference toward working in groups or teams. Individuals who prefer teamwork are more satisfied and effective in collaboration [16, 19]. On the other hand, people who prefer higher work autonomy provide less help to and learn less from their teammates [20]. One major difference between working alone and working in a team is the coordination and interaction involved in teamwork. Without proper coordination, a team will not be effective, nor will the process be enjoyable. Therefore, people who prefer teamwork will perceive the resources spent on coordination worthwhile. They will pay more attention to partners' activities and be more likely to use all available communication tools for awareness conducts. As a result, they should obtain a higher level of awareness. Thus,

> H3a: Higher group propensity will lead to higher communication frequency.
> H3b: Higher group propensity will lead to higher groupware usage.
> H4: Higher group propensity will lead to higher level of perceived awareness.

Media Use. Since all interaction among virtual team members is carried out through electronic media, both seeking and presenting awareness information have to be conducted via networking technologies. The use of communication channels is a necessary condition to obtain awareness. The more frequent participants use these media, the more opportunities they have to gather awareness information. Thus,

H5a: Higher communication frequency overall will lead to higher level of perceived awareness.
H5b: Higher groupware usage will lead to higher level of perceived awareness.

While counting of communication events can provide a general picture of communication pattern, it can be problematic to weigh an email message as the same as an hour-long videoconference due to their apparent differences in interaction intensity and media properties. To further explore the role of media in virtual teams, members' subjective reliance on each of the communication channels for awareness purposes is also investigated.

Research Question: What are the relationships between channel preferences and task interdependence, group propensity, and awareness level?

3   Method

Seven student engineering design teams participated in the study with six to eight members per team. Students were recruited from Mexico, Russia, and the United States. These teams were told to collaboratively complete a 13-week design project provided by industry partners. The projects included the design of automotive air conditioning system refinement, stair-climbing wheelchair, small animal intensive care unit, end effector storage unit, mechatronics laboratory demonstration system, CAD modeling of automotive structures, and automotive thermo electric units. All teams were provided with a range of communication tools including ISDN- and IP-based video conferencing systems, telephone, fax, email, and TeamSCOPE.

TeamSCOPE is a Web-based groupware application, which was designed to meet the needs of a common information repository and awareness support for distributed collaboration [2]. It features a shared file space, calendar, message board, and several awareness related functions. All activities occurred in TeamSCOPE are recorded. These records of activities are then presented in various ways, including activity summary, file activity history, viewing records, and login history, for users to conveniently acquire awareness information.

Data Collection and Measurement. Three sets of data were collected in this study. First, all participants submitted weekly communication logs. The average numbers of weekly communication incidents reported on these logs were calculated as participants' communication frequency. Second, TeamSCOPE usage was recorded by the system. Each user's average page requests per week were calculated to form the measurement of TeamSCOPE usage. Third, a survey administrated at the end of the project to acquire the following subjective measurements.

Task interdependence was measured as team members' perception on the degree to which they rely on each other to perform their own tasks. Three 5-point items were adapted from the literature [16, 20, 21] to form the scale of task interdependence.

Group propensity measures the individual's preference toward teamwork. Three 5-point items modified from Campion et al. [16] were used. One example item reads, "I prefer working alone to working in a group."

Awareness is defined as possessing knowledge about the current status and actions in a collaborative environment. Four 5-point items converged to form the awareness scale used in the analysis. One example item reads: "I usually had a good idea of what my distant teammates were working on."

Finally, regarding reliance on email, TeamSCOPE, and real-time distant meeting for awareness information, participants provided a single evaluation for each channel, ranging from 1 ("not at all") to 5 ("very much").

4 Results

Results of the correlation analysis are reported in Table 1. As hypothesized, perceived task interdependence ($r=.47$, $p<.001$) was a positive predictor of perceived awareness. H1 was supported. Task interdependence also had a positive relationship with communication frequency ($r=.28$, $p<.05$). H2a was supported. TeamSCOPE usage was positively associated with task interdependence; however, their relationship was not statistically significant. H2b was not supported. Group propensity was proposed to lead to higher communication frequency, TeamSCOPE usage, as well as higher level of awareness (H3a, H3b, H4). However, those hypotheses were not supported. The positive relationship of communication frequency and perceived awareness was confirmed ($r=.30$, $p<.05$). However, TeamSCOPE usage was not significantly related to perceived awareness level. H5a was supported, but H5b was not.

Correlation analysis also found that reliance scores on email and distant meetings were significantly associated with perceived task interdependence, but in opposite directions. Those who saw email as the main awareness conduit tended to report lower level of task interdependence ($r=-.34$, $p<.05$). On the contrary, the more participants relied on real-time meetings for awareness purposes, the more they felt they depended on distant teammates for the project (r=.57, p<.01). Reliance on distant meetings was also positively associated with perceived awareness (r=.47, p<.001). Reliance on TeamSCOPE, however, did not signify any particular trend.

Table 1. Bivariate Correlation Matrix

Heading level	TI	GP	CF	TU	RE	RT	RD
Perceived Task Interdependence (TI)							
Group Propensity (GP)	.41**						
Communication Frequency (CF)	.28*	.12					
TeamSCOPE Usage (TU)	.14	.27	.31*				
Reliance on Email (RE)	-.34*	.02	-.25	-.10			
Reliance on TeamSCOPE (RT)	.22	.07	-.03	.25	-.15		
Reliance on Distant Meeting (RD)	.57**	.25	.27	-.01	-.11	.16	
Perceived Awareness (AW)	.47***	.08	.30*	-.16	-.20	.02	.66**

*significant at .05 level, ** significant at .01 level, *** significant at .001 level

Since perceived task interdependence, communication frequency, and reliance on real-time meetings were all positively associated with perceived awareness, additional multiple regression analyses were conducted to test, based on the guidance offered in [22], whether the effect of task interdependence on awareness was mediated by communication frequency or reliance on synchronous meetings. The results were reported in Table 2. To assess the threat of collinearity, tolerance values were calculated and all tolerance values were larger than .66, well above the .10 level.

Table 2. Regression Analysis: Predicting Perceived Awareness

Model	1	2	3
Perceived task interdependence	.47***	.42**	.11
Communication Frequency		.18	.11
Reliance on Distant Meeting			.56***
R^2 (F)	.22 (13.82***)	.25 (7.97***)	.46 (13.22***)

** significant at .01 level, *** significant at .001 level

Models 1 and 2 above assessed the mediating role of communication frequency. As a mediator, communication frequency should be a significant predictor of perceived awareness and diminish the effect of task interdependence. However, it is not the case. In model 2, communication frequency did not contribute to awareness significantly, while the effect of task interdependence remained significant.

Reliance on real-time meetings was added to the regression analysis in Model 3. This newly added variable became the only significant contributor in the model (β=.56, p<.001). The result suggests that the subjective importance of distant meeting mediated the relationship between task structure and awareness level. Model 3 also explains more variance (R^2=.46) than Model 1 (R^2=.22) and Model 2 (R^2=.25).

5 Discussion and Conclusion

The focus of this study was to examine the antecedents of awareness. As expected, perceived task interdependence was positively associated with awareness level. However, individual preference for teamwork was not. Together, the results indicate that instrumental motivation, such as task requirements, has stronger dictating power on group process than intrinsic motivation, such as individual's teamwork preference. It suggests that to more effectively motivate team members to overcome communication barriers for awareness information, managers should carefully design task structure among different sites of a virtual team. Interestingly, though, people who prefer collaboration were more likely to see that they and their teammates were interdependent ($r=.41$, $p<.01$). It suggests that individual preferences may magnify the effect of task structure. An understanding of team members' teamwork propensity can help the managers to better their strategies in communicating task interdependence.

Higher degree of task interdependence also led to more communication incidents, which, too, was associated with the level of awareness. However, when both perceived task interdependence and communication frequency were entered into the same regression model, only task interdependence remained a statistically significant contributor. The number of communication incidents did not mediate the relationship between task structure and awareness level. It suggests that communication incident count was not a good enough predictor for awareness.

The lack of significant correlation between perceived task interdependence and TeamSCOPE usage, and between TeamSCOPE usage and the perceived level of awareness was not expected. In addition, the negative association, even though not significant, between TeamSCOPE usage and awareness level was surprising. These results may be explained by tool proficiency. Participants who knew where to find the information in TeamSCOPE would make fewer requests than those who did not. Therefore, high TeamSCOPE usage might indicate less effective search and, therefore, lower awareness level. Alternatively, the use of TeamSCOPE may indicate unsatisfied awareness demand. When members were not able to stay updated on distant teammates' progress, they were more likely to initiate awareness conducts. Since TeamSCOPE requires the least amount of efforts to acquire information comparing to other channels, logging on TeamSCOPE would be the easiest first step to take. However, due to its limitation in interactivity, TeamSCOPE may not be the most effective awareness channel and, therefore, users' awareness needs may left unaddressed. Overall, the lack of correspondence between the awareness level and groupware usage suggests a more fine-grained modeling of awareness [23].

The results related to participant's channel reliance also signify the importance of channels' interactive capability. Higher level of task interdependence was associated with perceived lower reliance on emails and higher dependence on real-time meetings. In addition, regression analyses showed that reliance on synchronous meetings completely mediated the relationship between task interdependence and awareness level. According to media richness theory, richer media allow people to better resolve ambiguity occurred in task communication [24]. Particularly, in the case of virtual teams, synchronicity provided by video or audio conferencing tools allows the awareness demand and supply to meet in real time. Even though setting up a meeting with distant members was much more difficult than writing an email or checking out records in TeamSCOPE, it was the preferred method for those who felt high level of interdependence embedded in their team's task structure. Future research should take subjective evaluation of communication media into consideration.

This research had several limitations. First, the use of student subjects, instead of real engineers, restricts the generalizability of the results. Additionally, the two-site formation with multiple-member subgroups was a specific team structure. Scheduling real-time meetings could be more difficult if more than two sites were involved. Also, the multiple-member subgroup arrangement allowed one to acquire project updates without interacting with distant teammates or using groupware. This arrangement made it difficult to determine the relationship between media use and awareness. The variation of team structure should be considered in the future studies.

In conclusion, despite its limitation, this study provided several insights on the relationships between awareness, task structure, media use, and media evaluation in virtual teams. For managing distributed collaboration, organizations should carefully design and communicate task interdependence among different sites. While task structure may serve as a motivator for virtual team members to overcome the communication barriers, it also posts higher awareness demand and may require rich media support to effectively meet the demand. Future research should continuously distill the demand and supply dimensions of awareness conducts and construct more fine-grained awareness model to enhance our understanding of awareness and offer better support to distant collaboration.

Acknowledgement

This study was part of the INTEnD project, which was led by Dr. Charles Steinfield.

References

1. Cramton, C.: Information problems in dispersed teams. In: Academy of Management Best Paper Proceedings, Southern University, Georgia, pp. 298–302 (1997)
2. Steinfield, C., Jang, C.-Y., Pfaff, B.: Supporting virtual team collaboration: the TeamSCOPE system. In: Group 1999. ACM Press, Phoenix (1999)
3. Tang, J.C.: Approaching and leave-taking: Negotiating contact in computer-mediated communication. ACM Transactions on Computer-Human Interaction 14(1) (2007)
4. Gutwin, C., Greenberg, S.: The Importance of awareness for team cognition in distributed collaboration. In: Salas, E., Fiore, S. (eds.) Team Cognition: Understanding the Factors that Drive Process and Performance, pp. 177–201. APA Press, Washington (2004)
5. Jang, C.-Y., Steinfield, C., Pfaff, B.: Virtual team awareness and groupware support: an evaluation of the TeamSCOPE system. International Journal of Human-Computer Interaction 56(1), 109–126 (2002)
6. Dourish, P., Bellotti, V.: Awareness and coordination in shared workspace. In: CSCW 1992. ACM Press, Toronto (1992)
7. Gutwin, C., Roseman, M., Greenberg, S.: A usability study of awareness widgets in a shared workspace groupware system. In: CSCW 1996. ACM Press, Cambridge (1996)

8. Cadiz, J.J., et al.: All ways aware: Designing and deploying an information awareness interface. In: The 2002 ACM conference on Computer supported cooperative work CSCW 2002, New Orleans, Louisiana, USA (2002)
9. Ganoe, C.H., et al.: Classroom BRIDGE: using collaborative public and desktop timelines to support activity awareness. In: The 16th annual ACM symposium on User interface software and technology, Vancouver, Canada (2003)
10. Biehl, J.T., et al.: Distributed coordination: FASTDash: a visual dashboard for fostering awareness in software teams. In: CHI 2007, San Jose, California, USA (2007)
11. Schmidt, K.: The problem with 'awareness'. Computer Supported Cooperative Work 11, 285–298 (2002)
12. Harper, R.H.R., Hughes, J.A., Shapiro, D.Z.: Working in harmony: An examination of computer technology in air traffic control. In: ECSCW 1989: the First European Conference on Computer Supported Cooperative Work. Gatwick, London (1989)
13. Heath, C.C., Luff, P.: Collaborative activity and technological design: Task coordination in London Underground control rooms. In: ECSCW 1991: the Second European Conference on Computer-Supported Cooperative Work. Kluwer Academic Publishers, Amsterdam (1991)
14. Thompson, J.D.: Organizations in Action. McGraw-Hill, New York (1967)
15. Van de Ven, A.H., Delbecq, A.L., Koenig, R.J.: Determinants of coordination modes within organizations. American Sociological Review 41(3), 322–338 (1976)
16. Campion, M.A., Medsker, G.A., Higgs, C.A.: Relations between work group characteristics and effectiveness: Implications for designing effective work groups. Personnel Psychology 46, 823–850 (1993)
17. Maznevski, M.L., Chudoba, K.M.: Bridging space over time: virtual team dynamics and effectiveness. Organization Science 11(5), 473–492 (2000)
18. Staples, D.S., Jarvenpaa, S.L.: Using electronic media for information sharing activities: a replication and extension. In: The Twenty-First International Conference on Information Systems, Brisbane, Queensland, Australia (2000)
19. Shaw, J.D., Duffy, M.K.: Interdependence and preference for group work: Main and congruence effects on the satisfaction and performance of group members. Journal of Management 26(2), 259–279 (2000)
20. Wageman, R.: Interdependence and group effectiveness. Administrative Science Quarterly 40, 145–180 (1995)
21. Van der Vegt, G., Emans, B., Van de Vliert, E.: Effects of interdependencies in project teams. The Journal of Social Psychology 139(2), 202–214 (1999)
22. Baron, R.M., Kenny, D.A.: The moderator-mediator variable distinction in social psychological research: Conceptual, strategic, and statistical considerations. Journal of Personality and Social Psychology 51(6), 1173–1182 (1986)
23. Ferreira, A., Antunes, P., Pino, J.A.: Evaluating shared workspace performance using human information processing models. Information Research 14(1) (2009)
24. Daft, R.L., Lengel, R.H.: Information richness: A new approach to managerial behavior and organization design. Research in Organizational Behavior 6, 191–233 (1984)

A Flexible Multi-mode Undo Mechanism for a Collaborative Modeling Environment

Tilman Göhnert, Nils Malzahn, and H. Ulrich Hoppe

Collide Group, Faculty of Engineering,
Department of Computational and Cognitive Sciences,
University Duisburg-Essen, Lotharstr. 63, 47057 Duisburg, Germany
{goehnert,malzahn,hoppe}@collide.info

Abstract. This paper presents a flexible multi-mode undo mechanism for a collaborative modeling environment supporting different types of graph representations (including Petri Nets and System Dynamics models) as well as free-hand annotations. The undo mechanism is first introduced on a formal basis. It is implemented as an extension of the underlying Match-Maker collaboration server and allows for selecting from and making use of several undo variants with minimal adaptation effort. This is a basis for future usability studies comparing different versions of undo and for better adapting the undo effects to the actual user goals.

1 Introduction

Undo and redo, i.e. the possibility to reverse the effect of an action and afterwards restore it, are functionalities that are especially important in collaborative scenarios. As Prakash and Knister state in [1]:

"Undo is important in collaborative applications because it provides freedom to interact and experiment in a shared workspace."

They explain that people are reluctant about changing shared documents as they fear to destroy the work of others.

Therefore being able to reverse changes makes it easier to explore ideas. This holds for synchronous and for asynchronous collaboration, since an undo functionality would be effective in both situations. Chen and Sun [2] describe undo as a useful functionality as it allows to reverse the effects of erroneous actions, to learn functions of software by trial-and-failure and to explore alternative solutions by backtracking. In addition they give three reasons, why undo is particularly valuable in collaborative applications:

1. "multi-user applications must have all the features in single user applications,
2. the potential cost of an individual user's mistake is multiplied many times since it may adversely affect the work of a large number of collaborative users,
3. the number of alternatives to be explored in a collaborative setting increases due to the presence of many users."

L. Carriço, N. Baloian, and B. Fonseca (Eds.): CRIWG 2009, LNCS 5784, pp. 142–157, 2009.

1.1   Undo/Redo vs. Undo/Repeat

Since undo and redo are not unambiguously defined we will introduce some terms and their meaning. We follow the definitions found in [3] and [4].

An *action* (or *operation*) is a change on a document like inserting or deleting characters in text documents or re-coloring objects in graphics. Actions like re-sizing a program window are not considered. In the description of algorithms actions will be denoted with lower-case letters: $a, b, c, \ldots$.

Undo is a function which allows the user to reverse the effect of an action. In the text editor example, adding the tailing character "n" to "actio" is an action and removing this "n" is the action caused by calling the undo function. The existence of an inverse action which reverses the effect of the original action is a precondition needed for the execution of undo. Eventually undo is offered only for a subset of actions a user may perform. The inverse action will be denoted with a bar over the letter of the original action: $\bar{a}, \bar{b}, \bar{c}, \ldots$.

Redo is the undo of an undo. In the above example it could be the restoration of the tailing "n" for "actio". While describing algorithms such actions will be denoted with a double bar over the letter as the inverse action of an inverse action: a for the action, $\bar{a}$ for the undo-action and $\bar{\bar{a}}$ for the redo-action. In many cases the redo-action is the same as the original action. In the given example the first addition of the tailing "n" is not different from the redo-action. This is not necessarily the case. It depends on the undo/redo algorithm.

Repeat is a function which is related to undo and redo. It allows doing an already performed action again. The main difference between redo and repeat is the dependency on the earlier actions: Redo is an undo of an undo and therefore depends on an earlier undo whereas repeat is performing an earlier action again. Hence repeat only depends on the original action.

For the description of some undo/redo algorithms also states are needed. Sab denotes the situation after performing the actions a and b in an environment with the initial state S.

The term "state" is used in this paper to denote a set of variables including their current values. For undo processes often only a subset of the system state is considered. Especially in collaborative systems this subset can be chosen by the subset of synchronized variables.

To be able to reconstruct the actions, which have been performed, a so called *history* is used. In this data structure information about type and order of the last actions performed on a document are stored as well as any further information needed for the undo/redo algorithm like e.g. the id of the user who performed an action. The history itself is not considered for undo.

There are two further terms which are interesting for discussing elaborated undo approaches. One is *undo algorithm* which describes the part of an undo mechanism responsible for actually performing the undo for a given set of actions. The other is *undo mode* which stands for the part responsible for selecting the actions to undo. One way to enable flexible undo mechanisms is to separate these two parts so that the algorithm supports multiple undo modes which can be exchanged.

1.2   Undo in Single User Systems

Single-step Undo. For single-step undo the history only contains one entry for the last action. This variant does not support a dedicated redo. Being a the last action, i.e. the single action stored in the history, $\bar{a}$ is the associated undo-action. Performing this undo-action replaces the action stored in the history. At this point another call of the undo function would cause the inverse of $\bar{a}$ to be performed, resulting in a redo of a.

History Undo. History undo [1] is an undo variant used in Emacs [5]. In this case the history has theoretically infinite length. For every stored action the history has to contain all information needed to perform the action in the system state right before its original execution plus all information needed to reverse the action in the system state directly after the original execution of the action. For every call of the undo function an undo of the latest action in the history is performed. Any call of a function that is not the undo function will result in another action to be added to the history. This action is now the latest action.

1.3   Challenges for Undo Mechanisms in Collaborative, Multi-representational Environments

Concurrency. There are many different solutions for concurrency control. Pessimistic approaches use instruments like locking of resources by floor-passing mechanisms with or without user interaction. Optimistic approaches do not lock resources. They recognize and handle conflicts. One example for such a mechanism is transformation of actions like it is used in GROVE (GRoup Outline Viewing Editor, [6]) in form of the dOPT-algorithm (Distributed OPerational Transformation, [7]).

An example for transformation of actions: Two users A and B edit concurrently the string "ace", with numbering of character positions starting with 1. User A decides to insert the character "b" between position 1 and 2, user B inserts d between 2 and 3. For user A this results in the string "abce", for user B in "acde". If the actions of the users are now propagated to the temporary results of the other one, after an execution without any transformation the string of user A will read "abdce", while user B gets "abcde". To ensure the same result for both users it is checked if the local action has changed the edited document in a way that influences the remote action. The local action of B has no such influence. On the other hand the local action of user A influences part of the document that is changed by user B and therefore B's action is transformed to inserting character "d" between 3 and 4.

Cooperation with Changing Participants. The intended use of the undo mechanism described in this paper is synchronous collaboration. As it is possible to enter and leave a collaborative session special consideration for the latecomer-problem is needed. There are different solutions to this problem like a complete replay of all actions that have been performed so far or receiving the complete current state as starting point for further synchronization. Thus it may not be assumed that every participant has received the same actions as all other users.

Local vs. Global Undo. For undo in shared systems the question which action is the last action is more difficult to answer than in single-user systems [8]. Thus, we distinguish between local and global actions. A local action is an action on the user's system. A global action may occur on any of the systems in sync.

If the shared systems make use of the WYSIWIS-pattern (What I See Is What You See) in the strict interpretation, all users have the same focus on the shared document and therefore should see the same action as the last one happening. Therefore the last global action is the one to be considered for undo. If on the other hand every user is allowed to have an individual view on the document the actions of other users may be invisible. In this case the last local action is an appropriate action to consider for undo.

One approach to implement local undo is to store only local actions in the history. However, this may cause problems in case of dependent actions because actions of global users cannot be considered for depency checks.

Another approach is to keep a global history but filter it for local actions. This allows to select only local actions for undo but consider all actions performed in the shared systems for dependency checking.

1.4   Undo in Shared Systems

Selective Undo. Prakash and Knister introduce an undo mechanism based on history undo enhanced by storing further information to every action-record in the history ([1], [3]) which allows to select actions from the history for undo. The most important criterion for selecting actions is a user-id, but also time-based or location-based (within the document) selection is possible.

Selective undo makes some assumptions about actions:

- *Existence of an inverse*: There is one and only one inverse for every action, which leads to a document state equal to the state before executing the action if executed directly after the action
- *Dependency recognition*: Given two actions it has to be decidable if one is dependent on the other in the way that it could not have been executed before the other. An example would be resizing of a circle which is dependent on the creation of the circle whereas the creation of two circles is an example for independent actions.
- *Transposed actions*: For every two actions that are not dependent there are two transposed actions which comply with the following conditions: Let a and b be the original actions, S a document state and a' and b' the transposed actions. Then Sab and $Sa'b'$ have to result in the same document state independent of the initial state S and b' has to be the action that had been executed instead of b if a had not been executed before.

The mechanism is based on the idea to transpose an action selected for undo with the currently next action in a copy of the history until the transposed version of the selected action is the latest action in the history. Then the inverse action for this transposed version of the selected action is determined, executed and appended to the end of the original history. If a dependency occurs between the selected action and one of the later actions in the history, the mechanism stops giving an error report and no undo is performed.

The exact inverse and transposed actions have to be defined together with the original actions. To ensure this being possible only a finite and fixed set of so called primitive actions are allowed. These may be used to build further actions.

Any Undo. Another undo mechanism developed for collaborative systems is any undo developed by Sun [4] for text editors. The name is based on the approach to be able to undo any action at any time using any undo mode.

The mechanism itself is based on the assumption that for any given action a an inverse action $\bar{a}$ exists, which will reverse the effect of a if executed directly after it. To be able to execute the inverse action at any later time than directly after the original action, the inverse action is treated as if it had been executed concurrently to all other actions executed after the original action. The inverse action is considered independent of later actions but it depends on those actions executed before the original action.

2   A Collaborative Modeling Environment

To test the undo-redo-framework it was implemented into a collaborative modeling environment consisting of the collaborative modeling tool FreeStyler [9], [10] using the MatchMaker [11] as collaboration server.

2.1   FreeStyler

The FreeStyler is a collaborative modeling environment based on the idea of shared workspaces offering freestyle-sketching as well as visual languages (Screenshot see Fig. 1) [9], [10]. FreeStyler can be used in single-user scenarios as well as in collaborative ones.

The support of visual languages is very flexible, as it based on a plugin framework for graph-based visual languages that can be implemented independently of the FreeStyler following a common interface.

The plugin framework is based on so called reference-frames [12], [13]. These reference-frames include information about the components of the visual languages, i.e. the nodes and edges, and may also include rules describing the syntax of the represented language. Currently about 32 different plugins are available.

2.2   MatchMaker

MatchMaker offers a combination of client-side data replication and server-side data storage [11]. On the one hand, every synchronized application has a local version of those parts of the shared data it needs, on the other hand, all shared data is stored on the server and can be retrieved from it. There is a special

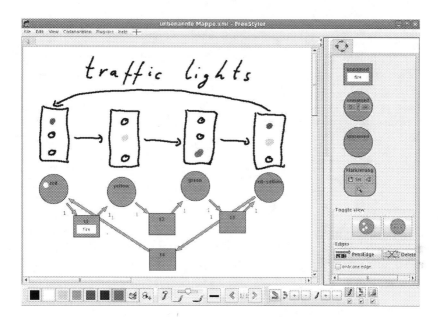

Fig. 1. Screenshot of the FreeStyler

mechanism to allow the server to keep its copy of the shared data in sync by applying the synchronized actions on the stored objects. So at any time a client, e.g. a latecomer, is able to fetch the current state from the server.

The shared data is organized in a tree structure which is especially useful for applications which use hierarchical data models. It is possible to address either a single object or a subtree by ids. Any message addressed to an object will be forwarded to the object itself and any ancestor objects of it. The server supports the following operations on the data: reading and writing of subtrees, writing of single objects, adding a child to an object and executing an action on an object.

The client acts as an interface between application and server. The application has to register itself as listener for any synchronized object it wants to get the synchronization messages for. The client passes the registration to the server and the messages coming from the server to the application. The client also allows accessing the operations on the synchronized data supported by the server in a slightly modified form. The client offers reading and writing of subtrees and executing actions on objects as the server does, the other operations are creating, changing and deleting of a single object.

Creating an object is an operation which needs the object to be shared and the id of the object to be the parent-object in the tree structure and returns the id for the new object.

Changing an object is performed by replacing it, so the new object and the id of the object to be replaced are needed as parameters.

Deleting is implemented as write-operation for a given id where the object to be written is a null-object. This causes the server to delete the whole subtree which has the given id as root.

Any of these operations is performed on the server to keep the copy of data on the server always in sync. The applications get the synchronization messages and have to handle them in a way that ensures their local copies of the shared data to be in sync.

The server offers an implicit serialization of the actions as it performs any modification locally and then propagates the action to all listening applications before processing the next operation-call.

3   A Flexible Undo Mechanism for a Collaborative Modeling Environment

In this paragraph the term *document* is used for the set of variables on which an undo functionality shall be provided. So a *(document-)state* is a set of specific values for those variables. Information which shall not be considered for the undo process is not part of the document state. An example for such information may be the name of the user who originally created the document as this data is not going to change afterwards. Therefore the term document state as it is used here does particularly not describe the set of variables that are externalized when saving a document.

This definition of the term state allows for a more precise definition of the terms used to describe the undo algorithm:

An *action* x is the changing of the values of a (sub-) set of variables. One action can be executable on a set of states and defines one state transition for every state it is valid on.

The *initial state* is the document state as it is before any action has been performed on the document.

A *history* H is a pair of a state S and a sequence of actions $x_1 \ldots x_n$ executed on that state: $\langle S, x_1 \ldots x_n \rangle$.

Applying an action x_1 to a state S results in a new state Sx_1. So the state Sx_1x_2 is defined as the resulting state of action x_2 applied to state Sx_1 and so on.

For every history $H = \langle S, x_1 \ldots x_n \rangle$ exists a *generated state* $S_H = Sx_1 \ldots x_n$. A state generated by a history with a finite number n of actions can be written as S_n.

A history $H = \langle S, x_1 \ldots x_i x_{i+1} \ldots x_n \rangle$ can be written as $H = \langle S_i, x_{i+1} \ldots x_n \rangle$.

An action b is called the *inverse action* of another action a, if for any state S_x which allows the execution of a holds $S_x ab = S_x$. The inverse action of a is written as $\overline{a}$.

Those terms being defined allow for a definition of two main elements of undo mechanisms. The following definitions follow the definitions which have been made by Chen and Sun [2].

Undo-effect: After executing $\overline{x_i}$ in the document state S_n ($i \leq n$) the following should be true:

- The effect of executing x_i is eliminated.
- The initial effects of executing $x_1, \ldots, x_{i-1}$ are preserved.
- The effects of executing $x_{i+1}, \ldots, x_n$ are preserved in the way they had been if x_i never had been executed.

This is called the undo-effect.

Undo-property: Given document state S_n the execution of $\overline{x_{i+1}}, \ldots, \overline{x_n}$ in appropriate order restores document state S_i.

This undo-property is notably weaker than the one postulated by Chen and Sun which can not be satisfied in this context. In the original version it has to be possible to execute the inverse actions in any order. That can not be guaranteed in this approach as dependencies between actions have to be considered and therefore a special order of executing the inverse actions may be necessary.

3.1   Requirements for the Implementation of Undo

The undo mechanism shall be action-based, therefore actions have to be logged and it has to be possible to propagate undo-actions into the system. In collaborative scenarios undo-actions must be globally effective instead of just locally.

User- vs. System-view. There is a break in FreeStyler and often in other interactive applications between the way a user observes actions and the way those actions are represented in the system.

Deleting nodes is an example for this. For the user it is one action to delete one node. For the system, this action is broken into a sequence of actions. At first any edge which is connected to the selected node is deleted. If there are no more edges connected to the selected node, the node itself is deleted. Therefore a mechanism is needed which allows to undo a user-action by only one undo-action independent of the number of system-actions the user-action is split into.

Flexibility in Terms of Undo Mode. As there are only a few proven statements about the ideal choice of undo modes depending on the system the undo shall be working in, the mechanism shall be flexible in so far that it allows using different undo modes. So on the one hand the mechanism can be used as basis for future research on that question and on the other hand it can be adapted to results in that area.

Flexibility in Terms of Integrity of the History. The undo mechanism shall be usable with an incomplete history. Incomplete means that the history does not begin with the first action ever executed in a session but it may begin with any action of a session. From that action on, all later actions have to be logged in the history. This is required for two reasons. It shall be possible for latecomers in collaborative scenarios to have a local history containing only those actions executed after the session was joined. The second reason is that it should be possible to delete a part of the history to free memory.

3.2   The Undo Mechanism

Based on *selective undo* by Prakash and Knister a flexible undo framework has been realized.

As the MatchMaker allows logging, local as well as global propagation of actions and provides a mapping of any system-action onto one out of a fixed set of synchronization-actions its client has been chosen as the interface used to add undo functionalities to. Therefore the four basic operations create, delete, change (replace) and execute (an action) are used to develop an undo mechanism.

To be able to handle dependencies, dependencies are described as follows:

Independence and Dependency of Actions. An action b is said to be independent of another action a if for all states S_x in which the execution of both actions is possible $S_x ab = S_x ba$ is true.

An action b is said to be dependent on another action a if and only if the actions are not independent.

The Actions and the Inverse Actions. There are four basic actions, which are listed with the parameters needed to describe a specific action. For any object synchronized exists one id which is assigned when creating the object.

- create(object): The given object is created and an id assigned to it.
- delete(ID): The object identified by ID is deleted.
- change(ID, object): The object identified by ID is replaced with the given object.
- exec(ID, action, parameter): The action "action" is executed with the given parameter on the object with the given ID.

For these actions a mapping to the inverse actions can be defined:

- create(object) → deleted(ID) with ID being the one assigned to the object created by the create to undo.
- delete(ID) → create(object) with object being the one the given ID has been assigned to.
- change(ID, newObject) → change(ID, oldObject) with oldObject being the one replaced by the change to undo.
- exec(ID, actionDo, parameterDo) → exec(ID, actionUndo, parameterUndo) with actionUndo and parameterUndo chosen in a way that the inverse action reverses the effects of the action to undo.

The completely instantiated actions containing the parameters needed for the original actions and those needed for undo are then put into the history.

Dependencies and Conflicts. There is no general way to provide the inverse action for the operation exec for every visual language. So it has to be programmed for every single language to comply with their particular requirements. This leads to the problem that those inverse actions have not yet been provided for all languages. To be able to handle this situation nonetheless actions for which no inverse action has been provided are treated as if they are in an irresolvable conflict. Therefore such actions and any further actions depending on these actions can not be undone.

With the abstract definition of dependency between actions, the described action-signatures and the knowledge about the use of a tree structure for organizing the shared the dependency between concrete actions can now be described.

In general it is impossible to say if any action could have been executed on a single object or would have lead to the same result if not all earlier actions on that object had been executed as well. Therefore any action is dependent on all earlier actions that have been executed on the same object.

For similar reasons an action can be said to be dependent on all earlier actions that have been executed on objects which are ancestors in the sense of the synchronization tree to the object the action is executed on.

As the described break between the user- and the system-view on actions is solved in this undo mechanism by grouping actions this leads to further dependencies. All actions belonging to one action-group are considered to be dependent on one another regardless of the dependencies that exist based on the objects involved. This allows to be sure either all or no actions of a group are undone.

Finally there is one form of dependency that comes from the application used to test the undo algorithm. As the visual languages are based on a graph structure dependencies coming from that structure have to be considered. The edges of the visual languages depend on the existence of the nodes they are attached to, so the creation of those nodes has to be done before the creation of an edge is executed.

Transposing of Actions. Actions considered in this undo mechanism are either independent and can be executed in any order without changing the resulting state or there is a dependency between them and then they can not be executed in any order other than the original one.

Logging of Actions into the History. The logging of executed actions is inspired by the selective undo mechanism by Prakash and Knister. Every system participating in the collaboration keeps a local history. This history is sorted in the same order the actions have been executed in. Caused by the implicit action serialization of the MatchMaker the actions in the history of all systems are sorted in the same order.

Do-Undo Pairs. The strict definition of dependency between actions has the advantage to prevent inconsistencies but also prevents lots of undo calls. Especially for later actions being dependent on the one to be undone, it is desirable to see if they have already been undone. If so they can be treated as if they never have been executed. This would allow more undo calls to be executed. Following the selective undo approach additional information is logged to allow the recognition of do-undo pairs of actions. A copy of the history is created when an undo is requested. The do-undo pairs are deleted from this copy before checking for dependencies so the actions which have already been undone are not treated as being dependent on an action for which the undo is requested. This results in less undo calls being unnecessarily prohibited by the dependency-recognition.

3.3   Handling the Challenges for Undo Mechanisms

Concurrency. As the developed undo mechanism relies on an underlying co-operation mechanism the concurrency control of that cooperation mechanism is used. Thus concurrency has not to be addressed further in the undo mechanism itself.

Cooperation with Changing Participants. The undo mechanism is de-signed to allow for the history to contain only those actions which were logged by any participant after joining a session according to the requirements formulated in 3.1. Therefore latecomers can use the undo mechanism with the information they receive during their participation in a collaborative session.

Local vs. Global Undo. As the decision for local or global undo is situation-dependent the undo mechanism itself is not limited to one of the two variants. For dependency checking always the complete history is considered whereas it may be filtered before selecting actions to undo. Thus local undo can be provided by an undo mode which filters the history list for actions of the local user.

4   Implementation

4.1   Architecture

To be as flexible as possible concerning the undo modes we designed the undo framework following the strategy pattern [14]. Using this pattern we introduced an abstract class UndoMode (see Fig. 2) representing the specific modes like HistoryUndo, SelectiveUndo or OfficeUndo. The UndoMode offers an interface to request an undo. The action(s) to be undone are determined by the concrete implementation of the UndoMode using the history of the UndoClient. After selecting actions for undo, the UndoClient is called to process the undo.

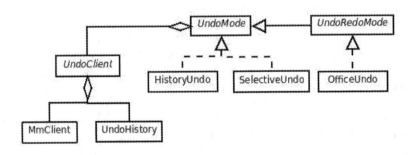

Fig. 2. Diagram of the architecture

In summary this architecture allows to exchange the concrete undo modes without having to change any other source. Furthermore this approach enables undo in a bunch of applications that use a MatchMaker as collaboration server. So presented framework is flexible both concerning the undo modes and, due to its collaboration server approach, concerning the application. Any application using the MatchMaker as collaboration server automatically may be enabled with the undo functionality by adding the appropriate button to the application.

The UndoMode uses the following algorithm to calculate if an undo is possible and which actions have to be undone:

Input:

1. List L_1 of actions to be undone.
2. Boolean variable B determining if dependent actions should also be undone.

Algorithm:

1. Check if all actions provide the necessary information to be undone. If not all information is available: abort undo.
2. Generate a list L_2 with all actions depending on the initial set of actions to be undone.
3. Remove all members of L_1 from L_2.
4. Remove all do-undo pairs from L_2.
5. Examine with the help of B if dependent actions shall be undone.
   (a) If they shall be undone, append L_2 to L_1.
   (b) If they shall not be undone and L_2 is not empty, abort undo.
6. Sort the resulting list L_1 by the reverse execution order of the actions.
7. Undo the actions in L_1 successively beginning with the first entry.

The proof of correctness of the algorithm may be found in [15].

Using this algorithm in our framework we have implemented several undo modes. One of these modes is *history undo* which shows the behavior as it would be expected from the history undo algorithm (see 1.2). Another one is a mode we call *"office undo"* as it is build to resemble the undo known from office suites. For undo always the latest action in the history which is either an original action or a redo is selected, for redo always the latest action which is an undo. Both of these modes exist in a global and a local variation. For local undo the history is filtered for actions originated by the local user before selecting an action for undo from it. For demonstration purposes also a *selective undo* is provided which allows to select any action from the history for undo.

5   Example – Modeling Petri Nets

FreeStyler provides (among other) a petri net plugin allowing to model and simulate petri nets. In computer science teaching this palette is often used to explain the basic characteristics of them. One of the examples we use to explain it is the example of two traffic lights at a crossing. One for cars and one for pedestrians (see Fig. 3).

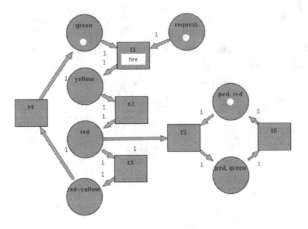

Fig. 3. Model of the crossing with traffic lights

In this example there are several problems. Two of them are caused by the fact that the network may die given a specific order of transitions:

1. t1, t2, *t3*, t4 (see Fig. 4)
2. t1, t2, *t5*, t6

The emphasized transitions are those that lead to dead ends.

To try alternative walkthroughs through the petri net the undo button may be used to backtrack the alternatives.

This small example contains several problems for the undo framework. Whenever the user wants to fire a transition he or she just presses the transition button.

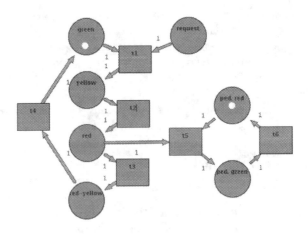

Fig. 4. Model of the crossing with traffic lights after firing t1, t2, t3, t4

This is just one user action, but the underlying simulation model generates several system actions from this. First of all the button is pressed, then for each of the nodes connected to this transition the particular change event - increasing or decreasing the token amount of the places - are executed. So instead of one user action a bunch of system actions are written into the history. If the user now wants to undo the changes he or she expects that all of the changes are undone at once. Especially for modeling languages with a formal semantics like petri nets it is important to undo all of the actions to maintain consistency of the model. If only a part of the actions is undone a model marking is reached that was never intended by the user and is wrong following the semantics of petri nets. This problem is solved with the action grouping approach mentioned above. Thus all of the actions belonging to the group created by the transition action are undone in reversed order of their execution, i.e. tokens removed are put back and generated tokens are removed.

Also some benefits of flexible undo in multiuser environments can be illustrated with this example. The task for a pair of students may be split into two parts: one for cars and one for pedestrians. In a first phase these types of traffic lights are modeled independently. One per student. In the second phase the two parts are combined collaboratively. In this scenario the first phase is best supported with local undo so the two students can work on their part of the traffic lights without interfering each other while maintaining a shared view. The second phase may be better supported with global undo since they also work on the same objects. Since our undo mechanism supports switching the undo mode during runtime we can offer the appropriate support for every phase.

6   Discussion and Future Work

Abowd and Dix [8] have stated that most current research on undo treats the problem from system-centered technical point of view, neglecting user intentions and goals. In this sense, they suggest a new perspective on undo moving from questions like "what does this button do?" to a more pragmatic or task-oriented view, e.g. by asking "what is this button for?". This would indeed address the user's intention. Switching the perspective in this way calls for an enhanced user support since the undo mechanism will then aim at the user's goal while pressing the "undo" button instead of just reversing the last action executed by the application. To obtain insight into the users' intention, the analysis of action logs using task models (cf., e.g., Hoppe's "task-oriented parsing" approach [16]) appears to be an adequate means. MatchMaker provides flexible tools for defining and using logging formats for the synchronized actions during a coupled session. The data derived from these logs may be used to gain insight into higher level task structures as a basis for a better adaptation to the user's needs.

1. As stated above, a single user action may result in several actions in the application's backend. By analyzing the co-occurrences of inverted actions in the log-file new groupings of actions may be proposed, which are better suited to the user's original purpose. This will narrow the gap between the user intended actions and the system actions.

2. Logging the actions will allow for a deeper understanding of the situations in which users tend to use undo and when they try to achieve the same effect by executing a set of other operations. On the one hand this opens another way of trying to capture the user's original intent to improve the undo mechanism and on the other hand we may be able to automatically offer hints at shorter workflows for the same effect by the system.

3. Currently we only provided a flexible framework for the implementation of undo/redo functionalities in collaborative applications. As a next step we will try to find out which undo mechanisms are best for different contexts. Cass, Fernandes and Polidore [17] have already worked on this issue, but they only used a pen and paper experiment and in a non-collaborative setting. Thus we want to know if collaborative software applications need a different undo support mechanism.

4. Prakash and Knister as well as Chen and Sun stated that undo allows the user to decrease the effort needed to fix errors and provides the opportunity to try out different approaches to a problem solution. In summary the user feels safer and thus may find solutions that otherwise would not have been tried out. We will investigate in which ways an undo functionality will influence the users' problem solving process and success.

References

1. Prakash, A., Knister, M.J.: Undoing actions in collaborative work. In: CSCW 1992: Proceedings of the 1992 ACM conference on Computer-supported cooperative work, pp. 273–280. ACM, New York (1992)
2. Chen, D., Sun, C.: Undoing any operation in collaborative graphics editing systems. In: GROUP 2001: Proceedings of the 2001 International ACM SIGGROUP Conference on Supporting Group Work, pp. 197–206. ACM, New York (2001)
3. Prakash, A., Knister, M.J.: A framework for undoing actions in collaborative systems. ACM Transactions on Computer-Human Interaction 1(4), 295–330 (1994)
4. Sun, C.: Undo any operation at any time in group editors. In: Computer Supported Cooperative Work, pp. 191–200 (2000)
5. Stallman, R.: GNU Emacs Manual, 16th edn. Free Software Foundation, 51 Franklin Street, Fifth Floor Boston, MA 02110-1301 USA (2007)
6. Ellis, C.A., Gibbs, S.J., Rein, G.L.: Design and use of a group editor. In: Cockton, G. (ed.) Engineering for Human-Computer Interaction, pp. 13–28 (1990)
7. Ellis, C.A., Gibbs, S.J.: Concurrency control in groupware systems. SIGMOD Rec. 18(2), 399–407 (1989)
8. Abowd, G.D., Dix, A.J.: Giving undo attention. Interacting with Computers 4(3), 317–342 (1992)
9. Hoppe, H.U., Gaßner, K.: Integrating collaborative concept mapping tools with group memory and retrieval functions. In: Stahl, G. (ed.) Proceedings of the International Conference on Computer Supported Collaborative Learning (CSCL 2002), New Jersey, USA, pp. 716–725. Lawrence Erlbaum Associates, Inc., Hillsdale (2002)
10. Gaßner, K.: Diskussionen als Szenario zur Ko-Konstruktion von Wissen mit visuellen Sprachen. PhD thesis, Universität Duisburg-Essen, Campus Duisburg, Fakultät für Ingenieurwissenschaften, Abteilung Informatik, Informations- und Medientechnik (2003)

11. Jansen, M.: Matchmaker - a framework to support collaborative java applications. In: Hoppe, H.U., Verdejo, F., Kay, J. (eds.) Shaping the Future of Learning through Intelligent Technologies, pp. 535–536. IOS Press, Amsterdam (2003)
12. Pinkwart, N.: A plug-in architecture for graph based collaborative modeling systems. In: Hoppe, H.U., Verdejo, F., Kay, J. (eds.) Shaping the Future of Learning through Intelligent Technologies. Proceedings of the 11th Conference on Artificial Intelligence in Education, pp. 535–536. IOS Press, Amsterdam (2003)
13. Pinkwart, N., Hoppe, H.U., Gaßner, K.: Integration of domain-specific elements into visual language based collaborative environments. In: Borges, M.R.S., Haake, J.M., Hoppe, H.U. (eds.) Proceedings of 7th International Workshop on Groupware (CRIWG 2001), pp. 142–147. IEEE Computer Society, Los Alamitos (2001)
14. Gamma, E., Helm, R., Johnson, R., Vlissides, J.: Design Patterns. In: Elements of Reusable Object-Oriented Software. Addison-Wesley Longman, Amsterdam (1995)
15. Göhnert, T.: Flexibel parametrisierbare aktions-synchronisation mit undo/redo in einer kooperativen modellierungsumgebung. Master's thesis, Universität Duisburg-Essen, Fachbereich Mathematik (2008)
16. Hoppe, H.U.: Task-oriented parsing - a diagnostic method to be used adaptive systems. In: CHI 1988: Proceedings of the SIGCHI conference on Human factors in computing systems, pp. 241–247. ACM, New York (1988)
17. Cass, A.G., Fernandes, C.S.T.: Using task models for cascading selective undo. In: Coninx, K., Luyten, K., Schneider, K.A. (eds.) TAMODIA 2006. LNCS, vol. 4385, pp. 186–201. Springer, Heidelberg (2007)

Forby: Providing Groupware Features Relying on Distributed File System Event Dissemination[*]

Pedro Sousa[1], Nuno Preguiça[1], and Carlos Baquero[2]

[1] CITI/DI, FCT, Universidade Nova de Lisboa
[2] DI/CCTC, Universidade do Minho

Abstract. Intensive research and development has been conducted in the design and creation of groupware systems for distributed users. While for some activities, these groupware tools are widely used, for other activities the impact in the groupware community has been smaller and can be improved. One reason for this fact is that the mostly common used applications do not support collaborative features and users are reluctant to change to a different application. In this paper we discuss how available file system mechanisms can help to address this problem. In this context, we present Forby, a system that allows to provide groupware features to distributed users by combining filesystem monitoring and distributed event dissemination. To demonstrate our solution, we present three systems that rely on Forby for providing groupware features to users running unmodified applications.

1   Introduction

Intensive research and development has been conducted in the design and creation of groupware systems. While for some activities, these groupware tools are widely used (e.g. instant-messaging, conferencing tools), for other activities the impact of the groupware community can be further improved. One example is the support of asynchronous cooperative editing, where version control systems (e.g. CVS, subversion) are widely used for data management but the proposed awareness mechanisms [5,17,14] are rarely used in practice.

One reason for this fact is that the mostly common used applications do not support collaborative features and users are reluctant to change to a different application. This fact has been pointed out before [18] and several approaches for providing groupware features in existing applications have been proposed in literature, as reviewed in the related work section.

In this paper we propose a different approach, relying on the combination of modern file system mechanisms and an event dissemination system. Current operating systems include lightweight file system mechanisms to monitor files accessed and modified by users, allowing to infer user activity based on monitored actions. By combining the information obtained from these mechanisms with

[*] This work was partially supported by FCT/MCES (POSC/FEDER #59064/2004 and PTDC/EIA/59064/2006).

an event dissemination system to disseminate this information among multiple users, it is possible to provide groupware features to users in a distributed environment. As exemplified in the applications presented in section 5, a wide range of functionalities can be provided, from awareness information to new groupware applications.

As far as we know, this approach has been previously neglected by the groupware community. However, when compared with other proposals, it has the advantage of being application-independent, allowing the developed solution to work with all applications, thus providing groupware features while users can continue using their preferred applications. A disadvantage of this approach is that it can only be used when file system activity occurs. This limits the scenarios where our approach can be used. Thus, we think that this approach is not intended to be a replacement of existing proposals, but rather a complement to such proposals.

The remainder of this paper is organized as follows. Section 2 discusses related work. Section 3 presents existing file system mechanisms and surveys their availability in the most popular operating systems. Section 4 presents Forby design. Section 5 discusses how to use Forby in the context of three different applications. Section 6 concludes the paper with some final remarks.

2   Related Work

Several approaches have been proposed to integrate or improve groupware features in existing applications. In this section, we survey some of these proposals, grouping them by their approach.

The first approach is to provide application sharing, by allowing a group of users to transparently share the same application. Systems like Microsoft Net-Meeting provide such functionality. However, as discussed in [2], this approach has a series of shortcomings, such as the lack of support for real concurrent work and large network bandwidth requirements.

The second approach consists in dynamically replacing single-user interface components of applications by multi-user versions. The replacement occurs at run-time and is transparent to the original application. Flexible JAMM [2] provides this functionality. This approach is very effective, but its applicability tends to be limited in the number of applications that can be supported.

The third approach consists in creating minimal modifications to existing applications or relying on information already provided by applications. For example, in [6] the authors present a generic mechanism for awareness applied to the BSCW shared workspace. In this context, events that are to be propagated are obtained by sensors that are integrated with the BSCW server. In [12], the authors use existing operations for getting and setting the state to provide synchronous editing functionalities in existing editors.

This approach is interesting but solutions are application specific. Thus, for providing the same functionality in different applications, the solution (or part of it) must be re-written, thus limiting its applicability. Additionally, when new versions of the software are released, it is necessary to verify if the developed solutions still work with the new version.

The fourth approach is based on the use of application plug-ins. In QnA [1], the authors present a plug-in for commercial instant-messaging clients that increase the salience of incoming messages that may deserve immediate attention. For the open-source Eclipse editor, plug-ins developed in Palantír [16], Jazz [7] provide different levels of awareness information for shared files.

This approach has the same shortcoming of the previous approach, with the difference that plug-ins are officially supported and it is expected that new versions do not break existing plug-ins.

Some applications do not directly support plug-ins, but some extension can still be achieved (e.g. by intercepting application events at the operating system level, such as mouse movements, keyboard activity, etc.) This approach has been used in several systems [18,11] for integrating groupware features in applications such as Microsoft Word. This approach has the same problems as the previous ones, but it is even worse as there is no supported API for the created extension.

In the Placeless Documents system [4], it is possible to associate with each document some code, in the form of active properties, that is run when the document is accessed. This approach can also allow to provide groupware features to existing applications. However, unlike our approach, for this code to be executed, the files must be accessed through the Placeless Documents interfaces, that includes a distributed file system interface. Additionally, unlike the solution proposed in our system, this system provides no support for combining applications running in different sites, thus making it harder to provide groupware features in distributed settings.

3   File System Mechanisms

File system support has been evolving, with the introduction of new functionalities. Although different operating systems provide different mechanisms, most modern operating systems tend to provide the same file system functionalities relying on different APIs. In this section we overview some of the new features that can be used when providing groupware features.

3.1   Notification Mechanisms

Notification mechanisms allow user level applications to be notified when some access to the file system is executed. This includes the execution of open, read, write and close operations on files and changes in directories.

Notification is performed asynchronously, out of the loop of the system call, thus imposing minimal overhead to the operation execution. This approach also alleviates the efficiency requirements for the functions that process notifications, as the execution time is not reflected in the original operation execution.

These mechanisms are currently used in several applications, such as indexing systems for tracking changes in files and directories without requiring expensive periodically polling of the file system.

Notification mechanisms are available in the three most popular desktop operating systems. In Windows, Win32 API supports this feature natively for some time. Additionally, the .NET framework also supports this feature using the *FileSystemWatcher* library class, making it available to all supported languages. In Mac OS, the File System Events API provides such functionality since Mac OS v. 10.5. In Linux, the Inotify subsystem provides such functionality and it is integrated in the kernel since version 2.6.13. Other subsystems provide similar functionality (e.g. FAM, dnotify).

These APIs are available to user level applications, making the implementation and debugging of applications that use these mechanisms simple. Additionally, for each of the proposed mechanisms, there are mappings to several programming languages allowing to access these features easily in different programming environments. In the Java language, the JNotify library provides similar support in both Windows and Linux systems.

The APIs for all these systems consists in a very reduced number of methods, that can be used in a very straightforward way. The example in Figure 1 shows how to use Inotify for requesting notification of accesses to a file. It also shows how the application can have access to the notifications.

```
//***** Initialize Inotify *************************
int fd = inotify_init ();
if (fd < 0) return -errno;
//***** Add a Watch *******************************
int wd = inotify_add_watch(fd,
    "/home/user/myfile", IN_ACCESS | IN_MODIFY);
if (wd < 0) return -errno;
//***** Read an inotify event *********************
static struct inotify_event buffer;
*nr = read (fd, &buffer, sizeof(buffer));
```

Fig. 1. Setting up a notification request for accesses on a file and obtaining notifications in Inotify

3.2   Interception Mechanisms

Interception mechanisms allow to intercept file system calls and execute code before returning the result. This code may execute any action the system developer wants, including executing direct access to the file systems, sending messages over the network or even denying file operations by returning an error. This mechanism is used, for example, by anti-virus software to avoid that users run compromised software.

Unlike the previous mechanism, file system interception is executed in the path of the system call. Thus, the time consumed in executing the additional code is directly reflected in the delay the application experiences when executing the file system operation. Thus, the implemented code must be efficient, especially when no additional functionality is being executed, or otherwise the system performance will be affected.

The existing interception mechanisms can be broadly divided in two classes: kernel-based and user-level based. The first, run fully inside the kernel of the operating system. This approach is more efficient, as no additional user/kernel transition is necessary. However, it is much more complex to develop applications using this approach, as debugging code that runs in the kernel is very difficult (an error may lead to a freeze in the system). The second class, only includes a small module running in the kernel that intercepts the file system calls and redirects them to user level, where the additional code runs. This is less efficient than the previous solution, but it is much simpler to develop. Additionally, it is always possible to use this approach during the development phase and migrate the developed code to the kernel after the development is completed. Concerning user level solutions, it is also possible to intercept file system calls executed using dynamic linked libraries (e.g. the C library in Unix-like systems) by replacing the code of these libraries. This approach is efficient but it is not generic, as it only works for applications that use dynamic libraries for executing file system calls.

Several approaches exist to implement file system call interception. *File System in Userspace* (Fuse) is available for both Linux and Mac OS operating systems. This system uses a small kernel module that redirects intercepted calls to user level. Application developed using Fuse run in user level and must specify, for each file system call, the code to be executed. Fuse is very popular and simple to use, and a large number of systems were developed on top of it. For Linux (FreeBSD and Solaris), FiST [19] provides a kernel level solution that includes a specific language for simplifying the creation of stackable file systems. In the Windows operating system, the *Installable File System* API allows to intercept file system calls at the kernel level. An application implemented using this API can either fully execute in the kernel or redirect the call to user level.

4   Forby

Forby is a middleware system for providing groupware features to distributed users. The system provides two main components, present on each node: a file system monitoring module and a distributed event dissemination module, as shown in Figure 2. The file system monitoring module is used to infer user actions, based on file system activity. The distributed event dissemination module is used to propagate information among nodes. Applications use these two low-level components for providing groupware features to users. Next, we detail the two main components of Forby and later discuss how to build an application relying on these components.

4.1   Monitoring Component

The file system monitoring module monitors the user's working area, capturing the actions made by that user, processing them and forwarding them to the application if those events are important. This component is divided in three sub-modules that perform the operations required to capture, pre-process and deliver the important file system events to the application. The sub-modules are the activity monitor, event filter and event manager (Figure 3).

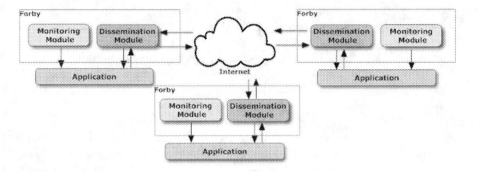

Fig. 2. Forby architecture

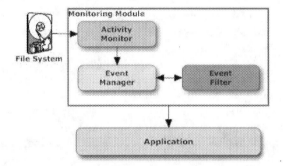

Fig. 3. Forby monitoring module

The activity monitor sub-module is responsible for monitoring the activity in the file system, obtaining the events through any existing tool or file system mechanism. Whenever a user performs an action in the monitored area, an event (*FSEvent*) is created. The FSEvent includes the information about the file, the access type and the current system time. The system can monitor any type of file access – create/delete, open file, read, write, etc.

This is the only module that is operating system dependent. Currently, there are two implementations based on *JNotify* [10] running on Windows and Linux systems and one implementation based on *Installable File Systems* (IFS) running on Windows systems. Forby is prepared to easily integrate other file system monitoring mechanisms by only creating a new activity monitor sub-module.

When using Forby, one activity monitor sub-module must be selected in each node, but different nodes of the same application may use different activity monitors.

As file system monitoring generates a large number of events, to simplify application development it is important to filter the messages propagated to the applications. Additionally, applications are often more interested in knowing that a given pattern of accesses have been executed, rather than having to process the all event sequence. For example, an application that just needs to know when a file has been changed, is not interested in processing the complete sequence of open, reads, writes, close events but rather prefers to receive a single event notifying that the file has been changed. The event filter sub-module is responsible to achieve this functionality by pre-processing the raw events.

As the necessary events are application-dependent, the application must define the event filter to use. Although the programmer can define a new event filter from scratch, a set of filters with basic functionalities is available for re-use. The basic functionality of these filters is to keep information about which files or directories should be monitored or ignored, updating the list of files/directories as the result of create/delete operations. All events on files that must be ignored are immediately discarded.

On top of this file management functionality, two basic filters were defined. The first, simply filters the events based on their type using some defined mask without further processing. The second, groups sequences of events into more complex events – for example, a sequence of write (resp. read) operations is notified as a file modification (resp. file access) event. These basic filters can either be used as is by the applications or be further extended. Our experience in developing applications using Forby shows that the basic filters can usually be used without changes, as application requirements tend to be similar.

The use of event filters makes sure that the application only receives the important events, therefore reducing the number of events that are processed. This approach simplifies the application's design and coding, since most of the effort of filtering the events is made at the event filter sub-module. Additionally, this is also important as it contributes for reducing the processing needs by filtering events as soon as possible, thus minimizing the overhead of Forby and improving system performance.

Finally, the event manager manages the events, linking the monitoring system to the application. It receives the events from the activity monitor, validates and pre-processes them using the event filter and ultimately, if that is the case, forwards the events to the application.

4.2   Dissemination Component

Although any event dissemination system could be used for this purpose, Forby dissemination component implements a system specially tailored for supporting users with minimal infrastructure requirements, thus allowing users to easily use the system. To this end, the event dissemination system manages the group participants, organizing them in a peer-to-peer dissemination tree that allows efficient event exchange among the peers.

Event dissemination is epidemic, where each participant that receives an event must forward the event through the tree, so that the event reaches all participants. Also, when an event is received, it is delivered to the system application in order to be processed.

For joining the dissemination tree, a new group member must obtain a list of participants already in the group, by contacting a *rendez vous point*. Two

approaches have been implemented. For minimizing the infrastructure requirements, Forby can rely on existing servers to maintain this information – either
a CVS/svn server used by the group or an email account supporting IMAP access (e.g. Gmail). If no remote server is available, peers can be organized into a
DHT (Pastry [15] in our current prototype), where one node is responsible for
maintaining the active participants of a given group.

After obtaining the list of participants, the new node enters the dissemination
tree and starts receiving, processing and propagating events, as any other node.

For small groups of users with no or minimal networking knowledge, private
networks are a problem as they reduce the connectivity among computers. This
is important as users' computers are increasingly in private networks, either because most organizations internally structure their network into private networks
or because access is usually mediated by a router (or firewall), even at home.

Forby addresses this problem by organizing private and public nodes in separate trees. Public nodes create a public dissemination tree as stated earlier.
Private nodes follow a different approach. The first node that connects to the
system becomes the private network leader and tries to establish a connection
with a public node. The other nodes that join that network connect to the leader,
forming a private dissemination tree. When an event is disseminated, it is forwarded through the tree and when it reaches the leader, it is sent to the public
peer, which will forward it in the public tree.

Any public node that receives an event, forwards to the other public nodes
and to all private nodes connected to it, so that every message reaches all group
nodes. Public nodes are used as a bridge between private networks, allowing
private networks to receive notifications. Figure 4 shows the organization of
public and private nodes.

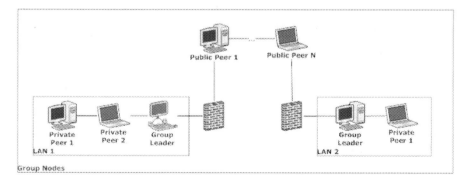

Fig. 4. Organization of private and public networks

All disseminated events are stored in event repositories for later use. Since
Forby follows a best effort dissemination policy (which offers no guarantee that
a message immediately reaches all nodes), this is necessary to provide fault-
tolerance. Additionally, it is also used to provide support for disconnected nodes.

Event repositories can be local to each peer or they can be set in a remote server. Once again, as for the *rendez vous* point, a CVS/svn or IMAP server can be used, thus imposing no specific infrastructure requirement for running Forby. Besides simplifying the system deployment, this approach allows each system/group of users to define the configuration that better suits their environment/requirements.

The use of a remote repository is also used to allow private networks to communicate with each other when no public node is available. In this case, the leader of each private sub-network periodically polls the server to check if new events are available and if they are, the leader forwards them to the other nodes in the network.

The dissemination system can be used to disseminate FSEvents or any other event that the application needs to propagate to other nodes. In any case, the disseminated event, *DisseminationEvent*, contains not only the event to be propagated but also all the information needed by the dissemination process (including the event sender and the sequence number). Each node keeps a summary of the received events (in a version vector) and periodically exchanges these summaries with neighbors (in the dissemination tree). A node can discover that it missed some event when exchanging these summaries or when it receives a new event (by discovering that some event is missing in the sequence of events from a node). In this case, the node can contact the remote repository — if one is being used — or disseminate an event request message, which will be replied by any node that has the requested event cached or locally stored.

For optimizing the event dissemination process, applications can define when an event makes a previous event obsolete – e.g. an event stating that a file has been modified usually makes a previous event with the same information from the same user obsolete. In this case, if the older event has not been propagated to all nodes, only the most recent event will be propagated.

Also, obsolete events can be deleted from the repositories and, when a node requests an event or a sequence of missing events, it needs only to receive the most up-to-date event that is not obsolete, optimizing the bandwidth and storage usage, and increasing the efficiency. Applications must define the rules used by the system to state that an event is obsolete. To simplify this process, the programmer can rely on base definitions or create a new one.

Finally, another feature offered by Forby's dissemination system is the possibility to send messages to a node using the reverse path followed by an event. This way, nodes can contact each other directly. As exemplified in the implemented applications, this is very useful for allowing a node to contact the original node to obtain additional information, when necessary.

5   Application Development

Forby can be easily used by any application that needs to provide groupware features relying on the information obtained by capturing and disseminating file system events. It allows programmers to focus on the application itself, rather then spending time dealing with the the details of efficient event capturing and dissemination. For using Forby, the application must set up the Monitoring and the Dissemination modules, by selecting the adequate parameters.

In the remaining of this section we detail three applications implemented using Forby. These applications have different goals – from providing awareness to a simple voting application, thus allowing to illustrate how Forby and its mechanisms can be used for supporting the development of different distributed groupware applications.

5.1   Awareness for Version-Control Systems

In client-server version control systems (e.g. CVS, subversion), users checkout each file they are interested on and create a local copy of the file. Then, they can modify the files in isolation for how long they want. Finally, when the user decides that she has finished her changes, she submits the new version back to the server. This approach may lead to concurrent changes, as more than one user may change the same file concurrently without knowing about the actions of other users.

This is a typical scenario where it would be interesting to provide additional awareness to users. Several solutions have been proposed to address this problem (e.g.[14,8,16,3,9]). However, the proposed solutions either require the user to use an additional application or are defined as a plug-in for some specific text editor, thus limiting its use.

Without changing the editor, a possible solution for providing additional awareness information is the following. Whenever a user accesses a file, by opening and reading it in any application, the user must be informed if a new version exists in the server or if some user has modified his local uncommitted copy of the file. Whenever a user is accessing a file and some other user concurrently modifies the file, the user is informed about the possible conflict. In any case, the user should have the possibility to request the new version either by obtaining it from the server or by requesting the other user to commit the new version.

We have implemented this application using Forby, relying on the dissemination of events for file change notification. Thus, whenever, a user modifies a file locally, our system automatically disseminates to all other users an event that represents that modification – including information on the version being modified and on the user. When receiving this event, each peer will store the event for possible future notification. If the file has been accessed recently, the user is also informed of the possible conflict.

Whenever a user accesses a file, by opening and reading it in any application, our system checks the events stored locally and determines if some user has modified her local copy (without propagating it to the server). Whenever this happens, the user is notified – figure 5 shows the notification presented when a user opens a file that is concurrently being modified (in this case, the user is using the *gedit* application, but this would work with any application).

When compared with the mentioned solutions that provide similar functionality, our approach has the advantage of working with any editor application, thus

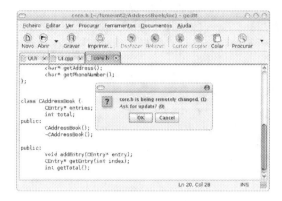

Fig. 5. User being alerted that the file she is opening has been modified by some other user

allowing users to continue using their preferred editors. Moreover, when compared with solutions that use additional applications, our solution allows the user to obtain the awareness information without having to change application – a simple pop-up is shown to users.

To implement this application in Forby, we have used the functionalities of the Monitoring and Dissemination modules.

The Monitoring module is used to monitor the file system activity and receive the needed events. In this case, we were interested in knowing, for a subset of the files in the file system, when a file had been accessed and when it had been modified. To this end, we have used the base filter that reports file changes and file accesses, with the change that file accesses are reported on the first read operation. On a file change, the application notifies other peers that a file has been locally modified. On a file access, the application uses this information to verify if the file has been remotely modified and to notify the user if necessary.

When deploying our solution, an adequate file system monitor should be used in each node, depending on the operating system.

The Dissemination module is mainly used to disseminate events about file modification by some user. To use it efficiently, the programmer must define when an event make others obsolete. In this case, when two file write events are on the same file and user, the older event is obsolete. Thus, in the worst case, for each file, the dissemination system will only have to keep one event for each user, even if a large number of updates are executed by that user.

When deploying our solution, it is necessary to configure the adequate repository for *rendez vous* and event storage. By default, *rendez vous* relies on the CVS server to store the needed information. For event storage, local storage is the default option but an external repository can be used if the network configuration requires that.

For implementing the requests for commit, our solution sends a message to the peer that has the new version using the reverse path for the update message.

We had previously implemented a similar solution without relying on Forby [13] and using the version control system to propagate events among users. When compared with that solution, our solution based on Forby is much simpler, as Forby handle the details related with file monitoring and event dissemination. Additionally, the system can be used without modification in any platform that runs Forby (Windows and Linux, at the current moment).

5.2   Automatic Directory Replication

In this section we describe an application that automatically maintains several replicas of a directory hierarchy up-to-date. Our intended use was to allow a family to cooperatively maintain the set of family photos, by allowing each family member to immediately access photos added by the other members[1]. However, our approach could also be used to maintain several replicas of a workspace up-to-date.

As in the previous example, we have implemented this application using Forby, relying on the dissemination of events for notifying workspace changes. Thus, whenever, a file is created, removed or updated, our system automatically disseminates this information to all other peers.

When receiving a file creation or file update event, the peer automatically downloads the file from the source peer[2]. If the peer has a public network address, a direct connection is used. Otherwise, the application relies on the dissemination system to communicate with the source peer. When receiving a file deletion event, the peer removes the file locally.

All the replication process is automatic, without need for the application to explicitly notify the users. For the targeted environment, we do not expect concurrent updates to occur, thus a simple conflict resolution mechanism is used to guarantee that all replicas eventually converge to the same state. When receiving a file update (or file delete), the correspondent action is executed only if no more recent event on the same file has been received (with the recent relation defined using the information used by the dissemination system). This guarantees that all peers will download the version that corresponds to the most recent event.

To implement this workspace replication mechanism using Forby, our application uses the Monitoring module to collect the information about file creation, deletion and modification. To this end, one of the basic event filter sub-modules has been used.

Regarding the use of the Dissemination module, it is necessary to define the obsoleteness rule. In this case, for some given file, a file update makes obsolete a file create and a file delete makes obsolete a file create and a file update from the same user.

---

[1] When compared with web-based services that provide similar functionality, our approach provides additional privacy as photos are not stored at some third-party server.

[2] For simplicity, we omit details related with configuration for delayed download or partial workspace replication.

Again, as in the previous example, when deploying the application it is necessary to decide which file system monitor, *rendez vous* and event storage mechanism will better fit the users environment.

5.3   Voting Application

The third application we present is a very simple voting system. In this application, a user in the group may start a new voting process. For each voting, a ballot document with some pre-defined text format is created in each peer involved in the voting process. The user can cast his vote by modifying the ballot document.

When implementing this application using Forby, the following approach is used. For starting a new voting process, a user must create a voting request file with a pre-defined extension. The file, with some specific pre-defined format, includes the question to be posed, voting possibilities and a flag stating that the voting process should be active. Currently, the voting request file is usually created by copy and paste from given templates, but this functionality could be implemented in existing or in a new applications.

When the application detects that a voting request file has been set to active (by changing the flag in the voting request file), it uses the dissemination system to propagate to other peers the voting request. Upon reception of a voting request, a ballot document is created and a notification is presented to the user. The user could later cast his vote by modifying the ballot document file stored in the file system. Upon completion of his vote – marked using a completed flag, the application propagates the ballot to the original site, where the final decision is computed when some minimum number of votes were received.

To implement this application using Forby, our application uses the Monitoring module to know when a voting request or ballot has been modified. By verifying the active (resp. completed) flag in the voting request (resp. ballot) file, the application knows that it is time to propagate a voting request event (resp. ballot event). The event filter sub-module used by this application only propagates to the application the notification that a file has been modified.

Regarding the use of the Dissemination module, the same obsoleteness rules of the replication application are used. Again, as in the previous examples, when deploying the application it is necessary to decide which file system monitor, *rendez vous* and event storage mechanism will better fit the users environment.

6   Performance Evaluation

Forby's event capturing relies on monitoring mechanisms that could possibly introduce some file system operation overhead. To analyze the relevance of that potential overhead, a simple experiment consisting on a set of file access operations was run. The experiment consisted on measuring the overhead, averaged over 500 repetitions, for read and write operations over different files and file sizes. Tables 1 and 2 show the resulting average execution time for the different monitoring tools.

Table 1. Average execution time for write operations

	Execution time (ms)		
File size	No monitoring	JNotify	Minifilter (IFS)
500 KB	17	18	17
1 MB	38	38	38
10 MB	485	485	491
20 MB	966	990	977

Table 2. Average execution time for read operations

	Execution time (ms)		
File size	No monitoring	JNotify	Minifilter (IFS)
500 KB	1	1	1
1 MB	3	3	3
10 MB	39	39	40
20 MB	80	81	84

The values on tables 1 and 2 show for both read and write operations a very low overhead. This low overhead is barely measurable and is probably more influenced by the increase in the overall system load than the potential delay, related to the registering of notifications, induced on the actual kernel operations. Moreover, the expected deployment environment consists on interactive user usage, in which this overhead is not observable in practice.

The low overhead is a noticeable benefit of kernel support for file system event notification, since, in contrast, file system call interception mechanisms induce observable delays [13].

7   Final Remarks

In this paper we present Forby, a system that leverages modern lightweight mechanisms for file system activity monitoring with distributed event dissemination to provide groupware features to distributed users. As such, the system is composed of two main modules. The monitoring module allows applications to obtain the file system events they are interested on using a high level, operating system independent, interface. The dissemination module allows to disseminate events among groups of peers, handling all details related with reliability (including persistence and disconnection support). The dissemination module was designed to impose minimal infrastructure requirements by allowing to use existing servers (CVS/svn servers or email servers – e.g. Gmail). This simplifies the use of the system by any user.

The examples presented in the previous section show three different uses of Forby. The first solution provides awareness information for users editing data managed by a version control system. The second solution provides automatic shared data replication. The third solution implements a voting system. These examples illustrate that Forby can be used to develop different groupware features to distributed users. One important feature of these solutions, and of solutions that rely on file system mechanisms, is that it allows providing groupware features to any application, allowing users to continue using their preferred applications.

However, it is important to understand that these mechanisms can only act when user applications access the file system. This makes these solutions better suited for providing groupware features in non-realtime (or asynchronous) settings. This limitation can be alleviated in some applications, by setting the auto-save period to a short time, thus allowing the system to have knowledge about user actions very quickly.

As mentioned earlier, we believe that combining the use of file system monitoring and distributed event dissemination has the potential to allow providing groupware features to users using unmodified applications. We see this approach not as a replacement of other alternatives, but rather as a complement.

References

1. Avrahami, D., Hudson, S.E.: Qna: augmenting an instant messaging client to balance user responsiveness and performance. In: CSCW 2004: Proc. of the 2004 ACM conference on Computer supported cooperative work, pp. 515–518. ACM, New York (2004)
2. Begole, J., Rosson, M.B., Shaffer, C.A.: Flexible collaboration transparency: supporting worker independence in replicated application-sharing systems. ACM Trans. Comput.-Hum. Interact. 6(2), 95–132 (1999)
3. Dewan, P., Hegde, R.: Semi-synchronous conflict detection and resolution in asynchronous software development. In: Proc. of the European Conference on Computer-Supported Cooperative Work - ECSCW 2007, pp. 159–178 (2007)
4. Dourish, P., Keith Edwards, W., LaMarca, A., Lamping, J., Petersen, K., Salisbury, M., Terry, D.B., Thornton, J.: Extending document management systems with user-specific active properties. ACM Trans. Inf. Syst. 18(2), 140–170 (2000)
5. Fitzpatrick, G., Marshall, P., Phillips, A.: CVS integration with notification and chat: lightweight software team collaboration. In: Proc. of the 2006 20th anniversary conference on Computer supported cooperative work, pp. 49–58 (2006)
6. Gross, T., Prinz, W.: Awareness in context: a light-weight approach. In: ECSCW 2003: Proc. of the eighth conference on European Conference on Computer Supported Cooperative Work, Norwell, MA, USA, pp. 295–314. Kluwer Academic Publishers, Dordrecht (2003)
7. Hupfer, S., Cheng, L.-T., Ross, S., Patterson, J.: Introducing collaboration into an application development environment. In: CSCW 2004: Proc. of the 2004 ACM conference on Computer supported cooperative work, pp. 21–24 (2004)
8. Ignat, C.-L., Oster, G.: Awareness of Concurrent Changes in Distributed Software Development. In: Proc. of the International Conference on Cooperative Information Systems (CoopIS 2008), Monterrey, Mexico, November 2008, pp. 456–464 (2008)
9. Ignat, C.-L., Papadopoulou, S., Oster, G., Norrie, M.C.: Providing Awareness in Multi-synchronous Collaboration Without Compromising Privacy. In: Proc. of the ACM Conference on Computer-Supported Cooperative Work - CSCW 2008, San Diego, California, USA, November 2008, pp. 659–668 (2008)

10. JNotify. Jnotify linux api (2007), http://jnotify.sourceforge.net/
11. Li, D., Li, R.: Transparent sharing and interoperation of heterogeneous single-user applications. In: CSCW 2002: Proc. of the 2002 ACM conference on Computer supported cooperative work, New York, NY, USA, pp. 246–255 (2002)
12. Li, D., Lu, J.: A lightweight approach to transparent sharing of familiar single-user editors. In: CSCW 2006: Proc. of the 2006 20th anniversary conference on Computer supported cooperative work, New York, NY, USA, pp. 139–148 (2006)
13. Machado, D., Preguiça, N., Baquero, C.: Vc2 - providing awareness in off-the-shelf versioncontrol systems. In: IWCES9: Proc. of the Nineth International Workshop on Collaborative Editing Systems. IEEE Computer Society, Los Alamitos (2007)
14. Molli, P., Skaf-Molli, H., Bouthier, C.: State treemap: an awareness widget for multi-synchronous groupware. In: 7th International Workshop on Groupware - CRIWG 2001, Darmstadt, Germany (September 2001)
15. Rowstron, A.I.T., Druschel, P.: Pastry: Scalable, decentralized object location, and routing for large-scale peer-to-peer systems. In: Middleware 2001: Proc. of the IFIP/ACM International Conference on Distributed Systems Platforms Heidelberg, London, UK, pp. 329–350. Springer, Heidelberg (2001)
16. Sarma, A., Noroozi, Z., van der Hoek, A.: Palantír: raising awareness among configuration management workspaces. In: ICSE 2003: Proc. of the 25th International Conference on Software Engineering, Washington, DC, USA, pp. 444–454. IEEE Computer Society, Los Alamitos (2003)
17. De Souza, C.R.B., Basaveswara, S.D., Redmiles, D.F.: Supporting global software development with event notification servers. In: Proc. of the ICSE 2002 International Workshop on Global Software Development (2002)
18. Sun, C., Xia, S., Sun, D., Chen, D., Shen, H., Cai, W.: Transparent adaptation of single-user applications for multi-user real-time collaboration. ACM Trans. Comput.-Hum. Interact. 13(4), 531–582 (2006)
19. Zadok, E., Nieh, J.: FiST: A language for stackable file systems. In: Proc. of the Annual USENIX Technical Conference, San Diego, CA, June 2000, pp. 55–70. USENIX Association (2000)

Extending a Shared Workspace Environment with Context-Based Adaptations

Dirk Veiel[1], Jörg M. Haake[1], and Stephan Lukosch[2]

[1] Department for Mathematics and Computer Science, FernUniversität in Hagen,
58084 Hagen, Germany
dirk.veiel@fernuni-hagen.de, joerg.haake@fernuni-hagen.de
[2] Faculty of Technology, Policy, and Management, Delft University of Technology,
PO box 5015, 2600 GA Delft, The Netherlands
s.g.lukosch@tudelft.nl

Abstract. Nowadays, many teams collaborate via shared workspace environments, which offer a suite of services supporting group interaction. The needs for an effective group interaction vary over time and are dependent of the current problem and group goal. An ideal shared workspace environment has to take this into account and offer means for tailoring its services to meet the current needs of the collaborating team. In this article, we propose a service-oriented architecture of shared workspaces, analyze this architecture to identify adaptation possibilities, introduce the *Context and Adaptation Framework* as a means to extend shared workspace environment for context-based adaptations and validate our approach by reporting on our prototype implementation.

Keywords: Shared workspaces, adaptation, group context.

1 Introduction

In today's global economy, many teams use shared workspace environments providing problem-specific and team-specific tools and artifacts. However, the types of artifacts and tools currently needed may vary over time, task, and phase of collaboration. Thus, shared workspace environments must deal with changing tool sets consisting of single user and cooperative tools, and provide an appropriate integration approach. In order to accommodate such changing needs, shared workspace environments need to be adapted. However, such adaptations are difficult and cause overhead. Thus, the collaboration environment should make adaptations as easy as possible. Furthermore, changing collaboration situations may require adaptations of multiple tools and artifacts.

Current approaches support manual tailoring of shared workspaces either in terms of artifacts and workspace structure or they focus on adaptation of single tools to the needs of the individual user or, in some cases, to the needs of a group. However, there is no support for adapting a shared workspace environment to the needs of a team.

We propose an extended shared workspace environment supporting context-based adaptation, which helps to minimize adaptation overhead, and which supports adaptations affecting multiple tools to better match the users' current needs. Our approach

L. Carriço, N. Baloian, and B. Fonseca (Eds.): CRIWG 2009, LNCS 5784, pp. 174–181, 2009.

consists of the *Context and Adaptation Framework*, which facilitates the conversion of tools into context-adaptive tools providing the necessary interfaces to our run-time adaptation environment, and a run-time environment executing adaptation rules, which lead to modified workspace configuration and tool behavior.

In the next section, we briefly examine related work. In section 3, we propose possible points of adaptation for service-oriented shared workspaces. Section 4 presents both, the *Context and Adaptation Framework* as a means to make tools ready for context-based adaptation, and the architecture of the run-time environment. Section 5 presents our conclusions, experiences, and ideas for future work.

2 Related Work

We start this section by reviewing relevant shared work environments and discuss how these systems deal with possible adaptations. After that, we take a look at context-adaptive systems in general. We review whether the taken approaches are suitable to support context-adaptive collaboration in shared work environments.

BSCW [1] and *CURE* [2] are web-based shared work environments that offer a variety of collaboration services, e.g. communication and document sharing. *CHIPS* [3] is a cooperative hypermedia system with integrated process support. *TeamSpace* [4] offers support for virtual meetings and integrates synchronous and asynchronous team interaction into a task-oriented collaboration space. *BRIDGE* [5] is a collaboratory that focuses on supporting creative processes and as such integrates a variety of collaboration services. All of the above examples focus on a specific application domain and only offer a fixed set of services to the user. Some of them, e.g. *CHIPS* or *CURE*, are highly tailorable, but they do not automatically adapt their offered functionality to improve the collaboration within a team.

The most prominent examples for context-based adaptation focus on single users and consider location as most relevant context information (e.g. [6, 7, 8] or focus on learner profiles (cf. ITS). Compared to single-user ITS, *COLER* [9] provides a software coach for improving collaboration. *CoBrA* [10] is an agent-based architecture that uses shared context knowledge represented as a ontology to adapt service agents according to a user's context. Gross and Prinz [11] introduce a context model and a collaborative system that supports context-adaptive awareness. The main restrictions of their approach are that the context representation is only used to update and visualize awareness information and that only one cooperative application can be used. The *ECOSPACE* project [12] aims at providing an integrated collaborative work environment and uses a service-oriented architecture and provides a series of collaboration services for orchestration and choreography. The orchestration and choreography is based on an ontology which still has to be described.

The above approaches focus on adaptations which are used in specific domains, e.g., single-user systems or ITS, or on sub-domains in the field of CSCW, e.g. awareness or knowledge management. Adaptation based on group context is intended only by *ECOSPACE*, but the required context model is still an open issue. In summary, current approaches do not provide a sufficient context model for adapting the collaboration in shared workspaces and do not use the context information to adapt the interaction of the users.

3 Possible Adaptations in Service-Oriented Shared Workspaces

A collaborative application typically uses the model-view-controller paradigm [13] and can be based on a client-server architecture. Nowadays, service-oriented client-server architectures are used to create collaborative services to be consumed by collaborative applications (i.e. clients). We address these kinds of architectures and split them up into four layers: *UI*, *Logic*, *Services*, and *Model Layer* (cf. Figure 2). The view and controller parts of the tools (*Application$_X$ UI*) are usually running on the client side (addressed by the *UI Layer*). The *Logic Layer* consists of the application specific business logic (*Application$_X$ Logic*) that can be built by using different services from the underlying *Services Layer*. The *Services Layer* contains *Application Services* as well as *Collaboration Services* and is used by the above layer to accesses the artifacts (*Artifact$_X$*) represented in the *Model Layer*. If more than one application is present, we talk about a shared workspace environment.

Following the layers one can distinguish four areas of adaptation within shared workspaces: (1) *Modification of the application user interface:* Usually, the UI can be modeled as a hierarchy of visual components supporting the following adaptations: (i) add, remove, or replace entire visual components in the hierarchy, and (ii) replace individual views or controllers. (2) *Modification of application logic:* Here, we may change the internal structure of the application logic by adding, removing, or replacing application services used by the application, or by changing the execution structure of the application services. (3) *Modification of services:* In this layer, application services as well as collaboration services can be added, removed or replaced. This impacts the way users may communicate, coordinate and share objects. (4) *Modification of shared model:* This layer could be adapted by manipulating artifacts and sessions. Artifacts respectively documents could be created or deleted and their attributes could be manipulated. Sessions could be created or closed, and members, tools, or artifacts could be added to or removed from a session.

In the following, we use the above adaptation possibilities to define an architecture for specifying and implementing such adaptations and to introduce a framework supporting developers in extending applications with adaptation functionality.

4 Adaptation in Service-Oriented Collaborative Applications

We now briefly introduce our context model necessary for defining and recognizing conditions for adaptation in our *Context and Adaptation Framework* (*CAF*) [14]. This context model addresses scenarios in which several actors collaborate to achieve a shared goal and captures basic concepts of co-located and distributed collaboration.

Figure 1 summarizes our context model and shows the basic context classes and their relations. An *Application* implements the model-view-controller (MVC) paradigm [13] and consists of *Views* and *Controllers* components. *Views* and *Controllers* use *Services* to access the *Artifacts*. *Artifacts* use *Services* to notify *Views* and *Controllers* about changes. Each *Application* is part of a *User Workspace* and is created by an *Application Factory* which specifies what *Applications* are available within a workspace and how these can be initialized. Each *Actor* has a *User Workspace* and belongs to at least one *Team*. The *User Workspace* defines the *Roles* of an *Actor* within this *Team*. Each *Role* allows an *Actor* to perform specific *Actions* within an *Application*. The available *Actions* within an *Application* are defined by its *Application Functionality*. *Actors* then interact with the *Application* by performing *Actions* allowed by their *Roles*. These *Actions* are received by the corresponding *Controller* components of the *Application*. The *Application Functionality* class has several subclasses not shown in Figure 1, e.g., *Communication, Shared Editing (SE), Awareness, Management* or *Workflow Management*. These classes are used to specify functionality (e.g., *Chat, Rich Text SE, Remote Field of Vision, Concurrency Control Management*) as well as supported action types (in case of *Chat*, e.g., *OpenChat* and *SendMsg*).

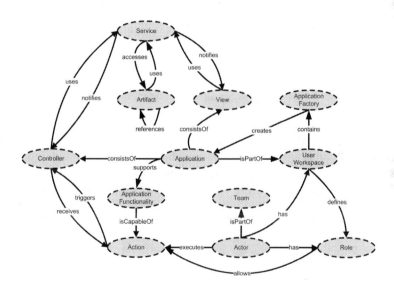

Fig. 1. Collaboration context model

All above classes are useful for modeling the configuration of shared workspaces and tools and for capturing the current state at runtime. As a sample scenario consider that a team consisting of Alice and Bob synchronously collaborates on a shared text document. We assume, that Alice and Bob created user workspaces sharing the design artifacts (i.e. documents). Alice then created a shared text document and opened a shared text editor to work on a design document. Bob later opened the same shared text document to review the current state of the design. Establishing a communication possibility among the two actors is an obvious adaptation possibility to improve their interaction in this state. The workspaces of both actors offer an application factory for a chat tool. The following pseudo code shows an adaptation rule which makes use of the current state and adapts the users' workspaces:

```
rule "open communication channel"
   when
       artifacts: getArtifactsInContext("OpenText:bob")
       actors: getActorsInContext(artifacts)
       communication: getApplicationsInContext(actors,
           "Synchronous Communication")
       selectedApplication: selectOneFrom(communication)
   then
       openForAll(selectedApplication, actors)
end
```

The above adaptation rule consists of a condition part and an action block. The action block may contain several adaptation actions in order to facilitate cross-application adaptations. By, e.g., adding a service to the current application configuration the adaptation action may change the current state. In the above example, the condition is triggered by *Actor:bob* opening the *Artifact:Text Document*. The action part of the above adaptation rule is executed if all conditions are valid. In this adaptation rule, `getArtifactsInContext` returns a set of artifacts which are in the context of the action *OpenText:bob*. The function `getActorsInContext` then calculates all actors which access the artifacts in the context of *OpenText:bob* (i.e. all objects in the state relevant to the action). The function `getApplicationsIn-Context` then determines in this case all applications which support the application functionality *Communication* and are connected to all actors accessing the same artifacts. The function `selectOneFrom` selects the context element which has been used most by the collaborating actors.

The above adaptation rule is only one example for a possible adaptation in the given context. There exist further adaptation possibilities to improve the interaction within a team, e.g., to provide additional awareness information or to enable concurrency control mechanisms. We use the current state information to recognize specific situations, which show potential for improvements as expressed in the rule's condition, and perform the adaptation described in the action part of the rule. In the above example we consider the situation where at least two users share the same artifact at the same time and assume that a communication channel would help to coordinate the users' work. Automated rule execution minimizes the users' adaptation efforts. Obviously, more generic rules (i.e. independent from specific users) are applicable in more cases and thus express general policies, while more specific rules can be used to express user or team specific preferences.

Next, we address the question of how to make either existing or new tools in a shared workspace environment ready for adaptation. From a developer's point of view, integrating an application into our *CAF* should be as simple as possible as well as minimize the implementation effort. These are the two requirements we have considered while designing the framework.

We propose an approach that extends usual client-server applications following the architecture as described in Section 3 by adding components of our *CAF* shown in Figure 2 in grey.

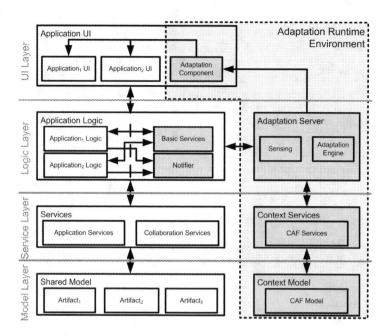

Fig. 2. Extended usual client-server application by *Context and Adaptation Framework (CAF)*

The *UI Layer* contains the *Application UI*s and the *Adaptation Component* of *CAF*. We use the *Adaptation Component* to start and stop *Application UI*s, or to use a specific interface an application offers to apply adaptation actions to the *Application UI*. Currently, this interface contains methods to show or hide a GUI component, to set the focus to a GUI component, to modify the content of a GUI component, to highlight a specific GUI component, to maximize or minimize the view, to set the read-only mode, to scroll to a certain position, or to lock the scrollbars.

The *Logic Layer* contains the *Application Logic* as well as the *Adaptation Server*. Our *CAF* adds the components *Basic Services* and *Notifier* to the *Application Logic*. The *Application Logic* can use *Basic Services* to integrate the application into the *CAF* or use the *Notifier* to send notifications to the client side. The developer can use *Basic Services* to integrate support for, e.g., sensing functionality, access right and user management, access to a database service etc. and thus, build a bridge to the *Adaptation Server*. Furthermore, the *Basic Services* can call services within the *Application Logic* of applications to, e.g., change the configuration. The *Adaptation Server* is based on a service-oriented architecture and hosts components like *Sensing*, and the *Adaptation Engine*. *Sensing* uses the information about service calls and user interactions to update the current state representation. The *Adaptation Engine* uses the current state representation to find corresponding adaption rules and to execute them. Adaptation actions can affect *Application UI*, *Application Logic*, *Services*, and the *Shared Model*.

Our *Adaptation Runtime Environment*, shown on the right in Figure 2, allows the execution of adaptation rules, which adapt the configuration of shared workspace environments and the behavior of applications offering our adaptation interface. It can be split up into the layers *Adaptation Server, Context Services* and *Context Model.*

A flexible adaptive system executes a cycle of: (1) User interaction; (2) Sensing user activities; (3) Adapting system behavior; (4) Modifying adaptation. The user interacts with the tools of the shared workspace environment. Usually, these interactions imply service calls that can be sensed to update the context representation of the current situation. The current state representation is used by the *Adaptation Engine* to find corresponding adaption rules and to execute them. An adaptation rule can modify the configuration of components in all layers.

Typical adaptations visible to a user affect the UI, e.g., show or hide a certain GUI component, set the focus to a specific GUI component. Adaptations that affect the *Logic Layer* may change the service composition (e.g., the concurrency control algorithm used). Starting or stopping a service (e.g., the chat service in the above example) can be a possible adaptation at the *Service Layer*. At the *Model Layer* the adaptations can range from changing attributes of an artifact to creating or deleting artifacts. All of these adaptations are supported by *CAF* and can be used in an adaptation rules' action part.

5   Conclusions

In this paper we proposed a service-oriented architecture of shared workspaces and analyzed possible points of adaptation on four layers (*UI, Logic, Service,* and *Model Layer*). We introduced a context model and adaptation rule syntax for expressing context-based adaptations of shared workspaces. Using *CAF*, we support the integration of service-oriented applications into our adaptation runtime environment.

Our approach exceeds the state of the art (cf. Section 2) in several ways: firstly, we provide a context model sufficient for adapting collaboration among users of a shared workspace and, secondly, our approach can exploit this context information to adapt the interaction among users and between users and their tools. Finally, our approach is open for inclusion of new services, applications, and adaptation rules. By extending the context model, new rules can be introduced and address new collaboration aspects and situations without rendering old rules meaningless.

We prototypically implemented *CAF* and our conceptual architecture (cf. Section 4) and used it to integrate a number of collaborative tools. Currently, our prototype implementation is used for functional testing and evaluation in pilot test cases. For the near future, we plan to include applications for adaptation rule tracing, editing and negotiation; testing of adaptation rules in real work situations with the goal of identifying good adaptation practice; and performance tuning of rule execution.

References

1. Appelt, W., Mambrey, P.: Experiences with the BSCW shared workspace system as the backbone of a virtual learning environment for students. In: Proceedings of ED-MEDIA 1999 (1999)
2. Haake, J.M., Schümmer, T., Haake, A., Bourimi, M., Landgraf, B.: Supporting flexible collaborative distance learning in the CURE platform. In: Proceedings of the Hawaii International Conference On System Sciences (HICSS-37). IEEE Press, Los Alamitos (2004)

3. Wang, W., Haake, J.M.: Tailoring groupware: The cooperative hypermedia approach. Computer Supported Cooperative Work 9(1), 123–146 (2000)
4. Geyer, W., Richter, H., Fuchs, L., Frauenhofer, T., Daijavad, S., Poltrock, S.: A team collaboration space supporting capture and access of virtual meetings, Boulder, Colorado, USA, pp. 188–196. ACM Press, New York (2001)
5. Farooq, U., Carroll, J.M., Ganoe, C.H.: Supporting creativity in distributed scientific communities. In: GROUP 2005: Proceedings of the 2005 international ACM SIGGROUP conference on Supporting group work, pp. 217–226. ACM, New York (2005)
6. Schilit, B., Adams, N., Want, R.: Context-aware computing applications. In: First Annual Workshop on Mobile Computing Systems and Applications (WMCSA) (December 1994)
7. Abowd, G.D., Atkeson, C.G., Hong, J., Long, S., Kooper, R., Pinkerton, M.: Cyberguide: a mobile context-aware tour guide. Wireless Networks 3(5), 421–433 (1997)
8. Kindberg, T., Barton, J., Morgan, J., Becker, G., Caswell, D., Debaty, P., Gopal, G., Frid, M., Krishnan, V., Morris, H., Schettino, J., Serra, B., Spasojevic, M.: People, places, things: web presence for the real world. Mobile Network Applications 7(5), 365–376 (2002)
9. de los Angeles Constantino-González, M., Suthers, D.D.: Automated coaching of collaboration based on workspace analysis: Evaluation and implications for future learning environments. In: Proceedings of the 36th Hawai'i International Conference on the System Sciences (HICSS-36). IEEE Press, Los Alamitos (2003)
10. Chen, H., Finin, T.W., Joshi, A.: Semantic web in the context broker architecture. In: Proceedings of the Second IEEE International Conference on Pervasive Computing and Communications (PerCom 2004), pp. 277–286. IEEE Computer Society, Los Alamitos (2004)
11. Gross, T., Prinz, W.: Modelling shared contexts in cooperative environments: Concept, implementation, and evaluation. Computer Supported Cooperative Work (CSCW) 13(3), 283–303 (2004)
12. Martínez-Carreras, M.A., Ruiz-Martínez, A., Gómez-Skarmeta, F., Prinz, W.: Designing a generic collaborative working environment. In: IEEE International Conference on Web Services (ICWS 2007), pp. 1080–1087 (2007)
13. Krasner, G.E., Pope, S.T.: A cookbook for using the model-view-controller user interface paradigm in Smalltalk-80. Journal of Object-Oriented Programming 1(3), 26–49 (1988)
14. Lukosch, S., Veiel, D., Haake, J.M.: Enabling context-adaptive collaboration for knowledge-intense processes. In: Proceedings: 11th International Conference on Enterprise Information Systems (ICEIS 2009). Springer, Heidelberg (2009)

An Evolutionary Platform for the Collaborative Contextual Composition of Services

João Paulo Sousa[1], Benjamim Fonseca[2], Eurico Carrapatoso[3], and Hugo Paredes[4]

[1] Instituto Politécnico de Bragança, Bragança, Portugal
jpaulo@ipb.pt
[2] CITAB/Universidade de Trás-os-Montes e Alto Douro, Vila Real, Portugal
benjaf@utad.pt
[3] Faculdade de Engenharia da Universidade do Porto, Porto, Portugal
emc@fe.up.pt
[4] Universidade de Trás-os-Montes e Alto Douro, Vila Real, Portugal
hparedes@utad.pt

Abstract. Besides services traditionally available in wireless networks, new ones may be offered that transparently adjust and adapt to the user context. The user would have more choice and flexibility if not only could he use platform and third-party services, but also compose his own services in an ad-hoc way, making it available to other users, involving them collaboratively in the construction of a wide set of services. Moreover, collaboration among users can be fostered by the availability of awareness services in mobile environments that enable them to execute joint tasks and activities. This paper presents iCas, an architecture to create context-aware services on the fly, and discusses its main modules. Also a collaborative application scenario is briefly described.

Keywords: Context-aware, Services composition, Semantic Web, CSCW.

1 Introduction

Mobile computation has an increasing community of users and is a natural field for cooperative applications. However, some mobile devices have important resources' limitations, so it is desirable that most applications can be used dynamically whenever available in the form of services. These services, if correctly specified, have the potential of being combined dynamically to produce richer services. In this scenario, the use of captured contextual information related to issues such as location, current activities, objects in the neighborhood and device features, plays a crucial role in the simplification of the interaction between humans and the digital world.

This paper describes an open infrastructure for a mobile network environment, in which a user can receive in his mobile device (e.g. PDA, notebook) context-aware information (location, neighborhood, user profile) and have a set of useful services sensitive to his current context. The user can also compose services dynamically in real time to create new highly personalized services with more features and use or share them as many times as he wants. All services are stored in a repository and have a Participatory Profile associated with each one. This profile is built based on the

L. Carriço, N. Baloian, and B. Fonseca (Eds.): CRIWG 2009, LNCS 5784, pp. 182–189, 2009.

contribution of all users. This contribution is both explicit, by classifying the services and thus helping other people on their utilization, and implicit, through the collection of usage data. Another collaborative and evolutionary feature is the ability to combine services from different users that resulted from previous compositions.

The remainder of this paper is structured as follows: section 2 presents major background issues behind the work described in this paper, introducing an ontology to describe context and discussing the several approaches for composing Web Services and the OWL-S ontology. Section 3 gives an overview of related work. Section 4 presents the iCas architecture and describes the details of each module. Section 5 describes a scenario for using iCas. Finally, we provide some conclusions and plans for future work, in section 6.

2 Background

Group awareness is an important issue in collaborative applications. In order to work together, users need information about the environment, namely who is present, current activities and shared artifacts [1]. With the emergence of mobile environments, which enable teams to cooperate while on the move, awareness information has become a hot topic in the scientific community leading to further developments in context-aware computing. Systems that are sensitive to context have the ability to "adapt according to the location of use, the collection of nearby people, hosts, and accessible devices, as well as to changes of such things over time" [2]. The fusion of context aware applications and mobile CSCW gains particular interest in the situation where teams are geographically distributed and on the move [3], such as activities in an academic campus and the management of emergencies [4]. In such environments, combined applications are able to deliver contextualized information and services, context based automatic execution of services and attach contextual information for later retrieval [5]. To do this, current CSCW infrastructures should use reusable components and services that can be dynamically composed in order to adapt their functionalities according to users' and groups' needs [6]. This composition can be achieved with the help of semantic information.

Contextual information models based on ontologies have been explored in several architectures that support context-aware services (e.g. [7, 8]). These models allow the cooperation among objects and the discovering, acquisition, inference, and distribution of contextual information.

The use of a semantic model brings several advantages:

- being based on Semantic Web standards, makes the exchange, reuse and, sharing of context information among context aware applications easier;
- it decouples the information context model from the programming model, unlike some architectures mentioned in the previous section;
- a high degree of expressiveness and formalism to represent concepts and relations in a context-awareness scenario, allowing reasoning about context.

To describe the context, we decided to use the semantic model SeCoM (Semantic Context Model), presented in [8]. SeCoM is composed of six main ontologies: Actor, Activity, Spatial, Spatial Event, Temporal Event, Device, Time; and six support ontologies: Contact, Relationship, Role, Project, Document, Knowledge.

Web Services have often been used for the composition of services. In [9] and [10] the authors compare several solutions, based on characteristics such as automatic composition, composition verification, scalability, goal satisfaction, connectivity and non-functional properties. When the purpose is to implement the composition of mobile services, we have to consider concerns such as the complexity of the services to be built. For this purpose, one must find a compromise between simplicity in service creation and flexibility. Flexibility requires more complex rules and probably specific technical knowledge. In this case the simplicity offered to end-users is lost.

To achieve this goal, we chose to compose services in an interactive way: the user gradually generates the composition with ad-hoc forward or backward selection of services. To use this approach for composing Web services requires that they can understand their features and how they interact together. WSDL specifies a standard way to describe the interfaces of a Web Service at the syntactic level. However, WSDL does not support the semantic description of services. OWL-S has appeared to fulfill this limitation and uses the OWL language to describe Web Services. OWL-S provides Web services with a set of markup language constructs for describing the properties and capabilities in an unambiguous interpretable form by the software agents. OWL-S is a framework that enables automatic discovery and matchmaking tasks, and composition and execution of Web Services. OWL-S consists of the following classes: ServiceProfile - specifies how the services are announced to the world; ServiceModel - specifies how to interact with the service; and ServiceGrounding - specifies the details of how an agent can access the service.

3   Related Work

A number of context-aware systems have been developed to demonstrate the usefulness of context-aware technology such as ParcTab [2], which was one of the first systems to offer a general context-aware framework; and ContextToolkit [11], which presents a modular context-aware framework with reusable components. This feature allows programmers to build more easily, interactive context-aware systems based on sensors. These systems do not have an open context model because often the context is described in an object-oriented base and so the information is strongly coupled with the programming model. More recently several studies appeared to support context-aware composition of services, one more generic and others dedicated to mobile environments [7, 12-14]. In [12] the authors propose a framework for dynamic composition of context-aware mobile services. The main features are service adaptation to the devices and the network, and service adaptation to the user preferences and user location. However the study does not specifies which approach is used to compose new services. The SOCAM [7] presents a middleware architecture for building rapid context-aware services. It provides support for discovering, acquiring, interpreting and accessing context information. It also presents one of the first ontologies that define the main classes of context: person, location, activity and computer entity. Nevertheless, this architecture does not allow the composition of services.

MyCampus [13] is a semantic web environment that uses agents able to find context information for enhancing everyday campus life. MyCampus architecture is composed by eWallets (knowledge containers), which support automated discovery and access to the context. The users can subscribe task-specific agents to assist them in different context tasks using the semantic information in eWallets. These agents are able to discover, execute and compose automatic semantic Web Services using the OWL ontology for services (OWL-S) [15].

In [14] the authors present CACS, a framework that enables context-aware composition of Web Services. This framework supports capability matches and goal-driven composition services flow. The CACS architecture uses software agents to discover, compose, select, and automatically execute Web Services using OWL-S.

In [12], [13], [14] we saw that these systems do not have an open model to describe context, which causes some lacks on sharing context knowledge and context reasoning with external systems. The [12], [14] studies present architectures that support the automatic composition of services. The user makes a request to the architecture, most of the times to a software agent that collects context information and tries to find the most suitable service, which agrees with the request description. If the agent does not find the service or it does not exist, then the software agent decomposes the request into multiple sub-goals in order to find the matching services.

In all the cases that use automatic composition, it is a hard task to maintain the details about the rules of services' invocation. These approaches also do not have an open model to describe context, which causes some limitations regarding the sharing of context knowledge and context reasoning with external systems.

4 Architecture and Implementation

To support the composition of context-aware services on the fly and provide context-aware information to the users, we propose a service oriented architecture (SOA) based on ontologies, named iCas. We divide the architecture into four essential engines to explore the potential of context, as shown in Fig. 1.

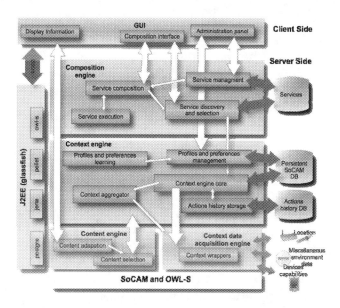

Fig. 1. Overview of iCas architecture

When a user selects the service composition interface, the service discovery module gets his preferences, parameters configuration and interests. With this information and the OWL-S services descriptions, the service discovery and selection module selects the services from the service repository to perform a context-based selection, and then delivers it as a list to the composition interface. The user may know clearly which tasks he wants to achieve with the composition or he can start to compose choosing compatible services that can suggest the creation of a new service.

Each time a service is selected to be part of the composition, the service discovery and selection module searches for services (Fig. 2) using data collected from the context engine core and returns further possibilities based on the current context and user policies. The search and selection are only possible due to the OWL-S service description, which allows creating relationships with other ontologies that can describe details about a service type and its features.

The search is performed using the description of the ontology's ServiceProfile class, which describes the service and specifies its input/output parameters, preconditions and effects. The first selection of services is performed using the ServiceProfile hierarchies, which choose the services from a particular category. Then a matching is performed, selecting the services whose input is syntactically compatible with the output of the current service (Filtering). Finally a scoring is carried out using the weights of the evaluation parameters defined in the ServiceProfile and a particular evaluation policy, which depends on the service category. The Participatory Profile shown in Fig. 2 is an extension of the ServiceProfile class, with new fields for storing information such as service's usage and user's evaluation. Furthermore, the user can add specific fields that help to classify it. This feature, along with the user's evaluation, makes possible for users to collaborate in the definition of information that helps in the selection of services. Each selection of a service is based on the matching between the service's Participatory Profile and the user's service preferences described in his profile. The services stored in the repository can be built by the system administrator or by the users. These services can be either atomic or the result of a prior composition of services. Their creators define the sharing permissions that are stored in the Participatory Profile to specify who can use them.

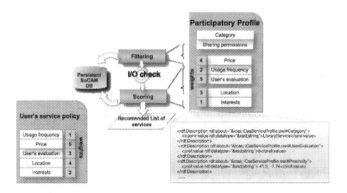

Fig. 2. Selection of services for composition

The ongoing user composition is supported by the service composition function, which generates a workflow of service calls. By the time that the user finishes the composition, the entity service composition has created a composite service that contains a workflow. This workflow is a composite service that has the three key descriptions of an OWL-S service: service profile, grounding and model, as mentioned in the end of section 2. This newly composed service can be saved, executed or used in another service composition task. To store the service, the service composer module uses the service management, and to execute the service service execution is called. When the service execution module executes a composition, it follows a workflow to call each individual service and exchange data between them, according to the flow constructed by the user.

The service management module deals with the services stored in the services container, providing operations such as adding, removing and sharing services using the policies properties. The service container only stores the OWL-S description of this service (service profile, model and grounding), but the service functionality is still provided by a third party.

The context engine is responsible for managing all related context data and for reasoning about context. All context information is stored in a permanent OWL ontology storage system. The context engine core uses the Jena API to store the RDF models of SeCoM using a PostgreSQL database. This engine is also responsible for extracting knowledge from the SeCoM model, using SPARQL queries and for making inferences to derive additional statements that are not described explicitly in the SeCoM (e.g. "if a user is located in the library, he is at the university campus", or if a user has interests in "ontologies" and context-awareness is related to "semantic web", hence the user is also interested in "semantic web"). Besides the usage of contextual information for a single user, this engine also enables sharing context among users, providing features of social networks by extracting information such as "who is near me?" or "what are my friends doing?".

The context aggregators keep in memory (non-persistent) highly changing dynamic data that is captured from various sources related to an entity (e.g. user, object). For each entity an instance is created that relates that entity with the data that come from the sources (e.g. user's location and data sensors), removing the computational charge caused by the frequent data updates into the persistent ontology.

The profiles and preferences management is responsible for managing the explicit user profile and interests information. Using the administration panel this module allows the user or administrator to manage explicit context such as insert, update and remove profiles parameters and user preferences.

The actions history storage captures each action performed by the context engine core and stores it in the actions history database. The main actions are search, insert, update and remove, and they are saved in the following format: Action + target Triplet (e.g. update: Bob isMemberOf the Sciences Students Group).

The profile and preferences learning module can change preferences and profile data through a learning algorithm (e.g. if a student queries many times for a particular book in the library services, the theme category of that book is added to the hasInterestesIn property of the knowledge ontology). It searches for particular actions stored in the actions history database, counts the occurrences of an action and, accordingly, changes specific parameters defined to be update. Although this is not an optimal approach, a good solution can only be achieved with a large-scale utilization of iCas architecture and the collecting of user feedback. In the future the algorithm may also evolve to an AI algorithm, searching for patterns in the database.

The context data acquisition engine collects data from several sources, such as location devices, sensors and external services, and prepares the data to be used by the content engine and context engine (e.g. convert units values from a data sensor, or transform the coordinates of the user location to a referential location (room 2.1)).

The content engine is composed by two modules: the content selection module is a timer function that periodically selects the user interests from the context engine and delivers it to the content adaptation module for transformation. This information is provided to the GUI as RSS feeds that are adapted by using XHTML Modularization.

All four engines are implemented in the Glassfish v2 application server, which provides the functions to the GUI client through HTTP, as Web Services. This configuration was chosen to support the ad-hoc composition of services in mobile devices, bringing the reasoner's computational requirements to the server side.

5 Application Scenario

We have chosen a university campus as a scenario for using iCas. The purpose is to support students and teachers in their campus life, helping them to keep updated and improve their social and pedagogical interaction. When a user arrives at the campus and connects his mobile device to the wireless network he will have to authenticate, to be identified in a wireless system and in the iCas architecture. This wireless network is also used as a campus location system, including other services that can provide useful information integrated to the iCas system, such as: an e-learning platform, library and administrative services. A possible composition could be the association of the e-learning services with the library service to generate the list of a course bibliography that is presently available in the library. A typical example of this usage scenario is the Friends' Awareness Service, in which the user combines the following set of services to get information about the activities of friends that are located in the campus: Friends Service – information gathered by analyzing the user's profile; Location Service – provides users locations based on the information gathered on the campus location system aforementioned; Calendar Service – provides information on user's appointments and tasks. By using the Friends Awareness Service a user has information that can be used to decide whether or not he can join a friend for achieving some specific task, thus fostering interaction and collaboration.

6 Conclusion

In this paper we have presented iCas, a service-oriented architecture that uses an ontological context model to provide personal and contextual information and to support the composition of context-aware services. The composition of services can be achieved collaboratively through the possibility of sharing the services between among users. By incorporating contextual information, the platform promotes interaction among users, thus enabling the collaborative execution of tasks and activities in a mobile environment.

A prototype of the iCas platform has been implemented and functional tests have been conducted in experimental setups. However, we intend to test the platform more accurately, by simulating critical conditions and submitting it to intensive real daily usage. For this purpose, it is our intention to implement an academic scenario, providing some basic services and promoting the usage of the platform by the academic community, stimulating the evolutionary production of more sophisticated services by users, taking advantage of the composition features. In such a scenario, collaboration will occur both in the construction of services and in the availability of collaborative services that users can benefit from to improve their self and collective performance in social and academic activities.

References

1. Gross, T., Prinz, W.: Modelling Shared Contexts in Cooperative Environments: Concept, Implementation, and Evaluation. Comput. Supported Coop. Work 13, 283–303 (2004)
2. Want, R., Schilit, B., Adams, N., Gold, R., Petersen, K., Goldberg, D., Ellis, J., Weiser, M.: The Parctab Ubiquitous Computing Experiment. In: Mobile Computing, pp. 45–101 (1996)
3. Jones, Q., Grandhi, S.A., Terveen, L., Whittaker, S.: People-to-People-to-Geographical-Places: The P3 Framework for Location-Based Community Systems. Comput. Supported Coop. Work 13, 249–282 (2004)
4. Papadopoulos, C.: Improving Awareness in Mobile CSCW. IEEE Transactions on Mobile Computing 5, 1331–1346 (2006)
5. Dey, A.K., Abowd, G.D., Salber, D.: A conceptual framework and a toolkit for supporting the rapid prototyping of context-aware applications. HCI 16, 97–166 (2001)
6. Neto, R.B., Jardim, C., Camacho-Guerrero, J., Pimentel, M.d.G.: A Web Service Approach for Providing Context Information to CSCW Applications. In: Proceedings of the WebMedia & LA-Web 2004. IEEE Computer Society, Los Alamitos (2004)
7. Gu, T., Pung, H., Zhang, D.: A service-oriented middleware for building context-aware services. Journal of Network and Computer Applications 28, 1–18 (2005)
8. Bulcao, R., Campos, M.: Toward a Domain-Independent Semantic Model for Context-Aware Computing. In: Third Latin American Web Congress (LA-Web 2005), vol. Proceedings of LA-WEB 2005, p. 61. IEEE Computer Society, Washington (2005)
9. Milanovic, N., Malek, M.: Current Solutions for Web Service Composition. IEEE Internet Computing 8, 51–59 (2004)
10. Srivastava, B., Koehler, J.: Web service composition - current solutions and open problems. In: ICAPS 2003 Workshop on Planning for Web Services (2003)
11. Salber, D., Dey, A., Abowd, G.: The Context Toolkit: Aiding the Development of Context-Enabled Applications. vol. CHI, pp. 434–441 (1999)
12. Panagiotakis, S., Alonistioti, A.: Context-Aware Composition of Mobile Services. IT Professional 08, 38–43 (2006)
13. Sheshagir, M., Sade, N., Gandon, F.: Using Semantic Web Services for Context-Aware Mobile Applications. In: MobiSys 2004 Workshop on Context Awareness, vol. Proceedings of MobiSys2004 Workshop on Context Awareness, Boston (2004)
14. Nan, L., Junwei, Y., Min, L., Yang, S.: Towards Context-Aware Composition of Web Services. In: Fifth International Conference on Grid and Cooperative Computing (GCC), vol. Proceedings of GCC 2006, pp. 494–499. IEEE Computer Society, Los Alamitos (2006)
15. W3C: OWL-S: Semantic Markup for Web Services (2004)

Gesture Interaction in Cooperation Scenarios

Carlos Duarte and António Neto

LaSIGE/Faculty of Sciences of the University of Lisbon,
Edifício C6, Campo Grande, 1749-016 Lisboa, Portugal
cad@di.fc.ul.pt, aneto@lasige.di.fc.ul.pt

Abstract. Several computer based tools have been developed to support cooperative work. The majority of these tools rely on the traditional input devices available on standard computer systems, i.e. keyboard and mouse. This paper focuses on the use of gestural interaction for cooperative scenarios, discussing how it is more suited for some tasks, and hypothesizing on how users cooperatively decide on which tasks to perform based on the available input modalities and task characteristics. An experiment design is presented to validate the proposed hypothesis. The preliminary evaluation results, also presented, support this hypothesis.

Keywords: Gestural Interaction, Collaboration, Shared Workspace, Evaluation.

1 Introduction

Gestures are a means of providing natural and intuitive ways to interact with computers across most, if not all computer application domains. This paper looks at ways in which gestural interaction can contribute to the design of collaboration scenarios and applications.

Gesture based interaction is becoming increasingly present, with the growing availability of devices and mechanisms capable of interpreting gestures. The first technological segment to generically adopt gesture based interaction was the mobile devices segment. Palm devices, followed by Windows Mobile based PDAs and Smartphones, were the first to offer gesture recognition capabilities, which were initially used for handwritten recognition. Gestures were performed by using a stylus against a capacitive screen. Later technology supported making gestures with fingers. The iPhone mass marketed the multi-touch technology, introducing gesture interaction to a wider audience.

Today, the introduction of new hardware, like the Microsoft Surface Table, complements older hardware, like touch enabled wall displays, and is making gesture interaction available in a whole new class of scenarios, capable of offering interaction and collaboration possibilities that are not feasible with personal mobile devices.

Our aim is to study the impact that gesture interaction can have on the ways people collaborate. In particular, since gesture interaction is better suited to certain tasks than the traditional keyboard and mouse input devices, we

L. Carriço, N. Baloian, and B. Fonseca (Eds.): CRIWG 2009, LNCS 5784, pp. 190–205, 2009.

hypothesize that people will adopt a collaborative decision process to assign tasks to each other based on a number of factors, which include the technological capabilities of their interaction devices.

This paper presents an experiment designed to assist in understanding how offering the possibility to use gestures impacts the collaboration and decision making processes. Initial results from the experiment are also presented.

The paper begins with a review of several works using gesture interaction. This fosters the identification of three factors that might impact the collaboration: the type of gestures available or required, the technology required and the application domain. The paper then describes how gestures can affect collaboration scenarios, and elaborates on the aforementioned hypothesis. The experiment designed to test the hypothesis is then described, and its initial results presented. Finally, the paper concludes discussing the results achieved so far, and what our next steps will be.

2   Gesture Interaction

This section will look at gesture interaction from three distinct, but complementary, viewpoints. First, we will discuss gestures by their type. Afterwards, we will focus on the gesture enabling technologies, with special attention given to touch screens. Lastly, we will consider the different application domains where gestures have been used, with emphasis on collaboration enabling domains.

2.1   Gesture Types

Although different classifications could be proposed, in this paper we distinguish between four different types of gestures: deictic, manipulative, semaphoric and gesticulating gestures.

Deictic gestures involve pointing to ascertain the identity or spatial location of an object within the context of the application's domain. The first example of the use of deictic gestures is Bolt's "Put that there" work [1]. In his work gestures are used in conjunction with speech enabling the user to point at a location on a large screen display in order to locate and move objects. Deictic gestures have also been used in the CSCW domain, like in Kuzuoka et al.'s work [2].

Manipulative gestures are said to happen when there is a tight relationship between the actual movements of the gesturing hand or arm and the entity being manipulated. These have been used in three interaction paradigms: on the desktop in a 2-dimensional interaction using a direct manipulation device such as a mouse or a stylus [3]; in virtual reality interfaces as a 3-dimensional interaction involving empty handed movements to mimic manipulations of physical objects [4,5]; in tangible interfaces by manipulating actual physical objects that map onto a virtual object [6,7].

Semaphoric gestures are defined in [8] as a gesturing system that employs a dictionary of static or dynamic gestures. This is one of the most widely applied styles of gestural interaction even though its concept has been a minuscule part of human interactions [8]. However, with the move towards more ubiquitous computing paradigms, the use of semaphoric gestures is seen as a practical method of providing distance computing in smart rooms and intelligent environments [9,10,11] and as a means of reducing distraction to a primary task when performing secondary task interactions [12]. These types of gestures can be performed using hands [13,5], fingers [14,15], head [16] or other objects such as passive or electronic devices, like a wand or a mouse [11,17].

Semaphoric gestures thus refer to strokes or marks which are mapped onto various interface commands. Examples include mouse strokes for back and forward control of a web browser [17], controlling avatars in virtual reality style applications using shorthand marks [18], for interacting with and issuing commands for desktop style applications [4,19,20] or for screen navigation or marking or pie menu selections [21,10,22].

The act of *gesticulating* is one of the most natural forms of gesturing and is commonly used in combination with conversational speech interfaces [8,23,24]. Gesticulations rely on the computational analysis of hand movements within the context of the user's speech topic and are not based on pre-recorded gesture mappings as with semaphores. Systems using gesticulating gestures are thus primarily multimodal. Typically they include combined speech and gesture interfaces which attempt to create a naturalistic, conversational style interaction without the need for electronic or passive devices to detract from the natural gesticulations people spontaneously perform.

2.2   Enabling Technology

We can classify the technology that enables using gestures as an input modality according to how the gestures are acquired by the computing device.

Non-perceptual input involves inputting the gesture through devices or objects that require physical contact to transmit location, spatial or temporal information to the computer. Examples of categories of non-perceptual technologies are mouse and pen input, touch and pressure input, electronic sensing devices (wearable, gloves, sensor-embedded objects and tangible interfaces, tracking devices) and audio input. Given that the technology employed in the experiments reported in this paper is based on touch screens, the next paragraph will further detail this technology.

Touch and pressure sensitive screens are a gesture enabling input technology seen in the literature since the mid 80's [25]. Touch based input is similar to gesturing with direct input devices. However one of its key benefits is to enable a more natural style of interaction that does not require intermediate devices like the mouse [26,27]. Touch sensors in particular have been widely discussed in the literature, primarily as input for gestures to mobile computing [28,16] and for tablet computer devices [29,19]. More recently, touch and pressure sensitive material is used to enable table top computing surfaces for gesture style interactions [5,4,15,30]. Touch sensors enable input for a wide variety of computer devices, ranging from desktop monitors [31] to small mobile screens [28] to large interactive surfaces [21]. The gesture interactions they enable are similar in nature to the contact style gestures that can be performed on a surface.

Perceptual input technologies are those which enable gestures to be recognized without requiring any physical contact with an input device or with any physical objects, allowing the user to communicate gestures without having to wear, hold or make physical contact with an intermediate device such as a glove or mouse for example. Perceptual input technology includes visual, audio or motion sensors that are capable of receiving sensory input data from the user through their actions, speech or physical location within their environment.

2.3   Application Domains

Gesture interaction has been applied in many application domains. In the next paragraphs we focus on those domains which have a bigger potential to be explored in collaborative scenarios.

In desktop computing applications, gestures are an alternative to the mouse and keyboard interactions, enabling more natural interactions with fingers, for example. Many gesture based desktop computing tasks involve manipulating graphics objects [32], annotating and editing documents [33], scrolling through documents [21], making menu selections [10] and navigating back and forward web browsers [17]. More novel style interactions that utilize natural gestures such as nodding have been explored as a means of interacting with responsive dialog boxes on a desktop computer [34].

One of the first application domains for which gestures were developed involved graphic style interfaces. The gestures used for this style of interaction involve manipulating the input device to draw strokes, lines, circles or to make movements in different directions for drawing, controlling the functionality provided by the application, switching modes and issuing commands [35]. Similar applications were also implemented on touch screens or tablets using fingers or pens for gesturing [25]. Pressure sensors were also used as a means of gesturing to manipulate and create graphical objects in a finger painting application [31]. Touch sensors were used to track the movement of the finger or input device on the screen to create the drawings, and the amount of pressure applied to the screen was used to determine the thickness of the lines being drawn. Graphic applications also use gestures for navigational and view purposes as well as to control the various functions associated with the application. 2D mouse based gestures have been used to control the view provided by a camera in a 3D graphics interface using a mouse or a pen for input [27].

Gestures have been used in the field of CSCW systems within a variety of computing domains including desktop computers or table top displays [4,5] and large screen displays [9,36]. Interactions such as sharing notes and annotations among groups of people are most common in the literature using stroke style gestures both within a single location or across a network for remote teams [37,26,38]. Annotations can be transmitted on live video streams for collaboration between an instructor and a student working on tasks involving physical objects [2,19]. In some works images from one party are transmitted to the second party, who in turn gestures to create annotations overlaid on the image which is in turn transmitted back to the original party.

Gesture based interactions have also been designed to enable users to control devices and displays at a distance or within smart rooms. As smart room technologies became more sophisticated, so did the use of gestures as a more natural mode for interacting with computers that were distributed throughout ones environment [39] and as a key mode of interaction with perceptual interfaces [40]. Gestures for smart room or perceptual style interactions are also used for interacting with large screen displays that can be distributed around a room or building [41,36]. Actual computer devices can be manipulated using gestures that involve bumping, tilting, squeezing or holding the device for interactions such as navigation within a book or document and for detecting handedness in users of such devices [42] or for transferring data and communication between tablet PC's for example [6].

In pervasive computing and mobile computing devices, gestures have been used to allow eyes-free style interactions, enabling users to focus the attention on their mobility rather than on having to visually interact with the devices directly [16,28,20]. This style of interaction typically involves PDA's with touch sensitive screens that accept gestures as input, providing audio output to the user and enabling eyes-free interactions with mobile devices.

3   Collaboration with Gestures

Gestures are considered to be one the most natural forms of interacting with computers. In the previous section we have shown how different technologies can be used to enable different gesture types for different application domains. Several of those domains are collaborative, which further motivates a study of how a naturally used modality for human-human communication can be explored to enhance collaborative applications.

To promote adoption of gesture based interaction in the CSCW field, we need to explore the naturalness of this modality, while, at the same time, it has to simplify interaction in order to justify its adoption. As such, one of the problems addressed through the use of gestures include developing simpler gestures sets and interactions [43,44].

With this in mind, it can be argued that gesticulating, although being the most natural and intuitive form of interaction, still poses technological challenges that detract from a fluid and efficient interaction, limiting its current usefulness for interactive applications, and, consequently, collaborative scenarios. Accordingly, deictic, manipulative and semaphoric gestures are the best options to introduce in a collaborative application, given the higher performance level attained in recognizing these types of gestures.

The selection of the type of gestures to include when developing the collaboration should be guided by two factors: the available technologies for gestural input and the tasks to be performed. Manipulative gestures have the potentially highest requirements in terms of technological investment, unless being used in a simpler 2D interaction controlled by mouse. Deictic and semaphoric gestures can be used effectively with both perceptual and non-perceptual input technologies. Touch-enabled surfaces are becoming common nowadays, supporting non-perceptual gesture recognition for semaphoric gestures and direct selection for deictic gestures. Camera-based systems with tracking algorithms are also becoming available, offering perceptual support for deictic and semaphoric gestures.

Thus, from a technological viewpoint, the development of a collaborative application resorting to deictic gesturing is, nowadays, something more easily achievable. The same can be said about the use of semaphoric gestures for such applications, although these pose another challenge: the definition of the gesture dictionary.

To foster the adoption of gestural interaction, one of the requirements guiding the definition of the gesture dictionary should be the intuitiveness of the gestures comprising the dictionary set. In other words, the mapping of a gesture to a command or action, in the context of the task being accomplished by a user, should be the most intuitive and natural as possible.

With this in mind, we have previously conducted a set of studies in order to understand which gestures are the most natural for a diverse set of tasks [45,46]. The tasks considered included:

– desktop objects manipulation - tasks involving window dragging, opening and closing, as well as files and folders cutting, copying, printing and compressing.
– other objects manipulation - tasks involving image browsing and simple image editing, as well as a variety of actions to be performed on the Google Earth application, comprising navigation, rotation, panning, zooming and place finding.
– simple applications control - tasks conducted in the context of controlling a multimedia player application, enabling the study of how gestures can be mapped to digital or analog controls.

The tasks were conducted with the assistance of a SMARTBOARD, a large touch-enabled display. In a first study [45], users were asked to perform the gestures that seemed more adequate to the requested tasks, with a single restriction that no multi-touch gestures were performed, since the SMARTBOARD does not support this technology.

Results showed that for some tasks, people intuitively perform the same gestures, while for other tasks this is not the case. Figure 1 presents some of the tasks and gestures where at least 75% of the test participants made a similar gesture.

A second study [46] tried to validate these results, by asking a different set of users to perform the requested tasks using gestures, without telling them what gestures mapped to what tasks. Results from this study demonstrated that, for those gestures with a high agreement in the first study, the new users were quickly able to reach the correct gesture to perform the action. Users took an

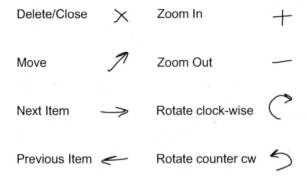

Fig. 1. Gestures performed by test participants with an agreement of at least 75%

average of 1.23 attempts and an average of 6.62 seconds to guess the correct gesture.

These results show the plausibility of designing a gesture dictionary that is simple and intuitive. Additionally, other results from our studies show that gestures are a pleasing and attractive alternative to users. For some of the requested tasks the test participants exhibited a preference to employ gestural interaction over traditional keyboard and mouse.

Based on these results, we proceed to study the use of gestures in collaboration scenarios. Given the greater adequacy of gestures to some types of tasks we hypothesize that users performing collaborative tasks, with different interaction modalities available, will collaboratively assign tasks to each other, based on the task features, the technology available and the types of gestures required. This hypothesis is summarized in figure 2.

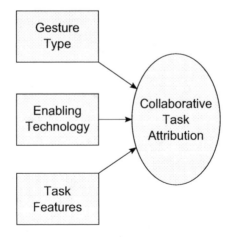

Fig. 2. A framework for task adoption in collaboration scenarios

4   Testing the Hypothesis

To test the hypothesis formulated in the previous section, we developed a simple collaborative whiteboard application. The whiteboard offers tools to draw geometric shapes (circles, rectangles, triangles), to perform freehand drawing, to paste pictures and to write text on the board. The application is capable of performing deictic and semaphoric gesture recognition, based on a gesture dictionary defined according to the results of our previous studies. This enables a user to interact using the traditional means, via a toolbar and menu with the available commands, or by employing gestural interaction. Figure 3 presents the visual interface of the whiteboard application.

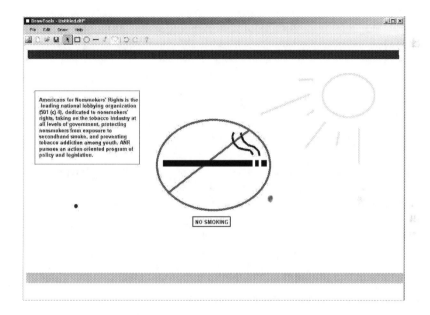

Fig. 3. The interface of the whiteboard application

The application is also endowed with communication capabilities. This allows several instances of the application, running in different platforms, to share the whiteboard between multiple users. Consequently, the application offers a shared workspace for synchronous interaction, with collocated or remote users.

To properly test the hypothesis that users will collaboratively assign tasks between themselves based on the tasks to perform and the technology available, we designed an experiment comprising two different interaction platforms.

One platform is a standard desktop or laptop PC, with standard interaction means: a keyboard and a mouse for input, a standard size output screen. This platform has no touch enabled interaction mechanisms, but gestures can still be performed using the mouse. The other platform is a wall sized display with a touch-enabled surface. Input is available through the touch-enabled surface, directly supporting gesture making. A virtual keyboard is available if required. Output is the result of a projection onto the same surface.

By installing the whiteboard application in such distinct interaction platforms, we expect to observe users taking advantage of the strong points of each platform. The PC platform is better suited for text input, for instance, while the wall sized display is better adapted to gesture making.

In the designed experiment, teams of two users, each using the application in one of the interactive platforms, have to replicate a document in the whiteboard. The document consists of text, images, geometric drawings and freehand drawings. Each member of the team is handed a paper copy of the document, and the task result is scored according to the whiteboard's resemblance to the original document. After concluding the document replication, a slightly different version of the document is handed to one of the team members. It is then the responsibility of that member to coordinate the team to perform the changes required to replicate the new version of the document. In each experiment two documents are replicated, in order to have both members of the team coordinating the correction stage.

During the document replication stage, users are free to communicate and decide on who does what task. They both have paper copies of the document and can negotiate which tasks are done in which platform. In this stage we expect to see if users decide collaboratively on which tasks to perform, and if the decision is influenced by the task's characteristics and the platform they will be using, in accordance to what was hypothesized.

During the correction stage only one user has access to the target document, but users are still free to communicate. We expect to gain some insights on the impact that having users with different knowledge levels about the task to perform has on the decision process.

In the experiment described above both users are collocated in the same room, even though they will be using different interaction platforms. We have also prepared an experiment where users are located in different rooms, thus resulting in a remote users setting. In this experiment, communication between users is achieved through a video conference application. With this new setting, we expect to understand the effects of collocation on the decision process.

4.1 Preliminary Results

We have conducted two preliminary runs of the experiment described above. Both the collocated and the remote settings have been tested with a pair of test participants.

The two platforms used in our experimental setting were the following:

– Standard interaction - one standard laptop PC, running Windows Vista SP1, with screen resolution of 1024 by 768 pixels, with the standard interaction means of a keyboard and mouse for input and an output screen with 15.4" active screen area.

- Large display interaction - one front projection SMARTBOARD, with 77"
  (195.6 cm) active screen area, connected to a PC running Windows XP SP3,
  with screen resolution of 1024 by 768 pixels.

Both platforms were networked, and both had a webcam and a microphone
attached for communication. The webcam for the large display platform was
attached to the top left corner of the SMARTBOARD, pointing towards the
area in front of the SMARTBOARD. With this set up, the camera was able to
show when the user was interacting with the SMARTBOARD even though it
could not show the SMARTBOARD's surface. The camera was also able to show
the user's face when looking at the video conference application window, which
was placed in the top left corner of the display.

In the first trial, both participants were collocated in the same room, with
one participant using the laptop and the other the SMARTBOARD. The laptop
participant had complete visual access to the large sized display, and could see
at anytime what the SMARTBOARD participant was doing. They engaged in
natural communication, turning to each other whenever needed, talking, looking
and gesturing freely. The video conference application was not used during this
trial.

In the second trial, the laptop participant was placed in a different room.
Communication was achieved through the use of the video conference application. This meant that, although the laptop participant could watch an almost
full body image of the SMARTBOARD participant, the inverse was not true,
with the SMARTBOARD participant being able to see only an image of the
laptop participant's face. This, obviously, results in a much more constrained
communication channel than the one available in the first trial.

In both trials the same set of documents was presented to the participants to
replicate. The first document was more text intensive. The version with changes
involved deleting some text and adding some freehand drawings. This version
was presented only to the laptop participant. The second document had a more
complex drawing to perform, and some text. The version with changes involved
adding more text to a different part of the document, as well as finishing the
drawing, but with very simple additions. This version was presented only to the
SMARTBOARD participant.

Although we are fully aware that the number of participants (four) involved is
not enough to draw significant conclusions, they have been, nevertheless, important to fine tune the experimental procedure, as well as to draw some preliminary
conclusions. The next sections present some insights about both trials and discuss their results.

4.2   The Collocated Trial

When presented with the first document, which was more text heavy, both participants immediately agreed that the laptop participant would be responsible
for text input, while the SMARTBOARD participant would be responsible for
drawing the requested pictures.

Even though the participants were told that time was not an issue, and only the quality of the replica would be evaluated, when the SMARTBOARD participant finished drawing, since the laptop participant was still entering text, he offered to enter some of the missing text. This offer was promptly accepted by the laptop participant, which meant both participants ended up entering text during this task, finishing it quicker than they would have otherwise.

When the laptop participant received the altered version of the document, he decided on deleting the text himself, and directing the SMARTBOARD participant to draw the new objects. To guide the SMARTBOARD participant in the task of placing the new objects, the laptop participant made extensive use of deictic gestures and expressions, like "place one to the right of the seagull and the other above it".

After seeing the second document, the test participants discussed thoroughly who would do what tasks. Based on the complexity of the requested drawing, they opted for the laptop participant to perform the drawing, while the SMART-BOARD participant would input the text. As both ended their tasks approximately at the same time, there was no need for any of them to perform additional tasks. Nevertheless, after finishing their tasks, they agreed that their task choice was not the better, with the SMARTBOARD participant strongly defending this position.

The behavior exhibited after receiving the changed version of the document supports this observation. The SMARTBOARD participant, who was responsible for the changes this time, performed the simple drawing additions, while dictating to the laptop participant the text he needed to enter. Once again, deictic pointing and references were used, this time by the SMARTBOARD participant, to specify where the laptop user needed to place the entered text.

4.3   The Remote Trial

For the first document, the behavior exhibited by this pair of participants was similar to the behavior that had been exhibited by the collocated trial participants: the laptop participant entered the text, while the SMARTBOARD participant made the drawings. These choices were kept when making the changes to the initial document. The laptop participant deleted the requested text and asked the SMARTBOARD participant to draw the new objects. However, one important difference to the first trial emerged. The laptop participant had difficulties in directing the SMARTBOARD participant to place the new objects correctly on the document. In the end, it was the laptop participant that moved the objects to the correct position.

The second document was dealt in the same fashion, that is, the laptop participant entered the text, and the SMARTBOARD participant made the drawings. Given that this document had a smaller amount of text and a more complex drawing, the laptop participant finished the task before. He then proceeded to help the SMARTBOARD participant, by making some precision adjustments to some of the previously drawn objects. This was a request from the SMARTBOARD participant, who was experiencing some difficulties in selecting lines and freehand drawings, in part due to a loss of accuracy of the SMARTBOARD[1].

When requested to direct the changes to the second document, the SMART-BOARD participant took it upon himself to enter the new text, and specified to the laptop participant what the requested additions to the drawing were. Contrarily to what had been decided by the SMARTBOARD participant of the collocated trial, this participant did not felt comfortable to dictate the text over the video conference application. From this decision resulted this pair of participants took longer to complete the task. Additionally, and more important in the context of the experiment, it was clear the difficulty felt by the SMARTBOARD participant to direct the laptop participant to place the additional drawings in the correct position.

4.4   Discussion

After observing both trials, the major conclusion that can be reached regards the collaborative assignment of tasks. It was clear that test participants agreed that text entering tasks should be performed by the laptop participant, while drawing tasks should be performed by the SMARTBOARD participant. This observation supports our hypotheses that users collaboratively decide task assignment based on the task characteristics and the different platform support.

The other major conclusion from this observation was the impact that the users being on the same room had on the communication and the coordination of the tasks. This was made clear by the difficulties felt by the remote pair in guiding each other to place objects on the correct position when only one the participants had that knowledge. In the collocated pair, this was quickly addressed by both users pointing to the large display to indicate where the object should be placed when in doubt. In the remote setting, even thought the workspace is shared, participants could not transmit deictic gestures, and the spoken deictic expressions proved to be insufficient.

Another effect of the collocation was a more vivid negotiation between test participants. Collocated users discussed to a greater extent the characteristics of the task to be performed. They explicitly considered, when negotiating, the amount of text and the complexity of the drawings before deciding on who performed what task. Remote users exhibited a more streamlined approach, quickly stating (and agreeing) who would do what. This difference in behavior reflects the different communication channels available in each trial.

Future versions of the application will need to include mechanisms capable of improving the awareness to deictic gestures performed by remote participants. One simple solution to this problem would be the inclusion of a proxy of the cursor of every remote participant. This would allow references to the position of the cursor and would ensure an easier guidance for this kind of tasks.

---

[1] The SMARTBOARD required frequent recalibration during the experiment.

5   Conclusions and Future Work

This paper examined how gestural interaction can contribute to the design of collaboration scenarios. It begun with a review of gestures according to their type, the technological requirements for performing them, and the application domains were they have been used. This review, together with our previous studies on gestural interaction, grounded the hypothesis that users of platforms with different interaction characteristics will, collaboratively, decide on who performs what task, given the type of gestures required, the technological platforms available and the task to be performed.

To validate this hypothesis we designed an experiment involving a touch enabled wall display and a standard laptop computer. A shared workspace application was developed to support the experiment. Tasks were devised that required different interaction mechanisms, made available on both platforms. Nevertheless, given their interaction characteristics, each platform was better suited to perform certain types of tasks.

Preliminary results of the experiments conducted so far seem to validate the hypothesis. Users collaboratively assigned tasks between themselves, and the task assignment is in accordance to the strengths of each platform.

Future developments will begin by considering the evolution of the shared workspace application, in order to include more advanced features requested by the participants, but also to include other features aimed at improving collaboration. These include the aforementioned cursor proxy for remote users, a tighter isolation between users through lock mechanisms able to prevent users from modifying the same object simultaneously, an improved awareness mechanism capable of informing users of the other users' current activity, and different tool selection modes, more suited to each platform.

After improving the application, a full scale study will be performed, in order to allow us to draw extended conclusions regarding the validity of the proposed hypothesis.

References

1. Bolt, R.A.: Put-that-there: Voice and gesture at the graphics interface. In: 7th annual conference on Computer graphics and interactive techniques, pp. 262–270. ACM Press, New York (1980)
2. Kuzuoka, H., Kosuge, T., Tanaka, M.: Gesturecam: a video communication system for sympathetic remote collaboration. In: 1994 ACM conference on Computer supported cooperative work, pp. 35–43. ACM Press, New York (1994)
3. Rubine, D.: Combining gestures and direct manipulation. In: SIGCHI conference on Human factors in computing systems, pp. 659–660. ACM Press, New York (1992)
4. Wu, M., Balakrishnan, R.: Multi-finger and whole hand gestural interaction techniques for multi-user tabletop displays. In: 16th annual ACM symposium on User interface software and technology, pp. 193–202. ACM Press, New York (2003)

5. Rekimoto, J.: Smartskin: an infrastructure for freehand manipulation on interactive surfaces. In: SIGCHI conference on Human factors in computing systems, pp. 113–120. ACM Press, New York (2002)
6. Hinckley, K.: Synchronous gestures for multiple persons and computers. In: 16th annual ACM symposium on User interface software and technology, pp. 149–158. ACM Press, New York (2003)
7. Patel, S.N., Pierce, J.S., Abowd, G.D.: A gesture-based authentication scheme for untrusted public terminals. In: 17th annual ACM symposium on User interface software and technology, pp. 157–160. ACM Press, New York (2004)
8. Quek, F., McNeill, D., Bryll, R., Duncan, S., Ma, X.-F., Kirbas, C., McCullough, K.E., Ansari, R.: Multimodal human discourse: gesture and speech. ACM Trans. Comput.-Hum. Interact. 9(3), 171–193 (2002)
9. Cao, X., Balakrishnan, R.: Visionwand: interaction techniques for large displays using a passive wand tracked in 3D. In: 16th annual ACM symposium on User interface software and technology, pp. 173–182. ACM Press, New York (2003)
10. Lenman, S., Bretzner, L., Thuresson, B.: Using marking menus to develop command sets for computer vision based hand gesture interfaces. In: 2nd Nordic conference on Human-computer interaction, pp. 239–242. ACM Press, New York (2002)
11. Wilson, A., Shafer, S.: Xwand: Ui for intelligent spaces. In: SIGCHI conference on Human factors in computing systems, pp. 545–552. ACM Press, New York (2003)
12. Karam, M., Schraefel, M.C.: A study on the use of semaphoric gestures to support secondary task interactions. In: CHI 2005 extended abstracts on Human factors in computing systems, pp. 1961–1964. ACM Press, New York (2005)
13. Alpern, M., Minardo, K.: Developing a car gesture interface for use as a secondary task. In: CHI 2003 extended abstracts on Human factors in computing systems, pp. 932–933. ACM Press, New York (2003)
14. Grossman, T., Wigdor, D., Balakrishnan, R.: Multi-finger gestural interaction with 3d volumetric displays. In: 17th annual ACM symposium on User interface software and technology, pp. 61–70. ACM Press, New York (2004)
15. Rekimoto, J., Ishizawa, T., Schwesig, C., Oba, H.: Presense: interaction techniques for finger sensing input devices. In: 16th annual ACM symposium on User interface software and technology, pp. 203–212. ACM Press, New York (2003)
16. Schmandt, C., Kim, J., Lee, K., Vallejo, G., Ackerman, M.: Mediated voice communication via mobile ip. In: 15th annual ACM symposium on User interface software and technology, pp. 141–150. ACM Press, New York (2002)
17. Moyle, M., Cockburn, A.: The design and evaluation of a flick gesture for 'back' and 'forward' in web browsers. In: 4th Australian user interface conference on User interfaces 2003, pp. 39–46. Australian Computer Society, Inc. (2003)
18. Barrientos, F.A., Canny, J.F.: Cursive: controlling expressive avatar gesture using pen gesture. In: 4th international conference on Collaborative virtual environments, pp. 113–119. ACM Press, New York (2002)
19. Ou, J., Fussell, S.R., Chen, X., Setlock, L.D., Yang, J.: Gestural communication over video stream: supporting multimodal interaction for remote collaborative physical tasks. In: 5th international conference on Multimodal interfaces, pp. 242–249. ACM Press, New York (2003)
20. Pastel, R., Skalsky, N.: Demonstrating information in simple gestures. In: 9th international conference on Intelligent user interfaces, pp. 360–361. ACM Press, New York (2004)
21. Smith, G.M., Schraefel, M.C.: The radial scroll tool: scrolling support for stylus or touch-based document navigation. In: 17th annual ACM symposium on User interface software and technology, pp. 53–56. ACM Press, New York (2004)

22. Zhao, S., Balakrishnan, R.: Simple vs. compound mark hierarchical marking menus. In: 17th annual ACM symposium on User interface software and technology, pp. 33–42. ACM Press, New York (2004)

23. Kettebekov, S.: Exploiting prosodic structuring of coverbal gesticulation. In: 6th international conference on Multimodal interfaces, pp. 105–112. ACM Press, New York (2004)

24. Eisenstein, J., Davis, R.: Visual and linguistic information in gesture classification. In: 6th international conference on Multimodal interfaces, pp. 113–120. ACM Press, New York (2004)

25. Buxton, W., Hill, R., Rowley, P.: Issues and techniques in touch-sensitive tablet input. In: 12th annual conference on Computer graphics and interactive techniques, pp. 215–224. ACM Press, New York (1985)

26. Gutwin, C., Penner, R.: Improving interpretation of remote gestures with tele-pointer traces. In: ACM conference on Computer supported cooperative work, pp. 49–57. ACM Press, New York (2002)

27. Zeleznik, R., Forsberg, A.: Unicam - 2d gestural camera controls for 3d environments. In: Symposium on Interactive 3D graphics, pp. 169–173. ACM Press, New York (1999)

28. Brewster, S., Lumsden, J., Bell, M., Hall, M., Tasker, S.: Multimodal 'eyes-free' interaction techniques for wearable devices. In: SIGCHI conference on Human factors in computing systems, pp. 473–480. ACM Press, New York (2003)

29. Jin, Y.K., Choi, S., Chung, A., Myung, I., Lee, J., Kim, M.C., Woo, J.: Gia: design of a gesture-based interaction photo album. Personal Ubiquitous Comput. 8(3-4), 227–233 (2004)

30. Schiphorst, T., Lovell, R., Jaffe, N.: Using a gestural interface toolkit for tactile input to a dynamic virtual space. In: CHI 2002 extended abstracts on Human factors in computing systems, pp. 754–755. ACM Press, New York (2002)

31. Minsky, M.R.: Manipulating simulated objects with real-world gestures using a force and position sensitive screen. In: 11th annual conference on Computer graphics and interactive techniques, pp. 195–203. ACM Press, New York (1984)

32. Bolt, R.A., Herranz, E.: Two-handed gesture in multi-modal natural dialog. In: 5th annual ACM symposium on User interface software and technology, pp. 7–14. ACM Press, New York (1992)

33. Cohen, P.R., Johnston, M., McGee, D., Oviatt, S., Pittman, J., Smith, I., Chen, L., Clow, J.: Quickset: multimodal interaction for distributed applications. In: 5th ACM international conference on Multimedia, pp. 31–40. ACM Press, New York (1997)

34. Davis, J.W., Vaks, S.: A perceptual user interface for recognizing head gesture acknowledgments. In: Workshop on Perceptive user interfaces, pp. 1–7. ACM Press, New York (2001)

35. Buxton, W., Fiume, E., Hill, R., Lee, A., Woo, C.: Continuous hand-gesture driven input. In: Graphics Interface 1983, pp. 191–195. Canadian Man-Computer Communications Society (1983)

36. von Hardenberg, C., Berard, F.: Bare-hand human-computer interaction. In: Workshop on Perceptive User Interfaces, pp. 113–120. ACM Press, New York (2001)

37. Wolf, C.G., Rhyne, J.R.: Gesturing with shared drawing tools. In: INTERACT 1993 and CHI 1993 conference companion on Human factors in computing systems, pp. 137–138. ACM Press, New York (1993)

38. Stotts, D., Smith, J.M., Gyllstrom, K.: Facespace: endo- and exo-spatial hypermedia in the transparent video facetop. In: 15th ACM conference on Hypertext & hypermedia, pp. 48–57. ACM Press, New York (2004)

39. Streitz, N.A., Geißler, J., Holmer, T., Konomi, S., Müller-Tomfelde, C., Reischl, W., Rexroth, P., Seitz, P., Steinmetz, R.: i-land: an interactive landscape for creativity and innovation. In: SIGCHI conference on Human factors in computing systems, pp. 120–127. ACM Press, New York (1999)
40. Crowley, J.L., Coutaz, J., Bérard, F.: Perceptual user interfaces: things that see. Commun. ACM 43(3), 54–59 (2000)
41. Paradiso, J.A., Hsiao, K., Strickon, J., Lifton, J., Adler, A.: Sensor systems for interactive surfaces. IBM Syst. J. 39(3-4), 892–914
42. Harrison, B.L., Fishkin, K.P., Gujar, A., Mochon, C., Want, R.: Squeeze me, hold me, tilt me! an exploration of manipulative user interfaces. In: SIGCHI conference on Human factors in computing systems, pp. 17–24. ACM Press, New York (1998)
43. Song, C.G., Kwak, N.J., Jeong, D.H.: Developing an efficient technique of selection and manipulation in immersive v.e. In: ACM symposium on Virtual reality software and technology, pp. 142–146. ACM Press, New York (2000)
44. Swindells, C., Inkpen, K.M., Dill, J.C., Tory, M.: That one there! pointing to establish device identity. In: 15th annual ACM symposium on User interface software and technology, pp. 151–160. ACM Press, New York (2002)
45. Neto, A., Duarte, C.: A Study on the Use of Gestures for Large Displays. In: 11th International Conference on Enterprise Information Systems - ICEIS (2009)
46. Neto, A., Duarte, C.: Comparing Gestures and Traditional Interaction Modalities on Large Displays. In: Gross, T., Gulliksen, J., Kotzé, P., Oestreicher, L., Palanque, P., Prates, R.O., Winckler, M. (eds.) INTERACT 2009, Part II. LNCS, vol. 5727, pp. 58–61. Springer, Heidelberg (2009)

Strategies and Taxonomy, Tailoring Your CSCW Evaluation

Kahina Hamadache and Luigi Lancieri

Orange Labs, 42 rue des Coutures, 14000 Caen, France
{kahina.hamadache,luigi.lancieri}@orange-ftgroup.com

Abstract. With the rapidly growing development of Computer Supported Collaborative Work technologies, the evaluation of these services becomes an essential aspect. This evaluation mixes technical, business, social, perceptive, ergonomic aspects which can't be considered independently. In this paper we propose a new taxonomy of CSCW evaluation methods based on previous works and on our analysis of current evaluation methods. With this new taxonomy and the description of CSCW development process and life-cycle we are able to propose evaluation strategies that can be adapted and tailored for most of systems.

Keywords: Evaluating CSCW, Groupware evaluation, evaluation strategy, evaluation adaptability.

1 Introduction

Many different techniques have been used to evaluate groupware technologies, applying approaches that range from engineering to the social kind. Such methodological variety is due to the CSCW field, and up to now no consensus has been reached on which methods are appropriate in which context. Trying to apply conventional evaluation techniques to groupware applications without adapting them can be impossible or lead to dubious results. Applying the method of cognitive walkthroughs to the evaluation groupware, modifying traditional techniques in order to be able to apply them to the study of groupware applications can be complicated and expensive.

Many researchers believe that groupware can only be evaluated by studying real collaborators in their real contexts, a process that tends to be expensive and time consuming.

Others believe that it is more practical to evaluate groupware through usability inspection methods which do not utilize a real work situation. Groupware usability evaluation is difficult to perform because the common tradeoffs provided by different evaluation methods are constrained by the complex multidisciplinary nature of groupware systems. Traditionally accepted methods for assessing usability, such as laboratory experiments and field studies, become increasingly unmanageable because they involve multiple persons, which can be hard to find with the required competencies, may be geographically distributed, or simply unavailable for the considerable time necessary to accomplish collaborative tasks. Traditional experimental and laboratory methods that

L. Carriço, N. Baloian, and B. Fonseca (Eds.): CRIWG 2009, LNCS 5784, pp. 206–221, 2009.
© Springer-Verlag Berlin Heidelberg 2009

remove the software from its context of use may obtain simplistic results that do not generalize well to real world situations.

As an alternative to the laboratory, many groupware researches advocate the use of ethnographic and sociologic methods that explicitly consider culture and context (e.g. Quick and dirty ethnography [1]). These methods have been successfully applied to real situations, but they tend to be expensive and somewhat limited. They demand considerable time and evaluator experience. They work best at the beginning of design to articulate existing work practices and at the end to evaluate how systems already deployed in the work setting are used.

These limitations led to the emergence of a collection of discount methods: Groupware Task analysis (GTA) [2], Collaboration Usability analysis (CUA) [3] and Heuristic Evaluation (HE) [4]. These methods lack the capability to quantitatively predict human performance.

New evaluation strategies are needed that uncover central issues associated with groupware success and failure, and they need to be more flexible than they currently are in order to adapt to a greater range of factors that need to be considered [5].

In this paper, we want to emphasize the necessity of a comprehensive evaluation taxonomy and strategy for applications in CSCW. In this perspective we will propose a taxonomy of evaluation methods helping us to characterize them. Exploring further this idea, we propose to organize evaluation of CSCW systems in Evaluation Strategies that provide the evaluation process for a given system, defining what kind of method to use at what time. With this solution we intend to be able to plan almost any evaluation of CSCW system.

The content of this paper is organized as follows: in section 2, we review the existing taxonomies of evaluation CSCW. In section 3 we present our own taxonomy. Section 4 presents how we can build a strategy and gives an example. Finally we outline some perspectives and discuss of our proposition.

2 Existing Taxonomies of Evaluation Methods

Due to a long history, relatively to computer sciences' one, many evaluation methods have been proposed and designed to evaluate CSCW systems. Despite this profusion of methods it's not always easy to know which method you should use for your evaluation. To solve this issue we need to elaborate evaluation strategies, composed of several methods combination. Elaborating such strategies is complex; it requires knowing what methods have to be employed for a given kind of system. Thus, the first step in this perspective is to construct an accurate taxonomy of evaluation methods, based on several aspects of CSCW. In this section we'll see the already existing taxonomies proposed on the literature. Then we will propose our own taxonomy based on the methods previously mentioned and CSCW features.

As a critical point and a bottleneck in term of CSCW evolutions, the evaluation has been an interesting but also one of the most complex research domain. In order to make it more understandable to mankind, several researchers decided to go above the evaluation domain and tried to organize the work that have been done by others. Among them, we can cite Randall [6]; they have identified four orthogonal dimensions to classify the kinds of evaluation in groupware:

- Summative X Formative;
- Quantitative X Qualitative;
- Controlled Experiments X Ethnographic Observations;
- Formal and rigorous X Informal and opportunistic.

The authors state that the most used types of evaluations are the summative controlled and experimental (considered a formal technique); and the formative qualitative-opportunistic approaches (considered an informal technique).

Usually a distinction is made between *formative* and *summative* evaluation. Formative evaluation is meant to inform designers and developers designing the service or application and getting user feedback about preliminary versions. Summative evaluation is meant to inform the client or the external world about the performance of the service or application in comparison to a situation where there is no such service available, or to a previous version, or to competing services; in brief, to demonstrate the usefulness of the system.

This first taxonomy is relevant as it allows considering any evaluation method. However, for a proper classification, we should be able to consider methods as intervals over the specified dimension. Indeed, as we can vary methods settings, their classification can't be limited to a precise point in Randall's space.

CSCW evaluation is a vast domain, and as such it can be fathomed from many perspectives. In [7] the author proposes to consider it from five aspects as shown on Table 1.

Table 1. Ramage evaluation methods taxonomy

Ethnography	Qualitative	Psychological	Systems Building	Taking Advice
• Ethno-methodology	• Interviews	• Lab Experiments	• Iterative Prototyping	• Consumer Reports
• Conversational Analysis	• Questionnaires	• Analytic Approaches	• Participatory design	• Consultancy Reports
• Interaction Analysis	• Group Discussion	• GOMS Approach	• Beta Testing	• Marketing literature
• Distributed Cognition			• Heuristic Evaluation	
• Activity Theory			• User Testing	
• Breakdown Analysis			• Semi-Situated Ethnography	
• Others				

- *Ethnography* is the study of an entire organization in its natural surroundings over a prolonged period of time.
- *Qualitative* methodologies ask people questions about their experiences and compare/ contrast the answers to other people surveyed.
- *Psychological* methods use either lab experiments that focus on the isolation and analysis of a very specific phenomenon or analytic approaches that attempt to describe human interaction using formal models.
- *Systems Building* focuses on the development of partial or complete systems with the goal of improving them based on the evaluation.
- *Taking Advice* uses oral, video, and written information about an application as an evaluation mechanism.

Ramage points out that the very nature of CSCW imposes his taxonomy to be imperfect. As this nature results of the intricate combination of various disciplines, it makes it even more complex to provide a unifying taxonomy of evaluation methods. Also, given the breadth of his taxonomy there is some overlap between the methodologies.

David Pinelle and Carl Gutwin [8] have an approach closer from Randall's as they consider CSCW systems from strictly unrelated aspects. They reviewed forty-five CSCW articles from 1990 to 1998 with the objective to evaluate the use of evaluations methods and their different categories. They classified the evaluations both in relation to the environment where they are accomplished (natural occurrence or simulation of the phenomenon), and the degree of the variables manipulation (rigorous or minimum control of variables). See Table 2.

Table 2. Evaluation classification [3]

		Manipulation	
		Rigorous	*Minimal/None*
Setting	*Naturalistic*	Field Experiment	Field Study
			Case Study
	Controlled	Laboratory Experiment	Exploratory

They report that most of the articles only include laboratory experiment or even no evaluation, only some articles provide an experiment in real settings. We can also point out the fact that the evaluation is not completely integrated into the development process, despite almost every book, article, teacher or expert laud it. Indeed, most of evaluation processes are held for a finished application, package or prototype, it is not yet a continuous task during the development process. Another part of this report states that the most classical mean for evaluation is the observation, with direct sight or videotapes. The second technique is composed of interviews and questionnaires. They point out that evaluations lack of interest for "*organizational work impact*" and that most of them only focus on "*Patterns of system use*", "*Support for specified task*", "*User Interaction through the system*", "*Specific Interface features*" and "*User Satisfaction*". They also point out the fact that the evaluation should include gradually more and more work settings during the development of the software. They stress the fact that it is really important to lead evaluation even at the beginning of the development, it can avoid serious problems or misunderstandings, allowing you to be sure of what you're doing and protect from "chain reactions", meaning that if you have created a part of the application without evaluation and when you finally test it in real conditions, users can tell you that "he didn't want to have a shared file storage, but a personal one", and then you can redevelop most of your application. Furthermore authors suggest that evaluations should be shifted around users and their organizations and those researchers should try to reduce time and cost of evaluations methods, making them more attractive for companies and researchers themselves. Their conclusion is that each work used different approaches, methodologies or techniques for conducting evaluations.

Recently, [9] have led an interesting survey on some evaluation methods. What they suggest is not to use only one evaluation method, but to divide it into three phases: the first one consists in formative lab-based methods which goal is to avoid main errors; the second are field methods where you have to consider users' context; finally the third ones are qualitative methods in real conditions.

In order to facilitate the planning of evaluation Herskovic et al propose a classification of evaluation methods depending on some simple, but still fundamental, characteristics with limited values:

- *People Participation*: can be users, developers, experts or any combination of them;
- *Time to Apply the Method*: the moment when the evaluation takes place (before, during or after the development of the application);
- *Evaluation Type*: describe if the evaluation is qualitative or quantitative;
- *Evaluation Place*: can be a laboratory or usual work place;
- *Time Span*: the time dedicated to the evaluation, it can be hours, days or weeks;
- *Evaluation Goal*: describes the purpose of the evaluation method, what it is aimed to. It can be the evaluation of the product functionality, the collaboration process of the system or the product functionality considering the collaboration context.

In addition to this first classification, Herskovic et al furnishes a second one to estimate the final cost of an evaluation method according to its characteristics. It is another tool facilitating the construction of the triple-phased evaluation process.

This work is particularly relevant as it is based on the analysis of existing methods. This characterization is a good step in the long walk to a better understanding and appreciation of evaluation.

3 Proposed Taxonomy

The first element we consider to build our taxonomy is the fact that it can't be composed of a simple dimension. On the contrary, it should be designed according to a complex space. However, in opposition to Randall, all the dimensions of this space are not mutually orthogonal, implying that some of them can partially overlap themselves.

The second step of this process is the identification of important aspects of evaluation methods. It gives us the list of characteristics that we'll be used in the taxonomy.

The third step consists in defining the "meta" aspects of CSCW evaluation; this step is done by analysing the previous list of characteristics and extracting the main categories.

Fourth, we sort the characteristics according to these categories with the possibility to have a given characteristics in several categories (but obviously not all the characteristics in all the categories).

Fifth, inside the categories we try to gather characteristics by discovering similarities between them and then building sub-categories.

These five steps have led us to the following taxonomy (Table 3):

Table 3. CSCW Evaluation Taxonomy

- *Development specific aspects:*
    - *Development process:*
        - ➤ *Type:*
            - *Iterative;*
            - *Waterfall;*
            - *Extreme Programming;*
            - *...*
        - ➤ *Development Step;*
        - ➤ *Goal:*
            - *Maintenance;*
            - *New System;*
            - *...*
    - *Final System:*
        - ➤ *Goal;*
        - ➤ *Scalability;*
        - ➤ *End-users type*
            - *Anyone;*
            - *Developers;*
            - *Scientists;*
            - *Government;*
            - *...*
    - *Current System:*
        - ➤ *Scalability;*
        - ➤ *Step in development process;*
        - ➤ *Automation capacity*
        - ➤ *Evaluation Cost:*
            - *Computational Cost;*
            - *Human Cost;*
            - *Time Cost;*
        - ➤ *Feature to evaluate:*
            - *Feature type;*
            - *Feature maturity;*
        - ➤ *Evaluation needs:*
            - *Modus operandi:*
                - o *Exploration;*
                - o *Evaluate some precise points;*
            - *Evaluators type:*
                - o *Experts;*
                - o *End-users;*
                - o *Developers;*
                - o *Diversified;*
                - o *Representative Sample.*

- *Method specific aspects:*
    - *Goal:*
        - ➤ *Focus:*
            - *Usability;*
            - *Quality;*
            - *Performance;*
            - *Sustainability;*
            - *Utility;*
            - *Coherence;*
            - *Extensibility;*
            - *Scalability;*
            - *...*
    - *Type:*
        - ➤ *Formality:*
            - *Formal;*
            - *Informal;*
        - ➤ *Business consideration:*
            - *None;*
            - *Weak;*
            - *Average;*
            - *High;*
            - *Full;*
        - ➤ *Users multiplicity:*
            - *Single;*
            - *Multiple;*
    - *Cost:*
        - ➤ *Time Cost;*
        - ➤ *Human Cost;*
        - ➤ *Computational Cost;*
    - *Evaluation context aspects:*
        - ➤ *Evaluators type:*
            - *Experts;*
            - *End-users;*
            - *Developers;*
            - *Diversified;*
            - *Representative Sample.*
        - ➤ *Evaluation Place:*
            - *Laboratory;*
            - *Real location;*
        - ➤ *Evaluation Step:*
            - *Preliminary;*
            - *Main;*

Table 3. (*continued*)

• *Collaboration specific aspects:*
   - *Collaboration Model:*
     ➤ *Mode:*
       • *Asynchronous;*
       • *Synchronous;*
       • *Mixed*
     ➤ *Group Structure:*
       • *Size;*
       • *Scalability;*
       • *Members Coupling;*
       • *Members type:*
         ○ *Scientists;*
         ○ *Developers;*
         ○ *Diversified*
         ○ *...*
     ➤ *Evaluation Cost:*
       • *Computational Cost;*
       • *Human Cost;*
       • *Time Cost;*
   - *Evaluation focus:*
     ➤ *Single user behaviour;*
     ➤ *Multiple user behaviour;*
     ➤ *Mixed.*

The proposed steps for building the taxonomy are not mandatory; their main goal is to provide guidance for us and users in the representation of evaluation context and is largely inspired by classical taxonomy and ontology construction methods such as Bachimont in [12].

Obviously this taxonomy isn't exhaustive; it does not intend to address every methodology with all their details in this presented form. However, it can be simply *extended* to support any new kind of method.

4   Strategies for Evaluation

Evaluation of CSCW project is, of course, very important from a managerial point of view. In the first place, evaluation before a project is started is a key element to decide whether the project is worthwhile. Afterwards, evaluation is also useful as a basis for rewarding participants, to justify financing similar projects, or to justify a second phase of the project [10].

All evaluations have common features: in all cases there is an object being evaluated, a process through which one or more attributes are judged and valued, and all evaluations have a purpose.

As we mentioned in the previous section [9] propose an interesting approach to CSCW evaluation by proposing a three-phased strategy. This strategy relies on the principle that you don't need the same method at each step of your development:

1. *Formative lab-based methods* (perform some pre-evaluation to avoid main errors).
2. *Field methods* (with the participation of users associated context).
3. *Qualitative methods in real work settings* (evaluate if it really works).

This work is one of the few we found to propose a real strategy of evaluation. It is even more valuable as it builds a frame for evaluation, meaning that instead of telling which method to use, is only give a more general category. Then you're free to use the best method for your system, picked-up in the right category.

Relying on this good idea, we decided to go deeper in the definition of strategies refining the description of evaluation methods according to the development strategy, processes and steps. Naturally this refinement takes also advantage of our previously presented taxonomy.

Another point has to be noticed before diving into the strategies. We think that the good evaluation of a CSCW system has to be organized in three phases:

1. Evaluate the collaborative aspect;
2. Evaluate the business aspect;
3. Evaluate the combination of collaborative and business aspects.

This separation is particularly relevant as it allows identifying quickly and efficiently lacks in collaborative and business aspects of the system. Thus is can also help finding problems emerging when you integrate collaboration into the business domain.

4.1 Building a Strategy

Choosing a specific method instead of another is a critical need. It determines if the evaluation you'll lead is relevant or not for your system.

Deciding and choosing between these methods is puzzling. The methods have their own weaknesses, and trade-offs, they can be complementary or exclusive. Because the methods found overlapping problems, we expect that they can be used in tandem benefiting from each other, e.g., applying the discount methods prior to a field study, with the expectation that the system deployed in the more expansive field study has a better chance of doing well because some pertinent usability problems will have already been addressed.

To be quite exhaustive our model lets you two possibilities: picking up an existing strategy related to an evaluation close to your own; or building your own strategy "from scratch". The first opportunity is only interesting in some rare cases where you really are in the hurry and every hour or even every minute count. We'll get back on this first approach in the discussion, for the moment we focus on the second one.

So, how can we efficiently choose a strategy to evaluate a given system? The main idea is to find the best matching between the definition of your current system and the definition of the context in which strategies take place. For instance, the strategy will not be the same if you are at the beginning or at the end of your development lifecycle. This process is done according to five steps:

1. Describe the context of the evaluation.
2. Define the different phases of your development.
3. Extract the Evaluation Strategy Outline from the development phases.

4. Refine Evaluation Strategy Outline's methods' description.
5. Select Strategy's methods.

The first step to build a strategy is describing the context of the evaluation. In order to do so efficiently and exhaustively we propose to take a top-down approach. By this we intend to start by defining high level categories of the taxonomy and then going deeper and deeper. For instance, one of your first elements to describe is the development process type: do you use a traditional iterative process, a waterfall one or do you prefer the Extreme Programming methodology. Obviously, describing the development process is not sufficient, you also need to specify other system's relative aspects, business specific and collaboration's ones. Indeed, as the evaluation process is split in three different phases: business, collaboration and business + collaboration, we need to describe them sufficiently to find appropriate methods for each of them. Moreover, we believe in the proposition made in [9] to separate the evaluation in three phases: short lab experiment to detect main problems, field method with users' context to evaluate deeper and real settings evaluation to gain qualitative feedbacks.

Once the first description phase is done the second step consist in the description of the different phases of your development. For this step you have to define the different steps needed in your process and define the order in which they appear. For instance you should define the order in which you develop the different features of your system and of what types they are, business specific, collaboration specific or else. This part of the specification is really important as it is the base of the evaluation's outline generation.

Third step is the extraction of the evaluation's outline. As we've just said, this step relies on the description of development phases. Thus, to build the outline, or skeleton, of the evaluation strategy we have to consider each phase of the development process and establish if it requires an evaluation phase, moreover, we have not only to consider the development phase alone but we need to consider it and its position in the whole process of development in order to refine the evaluation methods and correctly establish if additional evaluation phases are needed. For instance, if the precedent phase of development was to create storage module and the current one focus on the development of an event logging feature, we not only have to test the new feature, but also the interaction between the storage and logging parts; for example to see if events are correctly represented in the repository.

The fourth step to build the evaluation strategy consists in refining the evaluation methods types selected to form strategy's skeleton. This part of the building process relies on the skeleton and on the previously description made through the taxonomy. Hence, the previous step gave us a set of evaluation methods types described with some broad criteria related to the development process phases. To complete this we use the description of the evaluation context made in the first step. Thus, we only have to "complete" the description of each method with the evaluation context and then access to a refined description of the evaluation context of each evaluation phase of the skeleton, giving us a refined skeleton.

The last step to build the strategy is the final selection of adequate methods. The main principle of strategy building is to find a matching between the evaluation context and the context in which a method takes place. As the evaluation context is described, we just have to find the corresponding methods. The natural way to perform it is by describing such methods. Hence the "matching" we propose rely on the "comparison" of system's context against methods' intended situation. Thus the description of evaluation methods through the taxonomy is crucial. But it's also a heavy task requiring a long study of each method. Still, that's not an unfeasible work as we think the definition of methods can be enriched by all users, closing the loop of collaboration. Besides, the taxonomy approach of our work enables users to only describe some general aspects of methods, resulting in a broader range of selected methods of system evaluation but saving large resources in exchange. By doing so, you load your burden with an extra task: choosing between a set of methods.

Finally, you've got an evaluation process consisting in an ordered sequence of evaluation methods: an Evaluation Strategy. Figure 1 sums up how you can build your evaluation strategy.

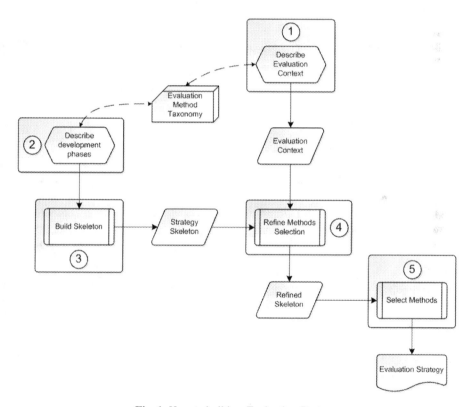

Fig. 1. How to build an Evaluation Strategy

To conclude this section we'd like to consider a critical point in the evaluation process which is often a source of conflicts: the lacks of adaptability of evaluation method to the evolution of development process. That is to say, if your development process suddenly speeds up, the heavy evaluation strategy you have chosen may not be able to make it. Thus, we think it is essential for a strategy to be able to be adapted to the changing context. In this perspective, the taxonomy we propose is central. Indeed, it provides a simple tool to find what methods have to be removed and which have to be used instead in your strategy to fit the new evaluation context.

4.2 Example of Evaluation Strategy

To illustrate the use of our approach, we'll now take an example of CSCW system evaluation. Let's consider a service we developed in a previous work [11]. This service is quite simple: it provides the capacity to automatically publish a questionnaire on a dedicated forum-like website. Thus it allows a team of users to efficiently communicate and collaborate by giving them to possibility to send questionnaires, answer to them and have a synthetic view of the responses even if they only have a low-resources device.

Table 4. Evaluation use case - Business and collaboration aspects

• Business aspects:	• Collaboration
- Questionnaire	- Questionnaire
➢ Editing	➢ Voting
➢ Sending	➢ Commenting
➢ Publishing	- Messaging
➢ Viewing	➢ Notification of publication
➢ Commenting	➢ Sending
➢ Voting	- Role management
➢ Synthesising	

As we'll obviously not write the full specification of the system we'll only focus on main aspects of this development.

Following the five steps we have defined, the first one is the description of evaluation context. In a first time we have to identify business specific aspects and collaboration ones as shown on Table 4. To complete the description of the context in which this development takes place, let's make a short description of the resources:

Table 5. Evaluation use case - Resources

• Human	• Time	• Hardware
o 20 Man-day	o 2 weeks (firm)	o All necessary
o 4 Peoples		

On the previous table (Table 5) we can see that the evaluation process have to be completed within two weeks. The hardware is not really a problem as required resources are relatively limited. Finally, the team has freed the equivalent of 20man/day to lead the evaluation to be distributed among 4 peoples.

Sticking with the taxonomy we can make the following assumptions:

- The development process relies on a fast iterative method;
- The goal of this process is to create a new feature for an existing system;

- End-users are accustomed to use communication means and web browsers;
- The evaluation have to explore the system in addition to validate the new features and check if it doesn't interfere with the normal behaviour of the system;
- Evaluators of the system need to be end-users for the final part of the evaluation;

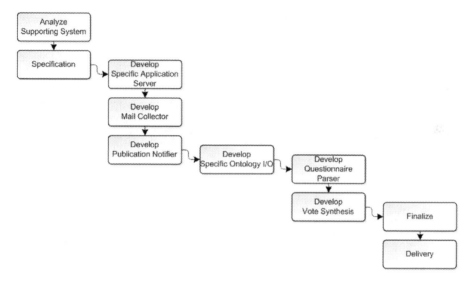

Fig. 2. Development Life Cycle

Considering the limited time granted for evaluation but the rather large amount of human and hardware resources, the proposed methods have to be adaptable for larger groups with intensive evaluation instead of small groups or loose evaluation sessions. Moreover, to be loyal with Herskovic's proposal ([9]) each evaluation step is subdivided into three phases, but in the case of iterative development process and even more when the development has to be fast, each loop on a same feature reduces the evaluation time and especially the time for lab experiment.

Secondly, we had to define the development lifecycle we use. Figure 2 shows how this development was organized: Analysis of supporting system in order to know if it was able to correctly support the new features, specification of the new features, multiple development phases, finalization and then delivery.

Based on this development life cycle we can extract the outline of the evaluation strategy. As we can see on Figure 3, to extract this skeleton we start by considering each phase that has been described earlier, for each one of them we figure out if it requires one or several evaluation stages. For instance in our example the "Analyze Supporting System" phase requires to evaluate if the system is suitable for the desired evolutions. As a direct consequence we deduce that we need three "Integration Feasibility Methods", one to evaluate if the system can handle the collaborative aspect, one to know if it correctly sustains the business part of the new features and finally a method to evaluate how the combination of these two aspects interacts with the existing system.

Fourth step of strategy building we have to refine the description made in the previous step with the help of the evaluation context defined in the first step. As it would be a little too long to describe all the methods evoked in Figure 3, we'll only refine one of the stipulated method. In order to have a relevant example, we consider the last evaluation step: "Full System Evaluation Method".

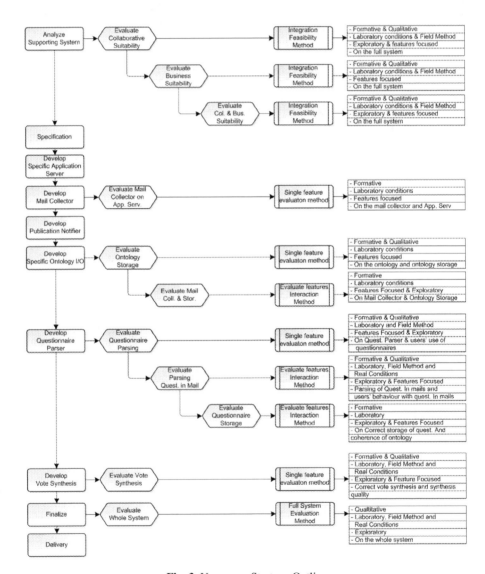

Fig. 3. Use case - Strategy Outline

Looking at the associated description we are able to see that this step has the following requirements: *Qualitative evaluation*, *Triple-phased* (laboratory, field study and Real Conditions), *Exploratory* (as all features has been tested previously we only have to figure out the overall quality of the system in its wholeness and if some unexpected challenges are leveraged) and obviously the evaluation concerns the *Full System*. Considering the evaluation context we know the selected method have to be rather fast than exhaustive as the evaluation time is short. Besides, we need to make end-users participate to the evaluation in order to have a real qualitative feedback but also to be able to efficiently explore the system with users' habits. Considering these requirements we can refine the method to the following ones: 1- Users' Exploration (let users use the new features without guidelines, just with the instructions to know how to use it and let them explore the system); 2- Scenario Based Evaluation (write scenarios to guide users in there walkthrough); and 3- Scenarios Refining Method (starting from some pre-written scenarios, users have to collaborate with peoples in charge or the development to refine scenarios through their own experience and desire). As we design it, to complete this step of strategy building you have to refine all the methods of your strategy skeleton.

The last step of this process is the final selection of methods. It has to be said that this step is not mandatory for all the evaluation steps of the skeleton. Indeed, for some of these steps you can have found only one method matching their evaluation context. In that case you obviously don't have to make a choice. Nevertheless, given the refinement level of an evaluation step (have it to desired or not), it can be matched by several methods. Finally, with all the required selections done, we've got our Evaluation Strategy.

5 Discussion

In this paper we tried to consider the barely dusted field of CSCW evaluation building. As we have seen in the past sections there have been several attempts to construct and propose a taxonomy that could help representing evaluation methods. If we refer to Randall [6] we can clearly say that even if this classification is relevant, it cannot efficiently help a user choose one method or another, it can surely give a trail, but no more. With more advanced taxonomies such as the one proposed by Pinelle and Gutwin in [8] users can find better ways to consider the evaluation they have to perform. In a different perspective, Herkovic et al in [9] has extended the representation of evaluation to several aspects leading to a finer representation. In addition they propose to organize evaluation in three steps. These papers are interesting and our work has roots in them, however there are some shortcomings: the granularity of the representation of evaluation is not sufficient to find a determine a precise method; another lack is the point of view of some taxonomies: most of them don't consider evaluation from the system point of view but only from "evaluation process" point of view. Our thought is that to consider evaluation properly you should see it from the system point of view, what you want to evaluate but also from the evaluation method point of view, how you have to evaluate.

We proposed a taxonomy whose goal is to provide a base to represent CSCW systems evaluation context. By a fine comparison of evaluation methods representation against evaluation context we can select relevant methods to be used. Moreover, following the development process of the system and keeping in mind the final result we are able to propose a complete evaluation strategy, indicating which method have to be used at what time of which step.

As we mentioned earlier, our approach offers the possibility for users to inspire themselves on existing strategies. This capacity is especially relevant in the case when people don't have enough resources to build their own strategy or when they need a strategy for a brand new system that doesn't match existing ones. In this last case, despite the fact that our approach should be able to propose a least a set of strategies, it can be interesting and relevant for the people in charge of the evaluation of this system to have a look at evaluation strategies related to closest already evaluated systems and maybe use one of them. In this kind of cases, where the taxonomy isn't sufficient because some specific parts of the system are not yet correctly represented, it is important that users can be allowed to complete the taxonomy by describing the missing parts. Besides this description, the feedbacks on the chosen evaluation strategy for this new system are even more significant and valuable.

Another point on which we want to insist is the possibility for users to choose an evaluation strategy and use some extra evaluation methods they think relevant. Once again, given the feedbacks of this extension, strategies can be updated and improved.

From the last points evoked we have to point out an emerging necessity of our system: the need for reasoning over the strategies. Thus, even if "Taxonomy" and "Ontology" are closely related parents, our model tends to become more "intelligent" and then our taxonomy tends to evolve to an ontology. Given that, we estimate than in a close future we'll be led to consider the design of inference rules, allowing us to naturally handle changes in our former taxonomy.

Continuing in the same prospect, our model will doubtlessly grow into some kind of expert system dedicated to the recommendation and knowledge representation and capitalization of CSCW systems evaluation. With further researches this systems should be able take into account feedbacks of evaluations to enhance and refine strategies.

Beyond these considerations, the very nature of our approach is based on the specification of the evaluated system. This particularity implies, once your specifications are finished, that you should have both your development design and your evaluation strategy. Furthermore, as evaluation can take place even at the beginning of the specification, the taxonomy we propose can help you lead some "pre-specification" evaluation.

Following the same perspective, the inherent flexibility of our model ensures you to be able to adapt your strategy according to the fluctuating requirements of the development and available resources.

Last but not least, the flexibility of our model provides an unusual but helpful advantage: the scalability in term of user's evaluation experience. Thus, our approach is suited for users with few or no experience in evaluation; in that case they can simply take one of the proposed strategies and apply it. But the model can also be used by expert in evaluation who will be able to rely on the obtained strategies and customize them the way they want. Such experts, via the feedbacks they can provide to our system, will confer it a part of their skills.

This paper doesn't pretend to solve all problems raised by the evaluation of CSCW systems. However, we think our contribution to this field of research can open a new perspective in the understanding and the way we consider the evaluation of complex systems. By rationalizing the evaluation and taking into account the limitations of available resources we tend to bring the evaluation processes more attractive and valuable for users. Besides, the foreseen automation of strategies evolution and recommendation, made through an expert system, will facilitate the job of many systems builders. Finally, defining evaluation strategies from the earliest steps of the development and by paradoxically keeping the possibility to make it evolve until the end of this process provides the flexibility that lacks in most of classical methods.

References

1. Hughes, J., King, V., Rodden, T., Andersen, H.: Moving out from Control Room: Ethnography in System Design. In: CSCW 1994, pp. 429–439. ACM Press, New York (1994)
2. Van der Veer, G., Van Wellie, M.: Task Based Groupware Design: Putting Theory into Practice. In: Proceedings of the conference on Designing Interactive Systems, pp. 326–337 (2000)
3. Pinelle, D., Gutwin, C., Greenberg, A.: Task Analysis for Groupware Usability Evaluation: Modelling Shared-Workspace Tasks with the Mechanics of Collaboration. Transactions on Computer Human Interaction, 281–311 (2003)
4. Baker, K., Greenberg, S., Gutwin, C.: Empirical Development of a Heuristic Evaluation Methodology for Shared Workspace Groupware. In: CSCW 2002, pp. 96–105 (2002)
5. Neale, D.C., Carroll, J.M., Rosson, M.B.: Evaluating Computer-Supported Cooperative Work: Models and Frameworks. In: CSCW 2004, Chicago, Illinois, USA (2004)
6. Randall, D., Twidale, M., Bentley, R.: Dealing with Uncertainly-Perspectives on the Evaluation Process. In: Thomas, P.J. (ed.) CSCW Requirements and Evaluation. Springer, Heidelberg (1996)
7. Ramage, M.: How to Evaluate Cooperative Systems. In: Computing Department, University of Lancaster, United Kingdom, Ph.D Thesis (1999)
8. Pinelle, D., Gutwin, C.: A Review of Groupware Evaluations. In: Proceedings of Ninth IEEE WETICE 2000 Workshop on Enabling Technologies: Infrastructure for Collaborative Enterprises, Gaithersburg, Maryland (2000)
9. Herskovic, V., Pino, J.A., Ochoa, S.F., Antunes, P.: Evaluation Methods for Groupware Systems. In: Haake, J.M., Ochoa, S.F., Cechich, A. (eds.) CRIWG 2007. LNCS, vol. 4715, pp. 328–336. Springer, Heidelberg (2007)
10. Baeza-Yates, R., Pino, J.: Towards Formal Evaluation of Collaborative Work and its Application to Information Retrieval (2006)
11. Hamadache, K., Manson, P., Lancieri, L.: Pervasive services, Brainstorming in situation of Mobility. In: 3rd International Conference on Pervasive Computing and Applications (ICPCA 2008), Alexandria, Egypt (2008)
12. Bachimont, B.: Engagement sémantique et engagement ontologique: conception et réalisation d'ontologies en Ingénierie des connaissances. In: Charlet, J., Zacklad, M., Kassel, G., Bourigault, D. (eds.) Ingénierie des connaissances, évolutions récentes et nouveaux défis. Eyrolles, Paris (French) (2000)

Analyzing Stakeholders' Satisfaction When Choosing Suitable Groupware Tools for Requirements Elicitation

Gabriela N. Aranda[1], Aurora Vizcaíno[2], Alejandra Cechich[1], and Mario Piattini[2]

[1] GIISCo Research Group, Universidad Nacional del Comahue
Computing Sciences Department, Buenos Aires 1400 - 8300 Neuquén, Argentina
{garanda,acechich}@uncoma.edu.ar
[2] ALARCOS Research Group, Information Systems and Technologies Department
UCLM-INDRA Research and Development Institute, Escuela de Informática,
Universidad de Castilla-La Mancha, Paseo de la Universidad 4 - 13071 Ciudad Real, Spain
{aurora.vizcaíno,mario.piattini}@uclm.es

Abstract. Global software development faces a series of problems related to various aspects of communication; for example, that people feel comfortable with the technology they use. In previous papers we have analyzed strategies to choose the most suitable technology for a group of stakeholders, taking advantages of information concerning stakeholders' cognitive characteristics. In this paper we present the preliminary results of an experiment in which our strategy was applied, and analyze stakeholders' satisfaction with regard to communication so as to discover if it is actually improved by our approach.

1 Introduction

Global software development (GSD) has become a common means to develop software [12]. However, in spite of the advantages that GSD offers [6, 16], the requirements elicitation process in such environments is particularly challenged by certain aspects. One critical point is the need to count on the best communication channels during the requirements elicitation process [5], while stakeholders' communication is challenged by the lack of face-to-face interaction, time difference between different sites and cultural diversity, among other factors [8].

Since communication in GSD projects takes place through groupware tools, it is quite interesting analyzing how those tools are chosen. As communication among people involves aspects of human processing mechanisms that are analyzed by the cognitive sciences, we have searched for references in Cognitive Informatics, an interdisciplinary research area that applies concepts from cognitive sciences to improve processes in engineering disciplines such as software engineering [17]. In such a direction, cognitive styles has been used as a mechanism to prove that heterogeneous software inspection teams perform better than homogeneous ones [15], where heterogeneity concerns the cognitive style of the participants. In our case, we have used cognitive styles as a means to select groupware tools and elicitation techniques in accordance with the stakeholders' cognitive style [14]. Although both works use cognitive styles to classify people, our approach differs from [15] because, rather than attempting to say which people seem to be more suitable to work together, our goal is to choose the best strategies to improve communication for an already given group of people.

L. Carriço, N. Baloian, and B. Fonseca (Eds.): CRIWG 2009, LNCS 5784, pp. 222–230, 2009.
© Springer-Verlag Berlin Heidelberg 2009

With such an idea in mind, this paper is structured as follows: First, we provide an introduction to some basic concepts concerning learning style models, and introduce a methodology for groupware selection based on concepts from fuzzy logic. We then present a controlled experiment carried out to validate our methodology, and we describe the preliminary results related to stakeholders' satisfaction concerning communication during a distributed elicitation process.

2 Supporting Stakeholders' Cognitive Preferences

Bearing in mind that elicitation is about learning the needs of the users [13], and it is also a scenario in which users and clients learn from analysts and developers [14], we consider that during the elicitation process everybody "learns" from others. We therefore focused our research on a special case of cognitive style models called learning style models (LSMs), which classify people according to a set of behavioural characteristics concerning the ways in which people receive and process information, and aim to improve the way that people learn a given task. The model chosen was the Felder-Silverman (F-S) model [9] since, according to our analysis, it covers the categories defined by the most famous LSMs (such as the Myers-Briggs Indicator Type, the Kolb model, the Herrmann Brain Dominance Instrument, etc.) and, additionally, the F-S model has been widely and successfully used with educational purposes in engineering fields [11]. There are four categories in the F-S Model (Perception, Input, Processing and Understanding), and each of them is further decomposed into two subcategories (Sensing/Intuitive; Visual/ Verbal; Active/Reflective; Sequential/ Global) [10]. The classification is carried out by means of a multiple-choice test[1], which returns a rank for each subcategory, and in which preferences for each category are measured as strong, moderate, or mild. According to the F-S model's authors, people with a *mild* preference are balanced on the two dimensions of that scale. On the other side, people with a *moderate* preference for one dimension are supposed to learn more easily in a teaching environment, which favours that dimension. Finally, people with a *strong* preference for one dimension of the scale may have difficulty learning in an environment, which does not support that preference.

Since our goal is to allow all those involved in the requirements elicitation process in a virtual environment to feel comfortable, we propose choosing the most suitable groupware tools and elicitation techniques according to each person's learning styles.

In order to obtain useful information before proposing our approach, we designed a survey to inquire into stakeholders' personal preferences and to look for behaviour patterns. The results of the first application of this survey, and a later replication, showed that people prefer using synchronous collaboration when their preference for the visual subcategory is stronger [3]. However, the result of analysing each category separately was not conclusive, so a combination of the preferences for the four categories had to be taken into account. To do so, we employed a methodology that uses fuzzy logic and fuzzy sets [1] to obtain rules from a set of representative examples, in the manner of behaviour patterns. Such a methodology comprehends two main stages (as is shown in Figure 1), which can be summarized as follows:

---

[1] http://www.engr.ncsu.edu/learningstyles/ilsweb.html

- **Stage 1. The Project Independent Stage:** The main goal of this stage is to obtain the set of preference rules. In order to accomplish such goal, first, many people are interviewed in order to obtain both their cognitive profile and two sets of examples (θ_1, θ_2), which are real data with regard to stakeholders' preferences in their daily use of groupware tools and requirements elicitation techniques. Second, data is analyzed by using a machine learning algorithm [7] so as to obtain a finite set of fuzzy rules. These obtained fuzzy rules, called *preference rules*, are project independent and can be improved as long as the set of examples and knowledge about the environment grow.
- **Stage 2. The project dependent stage:** This stage consists of the application of the preferences rules (obtained during the first stage) to a specific GSD project during a requirement elicitation process. Their application is carried out in two phases: First, we obtain the cognitive profile of every person in the virtual team and store this profile in a database. Second, the technology selection process is carried out by studying and confronting the personal preferences of people that need to work together. This is done by means of an automatic tool that chooses and suggests the most appropriate technology by using the fuzzy rules obtained in the first stage. The tool also takes into account other external factors that influence distributed communication such as the time difference between sites, the degree to which a common language is shared, and the current situation in the requirements elicitation process as it was explained in [2].

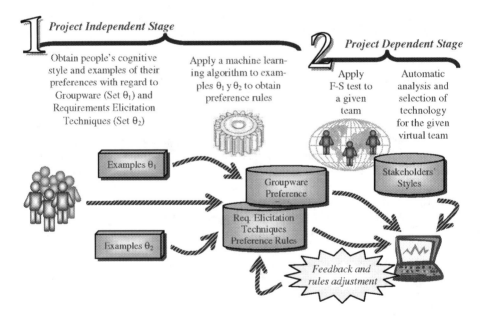

Fig. 1. Phases to define and analyze personal preferences to choose appropriate technology in Virtual Teams

3 Strategies for Cognitive Profile Combination

The previously obtained set of rules represents preferences according to people's cognitive styles, but they are used to discover the most suitable technology for only one person. This means that for each person in the virtual team, we obtain the groupware tool that is most suitable according to his or her cognitive style. However, since it is not expected that all the members of a team will be in agreement as to which groupware tool is the most suitable, it is necessary to provide strategies to combine the results.

According to the Felder and Silverman model, if some stakeholders' preferences are strong and the remaining stakeholders' preferences are moderate or mild, the choices that should be primarily considered are those of the people with strong preferences, since these people perform better when the technology is closer to the way they receive and process information [10]. Bearing this in mind, we have classified teams according to the occurrence of strong preferences, as follows:

- **Type 1:** There are no strong preferences in the team.
- **Type 2:** There are strong preferences but not on the opposite sides of the same category. For instance: if there are strongly visual people in the team, and there are no strongly verbal people, communication should be based on diagrams and written words, which would increase the involvement of visual people. People with slight and moderate preferences can easily become accustomed to them.
- **Type 3:** There are strong preferences on the opposite sides of the same category, so there is a conflict of preferences. For example, if there are one or more strongly visual people, and also some strongly verbal people, communication should support both kinds of styles, as we shall discuss later.

For each type of group we have proposed strategies for rules combinations. For example, the strategy for groups with a strong preference but no conflict (Type 2 groups), is represented as follows:

$$S_2 (\{g\}, (\{GS_1\}, ws_1), (\{GS_2\}, ws_2), \dots, (\{GS_n\}, ws_n))$$
$$\rightarrow g_i \in \{g\} \land g_i \in \{GS_j\} \land ws_j = \max(ws_1, ws_2, \dots, ws_n)$$

where GS_i represents the groupware tool that fits the i-th stakeholder's preferences, ws_i is the weight —meaning how strong the preferences are—, and the resulting g_i is a tool that is appropriate for the stakeholder whose personal preferences are the strongest.

An example of this strategy is shown in Figure 2: according to the preference rules, Chat is the groupware tool recommended for P1 and P2, while Email is recommended for P3. Since P3 has strong preferences, the recommended groupware tool for the group is Email, since this stakeholder will feel more comfortable with this groupware tool and the other stakeholders will not object because they have slight and moderate preferences.

As we explained, strategy S_2 is applicable in type 2 groups (with strong preferences but no conflict). In a similar way we have proposed a strategy S_1 for type 1 groups (without strong preferences) and a strategy S_3 for type 3 groups (with strong preferences on the opposite sides of the same category). Such strategies are widely explained by means of examples in [4].

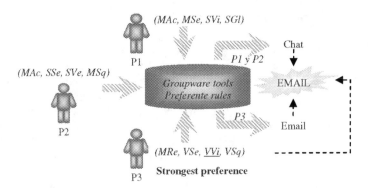

Fig. 2. Strategy for cognitive profile combination represented for 3 stakeholders with strong preferences without conflict

4   Applying Our Strategy S₂ in a Case Study

In order to validate our proposal, we carried out an experiment in which 24 computer science post-graduate students from Argentina and Spain took part. We attempted to simulate global development teams. The teams were therefore formed of three people in which two members played the role of analysts and the other played the role of client. The 'client' had to describe to the 'analysts' the requirements of a software product that the analysts would supposedly have to implement. The analysts then had to use the information obtained from the client's explanations to write a software requirements specification report. As the team members were geographically distributed they had to use a groupware tool to communicate.

After analysing the teams we realised that in each team there was at least one strongly visual person and there were no strongly verbal people; therefore, we applied the strategy for groupware selection for teams with strong preferences without conflict (S_2), explained in Section 3. Table 1 shows the most suitable tool for each team (second column).

Once obtained, each group was assigned a tool, in some cases according to their preferences and in other cases not, with the goal of testing whether there was any difference when they worked with the tool recommended by our approach.

Table 1. Groupware tools assigned to each team

Group	Team	Suitable GW Tool	Assigned GW tool	Suitability
0	G1	IM	Email	-
	G2	Audio	IM	-
	G5	IM	Email	-
	G7	Audio	IM	-
1	G3	Audio	Audio	+
	G4	IM	IM	+
	G6	IM	IM	+
	G8	Audio	Audio	+

The teams were later divided into two groups: those which used the best groupware tool according to our preference rules, and those which used a less suitable (according to our approach) groupware tool. The team that had to use the groupware tools that were not suitable for them (G1, G2, G5, G7), was referred to as Group 0; and the team that had to use the most suitable groupware tools according to our set of preference rules (G3, G4, G6, G8), was referred to as Group 1. The teams in Group 1 and Group 0 always had to use the tool assigned to their team. The resulting selection for each group is shown in the fourth column of Table 1.

5 Preliminary Results

Once the groupware tools had been assigned to each team, the team members were asked to simulate a requirements elicitation process for a given problem, using only the suggested groupware tool for analyst-client communication. As a result of this process they were asked to write an appropriate software requirement specification (SRS), and then they were asked to fill in a post-experiment questionnaire and rate their satisfaction with regard to communication with their partners during the requirements elicitation process. Satisfaction was scored by using a scale of 0-4 (0=very bad, 1=bad, 2=acceptable, 3=good, 4=very good).

According to the analysis of data collected by means of this post-experiment questionnaire, we obtained that most people in Group 1 ranked their satisfaction as 4="very good", while most people in Group 0 ranked their satisfaction as 3="good" (as it is shown in Figure 3). This difference between both groups would indicate that: *Stakeholders' satisfaction with regard to communication seems to be better in the groups that used the most suitable groupware tool according to our set of preference rules.*

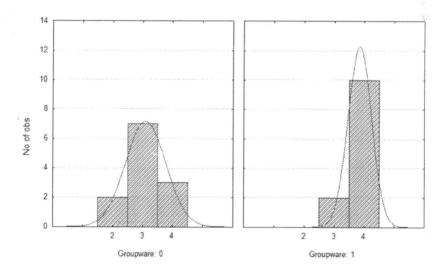

Fig. 3. Stakeholders' satisfaction with regard to communication during the requirements elicitation process

Finally, taking into consideration only the stakeholders with strong preferences (as it is shown in Figure 4), we noticed that satisfaction is clearly higher in the group that used the most suitable tool according to our proposal (Group 1).

This difference would indicate that: *Stakeholders' satisfaction with regard to communication seems to be better in the groups that used the most suitable group-ware tool according to our set of preference rules, especially when cognitive style preferences were stronger.*

The results obtained are close to our previous expectations, and we believe that they will assist us to evaluate the strengths and weakness of our proposal. We are currently working on the analysis of the quality of the written software requirements specifications, and its correlation with the use of the cognitive–based process for technology selection in order to discover whether the groups with the most suitable tools wrote a better requirements specification report. If this is proved to be so, we shall be able to state that working with a suitable groupware tool not only helps members to feel more comfortable but also helps to improve the results of their work.

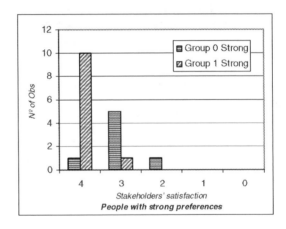

Fig. 4. Stakeholders' satisfaction with regard to communication according to cognitive style level of preferences

6   Conclusions and Future Work

When stakeholders are distributed throughout many distanced sites they must communicate with groupware tools. Choosing the appropriate technology for communication is thus crucial in such environments. We have therefore developed a methodology for technology selection based on the learning styles of the members of a virtual team.

In this paper we present the basis for the application of a strategy that combines the preferences of all the team members, searching for the best solution for the group as a whole, and detecting the strongest preferences in a team without conflicts. We also show the preliminary results of a controlled experiment in which this strategy was applied. As regards stakeholders' satisfaction with communication during the experiment, preliminary results indicate that the stakeholders from those teams that used the most suitable groupware tools suggested according to our proposal, perceived a better degree of communication. In other words, they felt more comfortable in the communication process than those who worked with another tool, especially in the case of those people with the strongest cognitive preferences. We plan to replicate the experiment in order to contrast these results in a similar environment.

Acknowledgements

This work is partially supported by the MELISA (PAC08-0142-3315), Junta de Comunidades de Castilla-La Mancha, Consejería de Educación y Ciencia, in Spain. It is also supported by the ESFINGE project (TIN2006-15175-C05-05) Ministerio de Educación y Ciencia (Dirección General de Investigación)/ Fondos Europeos de Desarrollo Regional (FEDER) and the FABRUM project (grant PPT-430000-2008-063), Ministerio de Ciencia e Innovación, in Spain; and the 04/E072 project, Universidad Nacional del Comahue, from Argentina.

References

1. Aranda, G., Cechich, A., Vizcaíno, A., Castro-Schez, J.J.: Using fuzzy sets to analyse personal preferences on groupware tools. In: Proc. X Congreso Argentino de Ciencias de la Computación, CACIC 2004, San Justo, Argentina, October 2004, pp. 549–560 (2004)
2. Aranda, G., Vizcaíno, A., Cechich, A., Piattini, M.: Towards a Cognitive-Based Approach to Distributed Requirement Elicitation Processes. In: Proc. WER 2005, VIII Workshop on Requirements Engineering. Porto, Portugal, June, pp. 75–86 (2005)
3. Aranda, G., Vizcaíno, A., Cechich, A., Piattini, M.: How to choose groupware tools considering stakeholders' preferences during requirements elicitation? In: Haake, J.M., Ochoa, S.F., Cechich, A. (eds.) CRIWG 2007. LNCS, vol. 4715, pp. 319–327. Springer, Heidelberg (2007)
4. Aranda, G., Vizcaíno, A., Cechich, A., Piattini, M.: Strategies to recommend Groupware Tools According to Virtual Team Characteristics. In: Proc. ICCI 2008, International Conference on Cognitive Informatics, Stanford, California, USA, pp. 68–174 (2008)
5. Brooks, F.P.: No Silver Bullet: Essence and accidents of Software Engineering. IEEE Computer 20, 10–19 (1987)
6. Carmel, E., Agarwal, R.: Tactical Approaches for Alleviating Distance in Global Software Development. IEEE Software 18, 22–29 (2001)
7. Castro, J.L., Castro-Schez, J.J., Zurita, J.M.: Learning Maximal Structure Rules in Fuzzy Logic for Knowledge Acquisition in Expert Systems. Fuzzy Sets and Systems 101, 331–342 (1999)
8. Damian, D., Zowghi, D.: The impact of stakeholder's geographical distribution on managing requirements in a multi-site organization. In: Proc. IEEE Joint International Conference on Requirements Engineering, RE 2002, Essen, Germany, September 2002, pp. 319–328 (2002)
9. Felder, R., Silverman, L.: Learning and Teaching Styles in Engineering Education. Engineering Education 78, 674–681 (1988)
10. Felder, R., Silverman, L.: Learning and Teaching Styles in Engineering Education. Engineering Education 78, 674–681 (1988) (and author preface written in 2002)
11. Felder, R., Spurlin, J.: Applications, Reliability and Validity of the Index of Learning Styles. International Journal of Engineering Education 21, 103–112 (2005)

12. Herbsleb, J.D., Moitra, D.: Guest Editors' Introduction: Global Software Development. IEEE Software 18, 16–20 (2001)
13. Hickey, A.M., Davis, A.: Elicitation Technique Selection: How do experts do it? In: Proc. International Joint Conference on Requirements Engineering (RE 2003), pp. 169–178. IEEE Computer Society Press, Los Alamitos (2003)
14. Martín, A., Martínez, C., Martínez Carod, N., Aranda, G., Cechich, A.: Classifying Groupware Tools to Improve Communication in Geographically Distributed Elicitation. In: Proc. IX Congreso Argentino de Ciencias de la Computación, CACIC 2003, La Plata, Argentina, October 2003, pp. 942–953 (2003)
15. Miller, J., Yin, Z.: A Cognitive-Based Mechanism for Constructing Software Inspection Teams. IEEE Transactions on Software Engineering 30(11), 811–825 (2004)
16. Richardson, I., Casey, V., Zage, D., Zage, W.: Global Software Development – the Challenges. University of Limerick, Ball State University: SERC Technical Report 278, p. 10 (2005)
17. Wang, Y.: On the Cognitive Informatics Foundations of Software Engineering. In: Proc. Third IEEE International Conference on Cognitive Informatics, ICCI 2004, Victoria, Canada, August 22-31 (2004)

Assessment of Facilitators' Design Thinking

Anni Karhumaa, Kalle Piirainen, Kalle Elfvengren, and Markku Tuominen

Lappeenranta University of Technology, Department of Industrial Management
firstname.lastname@lut.fi

Abstract. Meeting design is one of the most critical prerequisites of the success of facilitated meetings but how to achieve the success is not yet fully understood. This study presents a descriptive model of the design of facilitated meetings based on literature findings about the key factors contributing to the success of collaborative meetings, and links these factors to the meeting design steps by exploring how facilitators consider them in practice in their design process. The empirical part includes a case study conducted among 11 facilitators. Session goals, group composition, supporting technology, motivational aspects, physical constraints, and correct practices are found to outline the key factors in design thinking. Furthermore, the order of considering these factors in the design process is outlined. The results contribute to the discussion on how to improve the effectiveness of collaboration with better meeting design, providing also some insights into further development of design support tools.

Keywords: Collaboration engineering, meeting design, facilitation, group support systems.

1 Introduction

Facilitation and supportive technology, such as group support systems (GSS), can improve the efficiency and effectiveness of collaboration [1]. Collaboration researchers from different disciplines, e.g. [2–7], have found facilitation – especially the design task of facilitation – to be one of the most critical pre-requisites for the success of a meeting. However, the use of facilitation support does not automatically guarantee improved collaboration, but the success depends on how the support is applied [8, 9]. The problem is that the valuable expertise of applying collaboration support tends to remain as a tacit knowledge of experienced facilitators [10] and is lost as those facilitators are promoted to new tasks. Therefore, the facilitation expertise needs to be better documented. One of such approach is Collaboration Engineering (CE) which has been set up to develop and transfer guidelines and best practices for collaboration process design [11].

This paper was stimulated by the idea that the significant amount of research done on the factors contributing to the success of collaborative meetings would add to CE research if the role of those factors in meeting design was better understood. Thus far, structured design methods – addressed as the Way of Working in CE community [12, 13] – have been studied by presenting ThinkLets, codified facilitator interventions that aim to create desired patterns of collaboration [14–17], and by exploring the strategies and techniques that facilitators or collaboration engineers apply during the

L. Carriço, N. Baloian, and B. Fonseca (Eds.): CRIWG 2009, LNCS 5784, pp. 231–246, 2009.

design of collaboration processes [18, 19]. This understanding has also been applied to some (preliminary sketches of) tools, strategies, and techniques to support collaboration process design, e.g. [20], but more research is still needed in order to make the effective design practices wholly explicit and transferable [21, 22, 10]. Therefore, a deeper understanding of facilitators' design process would provide a valuable input.

This paper builds on the research question "how to design technology supported meetings effectively?" and presents a descriptive model of collaboration process design thinking. This is done through literature findings about the key factors contributing to the success of collaborative meetings, and by exploring how these factors are considered in meeting design by facilitators in practice. The model is meant to

- provide design support for (novice) collaboration engineers,
- increase our insight into the critical factors to be taken into account in the design of collaboration processes,
- increase our insight into how to emphasize the critical factors during different design steps,
- provide support for the creation of design support tools, and
- provide support for the training of collaboration engineers.

Our research follows a descriptive case study strategy. The empirical enquiry includes a case study conducted among 11 facilitators with semi-structured interviews and construct connection assignments. The case study concentrates on the GSS laboratory of Lappeenranta University of Technology (LUT), which has been working on facilitation and GSS for the last fifteen years. During that time, the laboratory has spawned one spin-off facility and has had a significant impact on Finnish GSS research and practice.

The rest of this paper will proceed as follows. Section 2 will first discuss the literature on meeting design and CE, and list literature findings about success factors in meetings. The research approach will be explained in Section 3, followed by the results of the case study in Section 4. Finally, conclusions and suggestions for further research are presented in Section 5.

2 Background

Facilitation has interested management researchers far before technical tools for facilitating group work existed. Lately, CE researchers such as Kolfschoten and Vreede have been studying the task of meeting design and the associated challenges. They have explained the trade-offs in the choice among facilitation techniques that the facilitator needs to take into account in meeting design [22], see Figure 1. As the figure shows, collaborative meetings and their intended deliverables are a function of several things: client goal and the stakeholders' expectations, available resources such as people and supportive technology, and the facilitator's ability to exploit these. Facilitator's task is to propose a meeting design which supports the group in completing the task in a way which is acceptable not only to the problem owner but also to the group that participates the meeting. Meeting design is thus a complex task as the often conflicting trade-offs need to be balanced to find a satisfying meeting design.

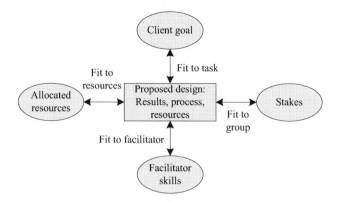

Fig. 1. Trade-offs in the choice among facilitation techniques

In GSS literature, meetings are usually described with an input-process-output model [23]. The model has three main parts as the name suggests: the inputs, process and outputs. The inputs are the resources, constraints and objectives of the meeting that set-up the problem space for the meeting design. The process includes the issues such as facilitation and meeting process structure that utilize the inputs in order to produce desired outcomes. From the design point of view, the process needs to be designed so that the problem space is met with the facilitation solution space in a way which results in satisfying outcomes. The outputs describe the task related outcomes such as quality and quantity of results and the social outcomes such as satisfaction and social group objectives. Again from the design point of view, the meeting output can be seen as not only an outcome but also as a measure of the fit between the problem space and the solution space. For instance, when the fit is good, the results are useful and meeting satisfaction is high. Thus, the trade-offs illustrated in Figure 1 can be seen as the process inputs which are, from our design point of view in this paper, the most interesting part of the input-process-output model. Therefore, we want to focus on the issues that make up a successful meeting.

Facilitation literature contains a considerable amount of discussion on meeting success and its measurement. In terms of meeting success and satisfaction; there is of course the technology adaption and acceptance literature, which is quite central to information systems research, including the widely accepted technology acceptance model (TAM) and its adaptations [24, 25]. Meeting satisfaction has often been suggested to be a key element in the acceptance of meeting results, thus facilitating continual use of GSS facilities [26, 33]. According to [26], there are at least two aspects of a meeting, which a participant could feel satisfaction about: the meeting outcomes and the process by which the outcomes are attained. To form a comprehensive illustration about the variables influencing GSS meeting success and the desired outcomes, we conducted a literature research on theory of meeting satisfaction and technology acceptance. The results are presented in Table 1. The measures and variables are roughly organized by the input-process-output model described above.

Table 1. Measurement variables used in evaluating GSS meetings [26–36]

Meeting outcomes	
Consensus	Ownership of results
Meeting productivity, decision time	Perceived quality of results/outcome
Organizational efficiency and effectiveness	Satisfaction with meeting outcome
Group performance	Satisfaction with meeting process

Meeting process factors	
Discussion quality/amount	Relative individual goal attainment (RIGA)
Faithfulness of appropriation	Shared/common purpose and team spirit
Group synergy	

Meeting inputs	
Adoption of correct practices	Individual goals
Cost of participantion	Motivation to participate
Facilitator expertice, guidance	Participants ability to assimilate and process
Facilitator influence	information
Group composition and homogeneity	Participants ability to exloit new information and
Group goals (implicit)	learn
Group interaction	Perceived value of goal attainment
Group norms	Place/time
Group size	Planning and organization of the sessions
Habituation to electronic communication	Session goals (explicit/out-spoken)
Habituation to group work	Support system features
Incentive alignment for participation	Task type
Individual ability to communicate	Technology

The meeting input factors shown in Table 1 represent the design trade-offs in the choice among collaboration techniques (see Figure 1). Facilitator's design effort, instead, can be seen as a process, a sequence of design steps ([18], see Figure 2), during which the facilitator encounters the problem of deciding and balancing between these design trade-offs. During the meeting design process, the designer diagnoses the task and the problem space, and tries to match them with the available facilitation techniques.

Fig. 2. Meeting design process

Based on this literature, we formed three *a priori* propositions about the role of the discussed trade-offs in effective meeting design. The propositions are presented in Table 2. We proposed that the meeting input factors in Table 1 formulate the content of meeting design and that these factors have a certain emphasis and order during the design process. In the following sections of this study, we examine how the design trade-offs and the meeting design process are interconnected and try thus to describe how facilitators design meetings. The propositions directed the data collection and analysis which will be explained in the next section.

Table 2. *A priori* propositions of the study

In effective meeting design...	
P1: Content	The main factors considered are the meeting input factors listed in Table 1.
P2: Importance	These factors have different weights.
P3: Order	These factors are taken into account in a certain order.

3   Method

To explore facilitators' design thinking, we carried out a case study among 11 collaboration process designers, with semi-structured interviews and structured construct connection assignments. The above presented propositions in the background, we developed the interviews and construct connection assignments based on (a) the exploratory questionnaire used in [10]; (b) the design steps described in [18]; and (c) the list of meeting inputs described in Table 1. The case-study approach was found to be appropriate for building a model to explain facilitators' design thinking [37]. By conducting semi-structured interviews and using connecting-constructs forms, deep knowledge was captured. While the number of interviewees was small, the facilitators possessed unquestionable expertise. Below, we present a summary of the research method.

3.1   Case Study

The case study explores the meeting design expertise accumulated in the GSS laboratory of Lappeenranta University of Technology (LUT) and its spinoff laboratory during the past 15 years. Both laboratories have a single meeting room designed to support up to 15-person electronic meetings. The laboratories apply numerous different software tools to support collaboration. The main software is GroupSystems, which contains all the general characteristics of GSS software. A more detailed description about the laboratory facilities can be found in [38].

Table 3. Summary of the case study interviewees

ID	Profession and degree	Experience (# of meetings)	Experience (years)
Experienced facilitators of LUT			
1	Professor, Ph.D.	20	1994-2008
2	Senior Manager, M.Sc.	20-30	2005-2006
3	Senior Assistant, Ph.D.	over 100	1999-2008
4	Professor, Ph.D.	30-40	1997-2001
5	Manager, Ph.D.	30-50	1994-1998
6	Assistant, M.Sc.	30-50	2004-2008
In-experienced facilitators of LUT			
7	Manager, M.Sc.	2	2008
8	Development Manager, M.Sc.	10	1998-2008
9	Research Assistant, M.Sc.	2	2002
10	Director, M.Sc.	10	1995-2008
Facilitator in the spinoff laboratory of LUT			
11	Trade Manager, M.Sc.	90	2005-2008

The GSS supported meetings at LUT are mainly arranged for teaching and research purposes. The laboratory's main partners in cooperation come from large size enterprises in forest, metal, and ICT industries. Some of the main topics of the meetings have been customer need assessment and idea generation and selection of product/service concepts. The spinoff laboratory of LUT is devoted to consulting small and medium size enterprises in their decision making problems such as planning for company strategies and marketing actions. Thus, the design expertise documented in this study focuses on designing collaborative meetings for the use of industrial companies. The facilitators interviewed for this case study and their expertise are summarized in Table 3. These facilitators comprise all the main facilitators of the laboratories, as well as some less experienced facilitators that were chosen as points of comparison.

3.2 Data Collection and Analysis

Following the case study strategy recommended by Yin [37], we developed a case study protocol[1] that contained the data collection instruments – a questionnaire and a construct connection assignment – as well as the procedures and general rules that should be followed in filling the questionnaire. The semi-structured interviews contained four parts:

1. Beginning, which contained a short introduction into the interview and the explanation of the interviewee's design experience.
2. Interviewee's own design process, where the interviewee drew/wrote down his or her own meeting design process.
3. Absorption in the themes, where the key themes were talked through by following the design process the interviewee had drawn.
4. Finishing, where the interviewee was asked to fill in the construct connection assignment and to give final comments on what the key issues in design are.

All the interviews were conducted by the first author. The questionnaire was used by the interviewer as a check-list of the key themes to be discussed during the interviews. Instead of following the order of the themes in the questionnaire, the discussion during the interviews was more directed by the interviewees and their own design processes. At the end of each interview, the interviewee was given the connecting constructs assignment where (s)he was asked to connect the design steps (see Figure 2) and meeting factors (see the meeting inputs in Table 1) according to which design step they thought each factor had to be paid special attention to.

All the interviews were audio recorded. The average duration of the recordings was 45 minutes, ranging from half an hour to one hour. 3 out of the 11 interviewees wanted to fill in the construct connection assignment outside the interview in order to be able to conduct the task at their leisure. The final number of separate links in the construct connection assignments was about 400.

The interview recordings were transcribed after which the texts were coded to assess how the meeting input factors (see Table 1) appear in the data. A coding scale of 1 to 4 was used. The idea was to understand the content of the facilitators' design. To identify unimportant factors as well as the factors over which the interviewees have highly divided opinions, the averages and standard deviations for the importance measures of different factors were calculated. In addition, the links between meeting factors and design steps in the connecting constructs assignments were entered into a database, where the number of links was presented in pivot tables with success factors in rows and design tasks in columns. The idea was to assess which design step(s) each success factor is especially important to consider in. These linkages were then presented in bar charts in order to make the patterns visually recognizable.

4 Results

This section presents the case study findings. The raw interview data was analyzed as described above. This section is structured around our positions (see Table 2). First, we discuss the design steps, i.e., the design process. Second, we examine the content of design, i.e., what are the defining factors in the design of collaboration processes. And finally, we connect the design process and content to each other by modeling when the meeting input factors are taken into account in the design process.

4.1 Design Steps

In the beginning of the interviews, the interviewees drew and/or wrote down their step-by-step views about the GSS meeting design process. The analysis of these process descriptions revealed that the interviewees' design steps cover relatively well the design steps discussed by Kolfschoten and Vreede [18]. Taking into consideration the iterative nature of design that appears in both the interviewees' and literature design descriptions, also the order of the steps can justifiably be argued to be similar. Thus, it was concluded that the results of this study support the design model presented in [18], and the model can be used to represent the design steps and their order.

4.2 Design Content

Table 4 shows which factors each interviewee takes into account during their meeting design process, and how these factors are emphasized. While the factors are based on the meeting inputs presented in Table 1, we organized them further into group, task, technology, and context related factors following the classification in [10], added by an input factor group of facilitation related factors thus highlighting also the meeting design trade-off of facilitator skills (see Figure 1) individually. As can be seen, task related factors are considered the most important in design. All the interviewed facilitators underlined a clearly formulated definition of the session goals as an essential requisite for a successful design. In addition to task related factors, also several other factors emerge from the list. Among group related factors, group composition is highly emphasized, except in the case of the spinoff laboratory. Technology-related support system features seem to be important in design, although not of key importance. The motivational aspects focus on customer commitment which is regarded as a vital pre-requisite for a successful meeting by the interviewees and thus as an important factor to be taken into account during the meeting design. Also facilitator expertise and adoption of correct practices are valued as critical success factors in design.

Table 4. Averages and standard deviations of the importance figures of different success factors considered in meeting design

Meeting factor	Average: 1 = no impact in design, 4 = one of the key factors considered in design				Std
	Exper.	In-exper.	Consult.	All	All
	(n = 6)	(n = 4)	(n = 1)	(n = 11)	(n = 11)
Group					
Group goals	2,5	2,0	1,0	2,2	1,2
Group composition	3,7	4,0	2,0	3,6	0,7
Group size	2,2	1,8	1,0	1,9	0,9
Individual goals	1,0	1,0	1,0	1,0	0,0
Ability to exploit new information and learn	1,5	1,0	1,0	1,3	0,5
Group interaction	3,2	2,8	1,0	2,8	1,0
Ability to assimilate and process information	2,3	2,3	3,0	2,4	0,7
Ability to communicate	1,7	1,3	1,0	1,5	0,7
Habituation to electronic communication	1,2	1,0	1,0	1,1	0,3
Habituation to group work	1,2	1,3	1,0	1,2	0,4
Task					
Session goals	4,0	4,0	4,0	4,0	0,0
Task type	3,5	3,5	4,0	3,5	0,5
Technology					
Technology	2,5	3,0	3,0	2,7	0,6
Support system features	3,2	3,0	4,0	3,2	0,4
Context					
Perceived value of goal attainment	3,5	3,0	3,0	3,3	0,6
Cost of participation	1,5	2,5	2,0	1,9	1,0
Motivation to participate	2,8	2,3	3,0	2,6	1,4
Incentive alignment for participation	3,2	2,8	3,0	3,0	1,1
Place and time	2,5	2,8	2,0	2,5	0,5
Facilitation					
Facilitator expertise	3,3	2,8	4,0	3,2	0,6
Adoption of correct practices	3,3	3,5	3,0	3,4	0,5
Facilitator influence	2,0	2,5	3,0	2,3	0,5

Were also found that the facilitators tend to ignore some meeting factors in their design: factors such as 'individual goals', 'ability to exploit new information and learn', 'ability to communicate', 'habituation to electronic communication', and 'habituation to group work' were rated very low on importance. Standard deviations being relatively low for each of these success factors, it can be concluded that this view is shared by all the interviewees. Interestingly, all these unimportant success factors belong to group-related factors and handle aspects from the *individual point of view*: an individual participant's goals, abilities, and habituations. Instead, the other group-related factors focus on *group as a whole*: group goals, composition, and interaction.

4.3 Design Order: Linking the Steps and Content

As we knew the design steps and content, the next step was to assess how these were connected to each other in order to fully understand facilitators' design thinking. The connections that the facilitators drew on the construct connection assignments are presented in Figure 3. These connections reveal how facilitators consider the design trade-offs (see Figure 1) by showing the decision making points in facilitators' design process for different meeting factors. The big chart in Figure 3 shows the average of the whole case while the small charts show the same data split into the three case groups. As can be seen, similar patterns across all the case groups can be recognized but some differences also exist. It is worth noting, that the bar heights in the figures of different case groups are not directly comparable, as the smoothness of importance figures between the groups differs significantly due to the varying number of interviewees (e.g. there are six experienced facilitators but only a single facilitator in the spinoff laboratory). According to our sensitivity analysis, using the two most important meeting factors from each success factor group in the importance calculations provides a good representation of facilitators' design thinking in general. Thus, we used the two most important factors from each group in the figures. The most important factors used in the calculations are the following:

- *Group*: Group composition; Group interaction
- *Task*: Session goals; Task type
- *Technology*: Technology; Support system features
- *Context*: Perceived value of goal attainment; Incentive alignment for participation
- *Facilitation*: Facilitator expertise; Adoption of correct practices

We draw the following conclusions about how the meeting design factors relate to the process of meeting design based on out interpretation of the weights of each factor in each design phase presented in Figure 3:

Step 1: In task diagnosis, the designers map out the situation: the most important success factors are related to the task, group, and context. Task-related factors have the overwhelmingly most important role during this design step. The designers also seem to have technology-related factors in mind. Facilitation-related factors, i.e. previous facilitation expertise, have a substantial effect in design-thinking in this step as well as in all the following steps.

Step 2: In activity decomposition, the facilitator's expertise is also emphasized. The designers use their expertise in order to decompose the task into subtasks and to build the solution process. Here, task- and group-related factors seem to have a substantial role in most cases. In-experienced designers rely less on facilitator expertise, which is quite obvious due to their lower level of experience.

Step 3: In technique choice, the role of facilitation-related factors gets even stronger. Keeping the task- and group-related factors still in mind, the designers focus on technology-related factors. It is worth noting that in spinoff laboratory the steps 'activity decomposition' and 'technique choice' are not separated and can therefore be interpreted together.

Step 4: In agenda building, the context-related factors with respect to motivational aspects seem to gain a considerable amount of designers' attention. Also the other factors are considered in order to build a good agenda, but they have gained more attention already in previous steps.

Step 5: In design validation, the designers use their expertise to evaluate the quality of the design. Experienced designers, in particular, seem to think about motivational aspects during design validation. In the spinoff laboratory, the group, the participants, are defined at this step, and the design is adjusted and validated according to the group needs.

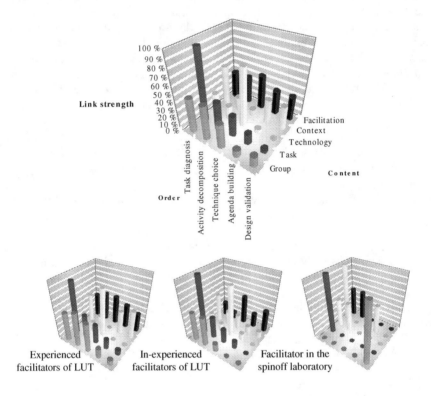

Fig. 3. Order of meeting factors in design: total average and case-group specific averages separately

5   Discussion and Conclusions

The study extends prior literature by highlighting the key meeting factors considered in collaboration process design and by linking these factors with the design activities conducted by facilitators. The study can form a basis for developing the theory of design thinking in collaboration engineering, for developing further support and training for collaboration engineers, and for studying the connection between the design and success of meetings.

When we look at the content of meeting design (see Proposition 1), it can be concluded that the factors discussed in prior literature are the same factors that facilitators take into account during their meeting design. The factors that emerged from the interview data correspond well to the input factors of meeting success that we listed based on the literature research (see Table 1) and further organized into five groups: group, task, technology, context, and facilitation related factors. The key design factors that arose from the content analysis of design thinking can be summarized with five factors as follows:

- *Session goals* give the start and provide the backbone for the whole design process. (Task)
- *Group composition* is considered to ensure correct information available in the meeting and to understand the demands of the group. (Group)
- *Supporting technology* is selected according to the demands of the task and group. (Technology)
- *Motivation* of the group is ensured in order to build commitment. (Context)
- *Correct practices* guide the designer in building a workable meeting agenda with right timeframes and content. (Facilitation)

However, we also made an interesting finding about some input factors that facilitators do not consider during the design: according to the case study results, the goals and desires of individual participants are ignored by the facilitators, but the participants are regarded collectively as a group instead. This may imply that the facilitators believe that the designed collaboration process together with technology will take care of the goal congruence, or that they believe considering individual participant needs to be too hard, even though it could be beneficial in building individual participants' commitment toward the group goal as proposed in [39].

When we examine the importance of different design factors (see Proposition 2), the facilitators emphasize some factors above others. Session goals are the number one design factor. However, when we consider different success factor groups, no group can be elevated to be significantly more important compared to the others. Physical constraints, such as time or group size were usually considered at least somewhat important, but their role in design thinking was smaller than that of more controllable factors. Through the semi-structured interviews it became clear to us that the design is always a resource-constrained assignment, and, if a design factor is predetermined, i.e. constraining, a facilitator rates it less important than a factor they can manipulate. These findings may imply that more understanding about the influence of design constraints is needed in order to wholly understand facilitators' design thinking.

When we look at the order of considering design factors during the meeting design (see Proposition 3), some tentative conclusions can be drawn. The design process which facilitators follow is consistent with the design approach presented in [18], comprising the steps of task diagnosis, activity decomposition, technique choice, agenda building, and design validation. Some factors, such as session goals or technology can rather easily be linked to a certain design step, but due to the iterative nature of design, the factors tend to have some impact on several design steps. Preliminary results of the order of different design factors are presented in the model of design thinking below.

Table 5. The model for meeting design thinking

Factor type	Key design factors	Design tasks				
		Task diagnosis	Activity decomposition	Technique choice	Agenda building	Design validation
Controllable *define and control*	Session goals *Goals, objectives, task type*	x	x			x
	Group composition *Group composition, interaction, goals, abilities*	x	x	x		
	Supporting technology *Tools, techniques, methods, technology features*		x	x		
	Motivation *Motivation to participate, commitment*	x			x	x
Constraining *understand and adapt*	Physical constraints *Place, time, other constraining factors*	x	x	x	x	
Guiding *exploit*	Correct practices *Best practices, facilitator expertise*	x	x	x	x	x

The resulting model of design thinking is shown on Table 5. By listing the key design factors and their role in design, the model outlines how to design technology supported meetings effectively which was the main research question. According to the results of this study, three types of design factors can be recognized in meeting design thinking: (1) controllable factors that the designer needs to define and modify; (2) constraining factors that set the boundaries of the design; and (3) guiding factors that the designer exploits across the whole design process in understanding, defining, and utilizing the other key design factors.

The controllable factors are in the center of design thinking, encompassing session goals, group composition, supporting technology, and motivational factors. As the 'x':s show in Table 5, each of these factors is emphasized in certain design tasks. For instance session goals, group composition, and motivation of participants are paid a special attention to during task diagnosis. The constraining factors influence the designer's decisions in the background. The case study results imply that every meeting has some preordained factors, such as meeting time-frame, which the designer just needs to adapt to. The constraining factors can be any factors that set limits to the design; some of the factors included in controllable factors in the model can actually be constraining if they have been predetermined before the design begins. Group composition, for instance, was regarded as a constraining factor by some of the designers interviewed in this study.

The controllable and constraining factors together formulate the contents of design thinking. The third type of design factors, the guiding factors, provide the designer with guidance in design decisions in the form of tacit knowledge and expertise of facilitators or of explicit best practise guides and design support tools. The role of guiding factors was ranked high across the whole design process, which implies that there is a real need for the support of the design process. Furthermore, the interviewed designers were found to use trusted design practices repeatedly, which implies that they are accustomed to the collaboration engineering type of design approach where design patterns are used repeatedly as building blocks of collaboration processes [40].

The most important contribution of this study is connecting the design factors and the design steps together, as explicitly presented in the model of design thinking. We propose that the model can provide valuable support for both the practicing facilitators (or collaboration engineers) and the creators of design support tools. First off, facilitators may use the model as a best practice guide about the design of collaboration processes. As it comes to the development of design support tools, the model provides valuable insights into the content, importance and order of the factors that need to be taken into account during the design process and that should therefore be implemented in the computer based expert tool which is mentioned to be critical to better support collaboration process design [18]. Furthermore, this would help CE community reach the optimized level of CE maturity where "there are management approaches and evaluation criteria to assess the quality of the [design] process" [19], and collaboration engineers could finally implement CE processes in organizations with success.

The study also raises a number of interesting directions for future research. First, the preliminary model of the linkages between design factors and steps could be validated. Second, this model could be used in developing supportive tools for meeting design as discussed. Third, the connection between effective design practices and meeting success could be studied. Fourth, an interesting topic for future research should also be to examine meeting design as resource-limited action with some resources, such as time and group size, pre-defined. As the results of this study propose, factor type, i.e. whether the factor is controllable or constraining, may have a significant impact in design thinking. Validating the model of design thinking could be done by conducting the same study for a bigger group of facilitators in order to make statistical analysis on the designers' views. The other questions that arise from this study could be studied through a combination of focused questionnaires, semi-structured interviews, and small case studies.

Note

1. The case study protocol can be requested from the first author.

Acknowledgements

We, the authors, would like to acknowledge the helpful comments from Prof. Gwendolyn Kolfschoten from TU Delft during the preparation of this paper. Kalle Piirainen acknowledges individually that the preparation of this publication has been conducted partly with funding from the Academy of Finland.

References

1. Fjermestad, J., Hiltz, S.R.: A descriptive evaluation of group support systems case and field studies. Journal of Management Information Systems 17, 115–159 (2001)
2. Antunes, P., Ho, T., Carriço, L.: A GDSS agenda builder for inexperienced facilitators. In: 10th Euro. GDSS Workshop, Copenhagen, Denmark (1999)
3. Clawson, V.K., Bostrom, R.P.: The importance of facilitator role behaviors in different face to face group support systems environments. In: HICSS, pp. 181–190. IEEE Press, Los Alamitos (1995)
4. Hayne, S.C.: The facilitators perspective on meetings and implications for group support systems design. ACM SIGMIS Database 30(3-4), 72–91 (1999)
5. Niederman, F., Beise, C.M., Beranek, P.M.: Issues and concerns about computer-supported meetings: the facilitators perspective. MIS Quarterly 20(1), 1–22 (1996)
6. Nunamaker, J.F., Briggs, R.O., Mittleman, D.D., Vogel, D., Balthazard, P.A.: Lessons from a dozen years of groups support systems research: a discussion of lab and field findings. Journal of Management Information Systems 13(3), 163–207 (1997)
7. de Vreede, G.J., Boonstra, J., Niederman, F.: What is effective GSS facilitation? A qualitative inquiry into participants' perceptions. In: HICSS, IEEE Press, Los Alamitos (2002)
8. Bostrom, R., Anson, R., Clawson, V.: Group facilitation and group support systems. In: Jessup, L., Valacich, J.S. (eds.) Group Support Systems: New Perspectives, pp. 146–168. Macmillan, New York (1993)
9. de Vreede, G.J., Davison, R., Briggs, R.O.: How a silver bullet may lose its shine – learning from failures with group support systems. Communications of the ACM 46(8), 96–101 (2003)
10. Kolfschoten, G.L., den Hengst-Bruggeling, M., de Vreede, G.J.: Issues in the design of facilitated collaboration processes. Group Decision and Negotiation 16(4), 347–361 (2007)
11. de Vreede, G.J., Briggs, R.O.: Collaboration engineering: designing repeatable processes for high-value collaborative tasks. In: HICSS. IEEE Press, Los Alamitos (2005)
12. Seligmann, P.S., Wijers, G.M., Sol, H.G.: Analyzing the structure of IS methodologies. In: Proceedings of the 1st Dutch Conference on Information Systems, Amersfoort, Netherlands (1989)
13. Briggs, R.O., Kolfschoten, G.L., de Vreede, G.J., Dean, D.L.: Defining key concepts for collaboration engineering. In: Proceedings of the Twelfth Americas Conference on Information Systems, Acapulco, Mexico (2006)
14. Briggs, R.O., de Vreede, G.J., Nunamaker Jr., J.F.: Collaboration engineering with ThinkLets to pursue sustained success with group support systems. Journal of Management Information Systems 19(4), 31–64 (2003)
15. Briggs, R.O., De Vreede, G.-J., Nunamaker Jr., J.F., Tobey, D.: ThinkLets: achieving predictable, repeatable patterns of group interaction with group support systems (GSS). In: HICSS. IEEE Press, Los Alamitos (2001)

16. de Vreede, G.J., Briggs, R.O.: Collaboration engineering: designing repeatable processes for high-value collaborative tasks. In: HICSS. IEEE Press, Los Alamitos (2005)
17. de Vreede, G.J., Briggs, R.O., Kolfschoten, G.L.: ThinkLets: a pattern language for facilitated and practitioner-guided collaboration processes. International Journal of Computer Applications in Technology 25, 140–154 (2006)
18. Kolfschoten, G.L., de Vreede, G.-J.: Collaboration engineering approach for designing collaboration processes. In: Haake, J.M., Ochoa, S.F., Cechich, A. (eds.) CRIWG 2007. LNCS, vol. 4715, pp. 95–110. Springer, Heidelberg (2007)
19. Santanen, E., Kolfschoten, G.L., Golla, K.: The collaboration engineering maturity model. In: HICSS. IEEE Press, Los Alamitos (2006)
20. Kolfschoten, G.L., Veen, W.: Tool support for GSS session design. In: HICSS. IEEE Press, Los Alamitos (2005)
21. Kolfschoten, G.L., Appelman, J.H., Briggs, R.O., de Vreede, G.-J.: Recurring patterns of facilitation interventions in GSS sessions. In: HICSS. IEEE Press, Los Alamitos (2004)
22. Kolfschoten, G.L., Rouwette, E.A.J.A.: Choice criteria for facilitation techniques. In: Briggs, R.O., Nunamaker Jr., F. (eds.) Monograph of the HICSS-39 Symposium on Case and Field Studies of Collaboration (2006)
23. Nunamaker, J.F., Dennis, A.R., Valacich, J.S., Vogel, D.R., George, J.F.: Electronic meeting systems to support group work. Communications of the ACM 34(7), 40–61 (1991)
24. Lee, Y., Kozar, K.A., Larsen, K.R.T.: The Technology acceptance model: past, present, and future. Communications of the Association for Information Systems 12(50), 752–780 (2003)
25. Briggs, R.O., Nunamaker Jr., J.F., Tobey, D.: The technology transition model: a key to self-susteining and growing communities of GSS users. In: HICSS. IEEE Press, Los Alamitos (2001)
26. Briggs, R.O., de Vreede, G.-J., Reinig, B.A.: A theory and measurement of meeting satisfaction. In: HICSS. IEEE Press, Los Alamitos (2003)
27. Briggs, R.O., Reinig, B.A., de Vreede, G.J.: Meeting satisfaction for technology-supported groups: an empirical validation of a goal-attainment model. Small Group Research 37, 585–611 (2006)
28. Brodbeck, Greitemeyer: A dynamic model of group performance. Group Process and Intergroup Relations 3(2) (2000)
29. Davison, R.: An instrument for measuring success. Information & Management 32(4), 163–213 (1997)
30. Griffith, T.L., Fuller, M.A., Northcraft, G.B.: Facilitator influence in GSS: intended and unintended effects. ISR 9(1) (1998)
31. Hackman, J.R.: The design of work teams. In: Lorsch, J.W. (ed.) Handbook of organizational behavior, pp. 315–342. Prentice-Hall, Englewood Cliffs (1987)
32. Limayem, M., Banerjee, P., Ma, L.: Impact of GDSS: Opening the black box. Decision Support Systems 42, 945–957 (2006)
33. Reinig, B.A.: Toward an understanding of satisfaction with the process and outcomes of teamwork. Journal of MIS 19(4) (2003)
34. Reinig, B.A., Briggs, R.O., Nunamaker Jr., J.F.: On the measurement of ideation quality. Journal of Management Information Systems 23(4), 143–161 (2007)
35. Shirani, A., Aiken, M., Paolillo, J.G.P.: Group decision support systems and incentive structures. Information & Management 33, 231–240 (1998)
36. Yuchtman, E., Seashore, S.E.: A system resource approach to organizational effectiveness. American Sociological Review 32(6), 891–903 (1967)

37. Yin, R.K.: Case study research: design and methods, 2nd edn. Sage publications, Thousand Oaks (1994)
38. Piirainen, K., Elfvengren, K., Kortelainen, S., Tuominen, M.: Enhancing learning with a group support system facility. In: Proceedings of NBE (2007)
39. Briggs, R.O., Kolfschoten, G.L.: Theories in Collaboration. In: Tutorial at the Hawaii International Conference on System Science, Waikoloa HI (2009)
40. Kolfschoten, G.L., Briggs, R.O., Appelman, J.H., de Vreede, G.J.: ThinkLets as building blocks for collaboration processes: a further conceptualization. In: de Vreede, G.-J., Guerrero, L.A., Marín Raventós, G. (eds.) CRIWG 2004. LNCS, vol. 3198, pp. 137–152. Springer, Heidelberg (2004)

Unraveling Challenges in Collaborative Design: A Literature Study

Kalle Piirainen[1,2], Gwendolyn Kolfschoten[1], and Stephan Lukosch[1]

[1] Delft University of Technology, Faculty of Technology, Policy and Management,
Section Systems Engineering, P.O. Box 5015, 2600 GA Delft, The Netherlands
{g.l.kolfschoten,s.g.lukosch}@tudelft.nl
[2] Lappeenranta University of Technology, Faculty of Technology Management,
Department of Industrial Management, P.O. Box 20, 53851 Lappeenranta, Finland
Kalle.Piirainen@lut.fi

Abstract. The complexities of modern business technology and policy are straining experts who aspire to design multi-actor systems to enhance existing organizations. Collaborative design is one approach to try and manage complexity in design activities. Still, collaboration in itself is not necessarily an easy mode of working. In this paper, we seek insight to challenges of collaborative design though a survey of design literature and qualitative content analysis. The literature reveals that the challenges can be condensed to five main challenges, creating shared understanding, balancing requirements of different stakeholders, balancing rigor and relevance in the process, organizing the collaboration effectively and creating ownership.

Keywords: Design science, design research, design methodology, collaborative design.

1 Introduction

As Herbert Simon [52] noted, "Design ... is the core of all professional training ..." (p. 111) Professionals and researchers alike solve problems daily, and many of the solutions involve developing or designing an artifact which is implemented to an existing system or organization to solve a problem in a complex context. Design is especially relevant to the information systems (IS) field and has been a subject of rising interest during the past few years. The discussion has been intertwined with the discussion on the nature of IS field, although similar strands can be found in e.g. management science, and demands for enhancing the relevance to practice while retaining academic rigor, e.g. [3, 5, 6, 53].

Increasingly, technology and systems tend to be complex and dynamic. Design has thus become a complex collaborative activity, moving beyond mere user involvement to become a highly interdisciplinary task of designing multi-actor systems. Often it is not feasible or even desirable for one professional to master all the knowledge required to design a complex system, and thus design requires collaboration between various domain-, process- and technical- experts. The rise of stakeholder-thinking [4]

L. Carriço, N. Baloian, and B. Fonseca (Eds.): CRIWG 2009, LNCS 5784, pp. 247–261, 2009.

has also propelled collaboration in design projects. In the interest of including different viewpoints, often policy officials, experts, as well as operators, management, economic experts, and investors can be included in the process.

Design itself as an activity is challenging, especially when considering large-scale projects [52]. In addition to design challenges, group work and collaboration themselves tend to be challenging. Nunamaker et al. [41] describe a myriad of challenges of collaboration and group work ranging from distraction to dominance and information overload. In addition to regular design challenges concerning e.g. stakeholder and requirement negotiations, collaborative design requires additional organization, negotiation and building of shared understanding on the issues concerning the design. Groups are unlikely to overcome the challenges of collaboration by themselves [48, 41] let alone the combined challenges of design and collaboration.

Design in general is about creating an understanding and definition of a problem and solving it through a process of finding a satisfying solution (for general discussion see [52]), or as Cross [17] puts it, it is about applying existing knowledge through a structured methodology to solve an existing problem. Design is therefore a purposeful act of creation. Collaboration on the other hand can be defined as "joint effort towards a goal" [13]. To integrate and to scope our study, we define collaborative design as *'Purposeful joint effort to create a solution'*.

Collaborative design and use of technology to support design activities are not new areas of research as such. Facilitation and technological support for e.g. requirement engineering and customer need assessment [28] and project planning [18] have been introduced. In practice, design is often supported with tools and environments, which have not been developed for collaborative work or tools that support one particular task or phase in the process. The present practice uses tools for design and modeling, as well as communication and project management, but somehow the results have not been what the technology has promised.

Design has been also researched, but while there is a wealth of system development methods and processes, everyday designers seem to be struggling to get results. It has been noted that especially professionals, i.e. practitioners, do not readily accept normative guidance and prescribed work methods [24]. Maher [37], for example, points out that there is a need to identify models of design processes that facilitate design rather than prescribe a design process. Even though the area has accumulated literature already, it seems that development of support would gain from a more holistic perspective to the design process. Which is why we focus in this paper on understanding the challenges of collaborative design, and it provides a base for developing guidelines to achieve better design results. The contribution is twofold; the results offer insight to the challenges of collaborative design as such, which is a step toward solving the problems. Uncovering and explicit formulation of these challenges serves also collaboration research community, as the insights to the process help facilitation of collaborative design and developing new ways to facilitate it.

The remainder of the paper is organized as follows. The following section presents the research design in more detail. The third section presents the analysis of design challenges. Subsequently, the fourth chapter summarizes and synthesizes the challenges of collaborative design. Finally, we conclude this paper and present an agenda for further research.

2   Research Approach

This section lays the foundation for identification of design challenges. First, we used bibliometric analysis to identify influential pieces of design literature, which we used to identify the challenges of collaborative design later in this paper. After the survey, we conducted a deeper analysis of the papers with qualitative content analysis. Next we describe the methods we used in the paper in more detail.

Bibliometric methods have been spreading to a wide variety of disciplines, and are considered helpful tools to uncover conceptual structures in a possibly fragmented field [44]. We aim to use bibliometric analysis as a means of explorative analysis to uncover the structures and important pieces of work in Design Research and Science (DSR and DS respectively).

To conduct our analysis we made a query to the ISI Web of Knowledge [31] database to retrieve a dataset for analysis. We chose the ISI Web of Knowledge as it represents a reasonable cross-section of 'high-quality' journals, and it interfaces with the tools we used for further analysis. The dataset was retrieved with the keywords "design science" OR "design research" OR "design theory" in topic, keyword or abstract. Before exporting the citation data, the dataset was narrowed down by concentrating on social, information and computer sciences, excluding sociological and political studies, as well as medical and life sciences. The resulting initial dataset was 189 articles with over 2000 references.

The main tool enabling the analysis was Henri Schildt's SITKIS toolset [45], which is able to import and manipulate Web of Knowledge datasets and to analyze the citation patterns. The dataset was exported from ISI and imported to SITKIS. During the inspection of the database, the list of cited articles was examined for misquoted author names and article names. We composed a co-citation network with a threshold of 5 citations, meaning those only publications which are cited 5 times or more together are included.

The main result we get from the survey is the sense that apparently design literature is gaining momentum at the moment, as citations to the main publications have been on the rise since 2003-2005. The co-citation statistics hint that design research builds a lot on few key articles. The strongest linkages are, not surprisingly, found between the ten most cited articles. The core of the literature seems to be a set of key articles published in MIS Quarterly [29, 39] and some other influential IS research journals [38, 56] and foundational books [49, 52].

Following the literature survey, we sought for greater understanding on the challenges in design and in particular collaborative design. We approached the issue by closer analysis of the contributions and coding them to find emerging patterns. In this meta-analysis we employ a basic qualitative content analysis technique, where we gather constructs from the literature to an existing classification to systematic reduce the textual material [34, 57]. We employ specifically a qualitative content analysis technique, where we code the text to existing classification by seeking expressions which answer to our question. The question we posed to analyze the data, was "what are the challenges of collaborative design?"

We started the analysis by examining the core literature identified in the survey, although because of research economies, we devoted our attention to articles and excluded the books with one exception, Simon's "Sciences of the Artificial" [52]. From there forward, we continued to snowball, identifying additional literature, through conventional searches and by examining citations in the identified articles as well as finding articles which cite the identified literature. The logic in compiling the table was that we continued to examine and search more literature until the same challenges started to appear over again, and thus a certain degree of theoretical saturation was reached. The basic coding of the literature was done by one of the authors, inspected by others, and the subsequent analysis and conclusions were based on the coding and discussions between the authors.

3   Analysis of Design Challenges

3.1   Framework for Analysis

As explained, we used an existing classification to code the challenges of collaborative design. The first dimension relates the challenges to different phases of a general design process. The most basic design process as outlined by Simon [52] for example is a simple generate-test cycle. That is to say, a designer finds and defines problem, generates a solution and tests it. Beside the most basic process, the literature describes or prescribes any number of design processes [50, 11, 22, 51, 40, 1, 14, 16]. The gist of the processes can be summarized to the following phases (1) problem definition, (2) generating an artifact or multiple artifacts, for (3) testing of the artifact, from where the approved artifact will pass to (4) validation and there to (5) implementation and (6) sustained use in the target system or organization. The problem definition could be also broken to two parts, to development of specification and representation of the present state, as commonly used in system engineering practice. We supplemented the sequential division to design process phases with classification built on the levels of design presented by Dorst [20], who discussed the different perspectives to design and outlined the basic levels of analysis concerning the science of design. The levels are (1) the object or artifact that is designed, (2) the context to which the object is designed and where it is supposed to function, (3) the actor or actors who design, and finally (4) the process and methodology for design.

Next, we discuss each phase on the design process separately and list the coded constructs on different levels of analysis. The challenges of collaboration we associate with respective design challenges are included in the discussion to form a basis for synthesis in next section.

3.2   Problem Definition

The phase of problem definition defines the design process and the subsequent content. Many of the challenges pertain to understanding the problem and developing a representation upon which all involved parties can agree. In fact, the main theme even at this stage of the process is to develop a shared understanding and a common language for the rest of the process.

> *1) Object of design*: The major challenges in relation to the object of design are to find a joint and mutually acceptable problem definition. For this, it is necessary to create a shared understanding about the problem

and integrate multiple perspectives. These challenges become visible in shifting, developing goals [52], the notion of a holistic outlook on the problem [4] and the problem in finding the users as well as identifying their needs [39].

2) Context of design: A great challenge which strongly relates to the context is the ambiguous, wicked, nature of problems [43, 52]. The complexity of the situation makes it hard to identify the real problem owner [52] and often results in conflicts of interest, difficulties in developing a consensus on the problem and goals [15, 43]. Finding the limiting resource and other design constraint [52] is a challenge which associates with framing the problem and finding the stakeholders.

3) Actors who design: Especially when considering collaborative design, developing a strong client/stakeholder orientation [4] plays a key role on later commitment and ownership to for the design [10]. Different interpretations of the design problem or specification [21] can be linked with misunderstandings because of insufficient overlap in experience and lack of shared context and language.

4) Design process and methodology: Considering the challenges above, a key issue in the process is focus in establishing the right specification [4] from the start. Developing the specification may require developing representations for design object and methodology [29, 42, 52, 54]. While developing the problem representation, there is also the trade-off between achieving representational simplicity and usability, and oversimplifying things, „assuming away" aspects of the problem [29]. Beside the conceptual challenges, there is the practical challenge of involving stakeholders in the design and problem definition [10, 4]. Following the specification there is a need for design strategy for the rest of the process and criteria for design iterations [33].

3.3 Artifact Construction

During construction of a solution to the previously defined problem, the challenges continue on the same track. Many challenges encompass the need to balance the different needs expressed by the stakeholders, and to develop a solution which can be implemented with confidence that the intended effect is achieved.

1) Object of design: An important challenge is addressing the effect of the design to the behavior of the user [10] and the ripple effect of the design to surrounding systems [52] through interfaces and interdependencies between systems and components. Finger and Dixon [25] also mention the issue while discussing the gap between system behavior and designed function. Also orchestrating the design of a large scale system with sub-systems [52] is a demanding task.

2) Context of design: The context present concrete challenges when abstracting the environment and finding criteria for satisficing solutions [29] and priorization of different stakeholders' needs [21]. Finding proper kernel theories and other foundations for the design [29, 39] is an important part of design where context also has a role. Goal shift was mentioned above, but often the shift continues after the initial specification is locked, and poses challenges for constructing the artifact.

3) Actors who design: The most apparent challenge is to manage coevolution of the problem (interpretation) and artifact [21] while maintaining adherence to the original specification. Also the human condition poses challenges in finding the solution, and often designers failure to develop multiple alternatives, and stick to favorite theories [4, 24] or "Marry the first option" [4].

4) Design process and methodology: Methodological challenges mirror the other levels of analysis, as the method is supposed to enable establising logical linkage between design features and requirements [33]. Proper decomposition of the problem representation [29] and choice of proper means to support the design [29]. The means can include development of sufficient representation to implement the artifact as intended [4] and effective (ad-hoc) search heuristics for solutions [21, 29].

3.4 Testing and Validation

Testing phase is for testing the solution or artifact for completeness and ability to solve the problem. The challenges in this phase are more technical in nature when compared to the conceptual challenges in previous phases. The challenge amounts largely to collecting enough data to pre-validate the artifact and to do the testing in a rational manner without playing favorites between possible alternative solutions. Compared to testing, validation is a more formal and stringent evaluation, to ensure the artifact lives up to the specification in terms of features and performance. Here the conceptual challenge are rare, and concern mainly the question whether that artifact represents the design accurately or not. The practical challenges deal with finding proper techniques for validation as well as finding criteria and metrics to judge whether the artifact is valid.

1) & 2) Object and context of design: The first challenge is ensuring that the artifact embodies the problem representation and the chosen kernel and design theories sufficiently and that the context for validation is realistic enough [29]. A large challenge in testing is to encourage users to use the system long enough to gather data for testing and validation [39]. Equally important is to create consensus on the validation before the fact and to choose proper metrics and methods to evaluate/validate the artifact [29].

3) Actors who design: Just as well as rationality may fail in elucidating and interpreting the requirements, a failure may occur while evaluating the solution [4] especially if they have some favorite feats of some of the collaborators. The same need to overcome bounded rationality is associated with assessing the effects of the design [52].

4) Design process and methodology: On methodology level, the main challenge is the selection and application of proper methodologies to validation [29], as well as developing a set of performance and evaluation criteria to judge the success of the design [4] in a way the collaborators can agree upon. The methodological challenges again are associated with the other levels, like developing performance and evaluation criteria to judge the success of the design [4] to support rational evaluation of the solution. Also ordering design iteration if performance objectives or behavioral patterns shift from expected [33] may prove to be a challenge.

3.5 Implementation and Sustained Use

The last two phases are the least documented in the examined literature. The challenges are nevertheless quite important considering the sustainability of the solution or artifact. Starting from the challenge of implementing the artifact in a way that ensures the intended features will emerge might be a major challenge in practice. Overcoming resistance to change may also connect to other challenges, as ownership and general willingness to use the artifact can conceivably influence the behavior of the artifact.

> 1) *Object of design*: In implementation and use the main challenge is to specify the artifact and its implementation in such a representation and with a degree of detail in design specification for the intended features to emerge [4, 29].
>
> 2) *Context of design*: Here, the challenge is to ensure that the realized artifact resembles the design [4]. Traditionally, resistance to change in the target organization [15, 39] has been a major challenge and important aspect is getting commitment to the artifact so that the organization can "learn to perform" [4]. Resistance to change and inaccuracies in representation can also cause deviation between designed function and the systems behavior [25]. The interface issues mentioned above can be realized in implementation, when the artifact has to interact with the existing organizations and systems.
>
> 3) *Design process and methodology*: The challenge of building ownership for the design [10] can be also seen as a challenge for the process. Also creating consensus behind implementation decision and methods is an important challenge.

3.6 Observations on the Different Levels of Analysis

We recorded additional viewpoints from the perspective of the four levels of design, presented above, while considering the challenges. Much of the examined literature was about design methodology from varied fields, so the objects of design which fall under the examined domain are quite varied. The main challenge concerning the problem seems to be "finding a reasonable and satisfying problem definition or representation". When moving on to compiling the artifact, the challenges on the interface of the artifact and context start to emerge, where many of the challenge seem to concern the interplay between the artifact and its surroundings. Similarly, the challenges in validation can be positioned to the interface between the object and context and the same can be said for implementation. One practical challenge rarely discussed here can probably also extend to implementation, namely the need to engage the user, and to supply a user with enough value to ensure sustained use.

Arriving to contextual challenges, the problem definition gathers even more attention. Much of the challenge boils down to assessing the situation and developing a representation for the problem which is robust to change and fluctuation in the immediate context as well as the environment. Following the challenges in defining the problem, the artifact can also be problematic to construct. The challenge of finding representations re-emerges in implementation when there is a need to convey the intended properties of the artifact and follows to implementation. Also challenges in implementing the artifact to an existing system are raised at this level. What is scarcely discussed, but we would like to highlight, is the effect of possible goal shift during the design phase. In modern fast developing world, requirements and regulations can develop quite quickly, and these changes can have surprising effect on design requirements.

The actor level should be particularly interesting to collaborative design. Much of the initial challenges in this level seem to be about committing the stakeholders to design. When the process proceeds, the challenges are more associated with overcoming limitation of rationality and being thorough and unbiased in the design. The literature mentions different and conflicting interest of the stakeholders and shift of design goals, both challenges can be traced back to the actors who participate in the collaboration. In organizational settings, actors might be inclined to behave according to their own best interest instead of the group objective if it seems that the artifact brings changes that threat their position. Another thing which has been left to little consideration is the values attached and conveyed in the design and artifact [30], which have a connection to the actors.

Perhaps not surprisingly, considering the literature, the process and methodology challenges are well represented in the examined texts. Finding problem representation is also present at this level, as the problem definition and the associated representation are an important topic in the process as it later affects methodological choices. Moving on, the challenges in the artifact construction pertain to interpreting the problem right and interfacing the facets of the problem to the design choices in a rational way. The same theme continues throughout the process as testing and validation is ridden with challenges in measuring the success in a valid manner to gain reliable information about the artifacts performance.

4   Discussion: Challenges of Collaborative Design

An interesting picture starts to form after review of the design literature. First, the reviewed literature, especially on Design Science, does not discuss collaborative design explicitly, although the same themes found in collaboration literature are present in many places. We would think that this is due to the implicit assumption that the designer is involved in the process as a rational agent who fulfills the stakeholder needs given the constraints, resources and available information. Although, nowadays the trend seems to be more toward collaborative design in practice at least in large scale projects. This view could also be a result of the rear-view mirror perspective often associated with literature studies.

Aside from the first fundamental realization, the second one is that the challenges described in the literature seem to pile up on the first phases of the process. The analogy between technology management is remarkable, where it is commonly attributed that the front-end, that is the problem definition and stakeholder negotiations, is an important but fuzzy phase, where the quality and subsequent success of the new product concept is largely determined [35, 36]. If we translate the management-of-technology lingo to design terms, the products are not co-developed, and the problem owner interfaces in the design process only in the problem definition and testing before the production run. What follows from this front heavy outlook on design is similar handling of designs as products. Namely, when we write that challenges pile up on the first phases, we do not mean to imply that the rest of the phases would be necessarily any less challenging, but they are or seem to be less studied and less discussed. In fact, one might suggest that one of most challenging aspects would be to ensure the organization adopts the artifact, learns to perform quickly, and develops ownership. This can be built also in the design features and during the front-end phases, but arguably achieving adoption and sustained use is one of most demanding test for an artifact.

Taking the discussion back to challenges of collaboration, collaboration was defined in the introduction as joint effort toward a goal, which is quite different from traditional design where a problem owner hires a designer who will elucidate the problem and will design an artifact without interference from outside, until the artifact is ready for trials and validation. "True" collaboration by definition would mean a co-evolution of the artifact by the stakeholders with a shared understanding of the goals and means of achieving them. When comparing the challenges of collaboration to the challenges of design, it might in fact be that the challenges of collaborative design do not differ much from other collaborative tasks. On other way around, it is probably safe to assume that normal challenges in collaboration and organization of group effort concern collaborative design in addition to the challenges presented above. Starting from the problem definition, it is obvious that it can be challenging to extract the real stakeholder demands from a group of people [4, 10, 21], whether there are real conflicts of interest or just challenges in collaboration. In collaborative design, different participants may have radically different stakes and interests, as they often represent different stakeholders. This difference in perspectives may result in conflict, and trust can be broken. People show different behavior in groups, some are dominant, and others are shy and in the background. When all experts represent important stakes within the organization, it is important that the opinion and knowledge of each participant is taken into account. On the other side of the spectrum, groups with a shared objective can get a tunnel vision of a problem when people are not critical but confirmative, resulting in a phenomenon described by Janis [32] as 'Group think'. Groupthink may then result on sub-optimal results or poor reception of the design artifact, if people compromise too much.

Another key challenge in collaboration is information overload. When more expertise is added to the collaborative task, more perspectives are accommodated; and more information needs to be assimilated by the participants. Information overload can be seen as a real danger when considering large scale designs, which can influence their environment as much as the environment influences the design [52] or otherwise ambiguous or wicked problems [43]. While structure and summarizing can resolve such overload, groups do not always have sufficient shared understanding to agree easily on the use of a framework or structure for their collaborative effort. Further, shared understanding in general is a problem when expertise among participants is non-overlapping [55].

An additional set of challenges occurs when groups do not only have to perform a task jointly but also have to make decisions or choices. The collaborators or designers have to make a choice on design features [29], have to agree upon the set of evaluation and validation criteria for the artifact [4] and have to decide whether further design is needed [33] and whether the artifact passes validation and is ready to implementation. Decision making in groups can be challenging, as both rational arguments and preferences of stakeholders can be different given their different stakes, expertise and role in the collaborative task. Consensus building in groups is challenging as actual disagreements can be obscured by differences in understanding, knowledge and mental models around the different proposals that the group has to consider [12].

Finally, the design of computer-mediated interaction, i.e. supporting the collaboration among multiple actors with adequate tools, is a challenge on its own [46]. Apart from just addressing the interaction of one actor with a computer system, the developer has to take the interaction between all actors into account. The interaction is mediated by the tool which implies that it has to fit the requirements and preferences of all actors.

Even though the traditional perspective design in is mostly that of professional designer, consultant or researcher, not that of a problem co-owner, or a designer who works in collaboration to co-evolve the artifact through dialog between different stakeholders. While user involvement is gaining increasing attention in requirement negotiation and problem definition, and more iterative design approaches are proposed [2, 9, 27], few examples exist where all critical stakeholders are involved throughout the design process. Regarding this finding, one of the real challenges might well be recognizing collaborative design as a class of design activity of its own. Exceptions include the work of Markus et al. [39] and their dialectic development, the spiral approach of Boehm [7, 8], which can be regarded as an example of collaborative development, and the Oregon Software development process which includes prospective end-users in the design process and fosters communication based on design patterns [46].

Regarding existing tools to support group work and collaboration, the first implication is that the challenges are not radically different from challenges of collaboration which are well known and exhaustively researched. The main difference is perhaps all the more important task of creating a shared understanding on the issues. More challenging from groupware perspective is that the design project may need support over longer periods of time and in a distributed setting. Conceptually more challenging task is to accommodate design tools and methodologies to the support system, to ensure rigorous and manageable design.

To conclude the discussion, we would like to propose five main challenges which summarize the content analysis. The first challenge can be labeled as creating an understanding between the actors. The challenge is to create a common language and shared understanding between people with different backgrounds and their own vested interests. A common language and representations is needed to support the co-evolution of the artifact throughout the process as per definition of collaborative design. The second critical aspect is linked to the first, as after finding the common language, the actors need to forge a satisfying compromise between different needs expressed by stakeholders. The third challenge emerges when balancing between rigor and relevance [29]. The purpose of methodology (rigor) is to ensure that the design process is foreseeable and that the right results are achieved for the right reasons, not just by chance or coincidence. Subsequently, relevance is ultimately about satisfying stakeholders, which requires their involvement, and consequently collaboration in the design process. Together, rigor and relevance support achieving satisfying quality. The fourth challenge is orchestrating the design effort in a way which supports answering the other challenges. Organizing the design and interaction is about overcoming the challenges of group work and bounded rationality of the actors. The fifth and last challenge is ensuring ownership, what happens to the artifact in the target organization or environment after the designer has left, and handed over the instructions for use?

Summarizing the challenges:

1) Creating understanding: ensuring shared understanding and mental models of the problem, current state of the system, and envisioned solution.
2) Satisfying quality: balancing individual requirements and joint quality constraints while making design choices
3) Balancing rigor and relevance: using methods that support both stakeholder involvement and design discipline.
4) Organizing interaction: effective organization beyond project management, facilitating interaction between the actors, achieving rationality in the process and finding ways and means to work effectively
5) Ensuring ownership: implementation of the design to an organization, and transfer of ownership.

5 Conclusions

The paper set out to answer the question "what are the challenges in collaborative design?" The first step was to conduct a literature review, providing a starting point for the identification of challenges. After identification of the core literature and analysis, we identified challenges of collaboration and challenges of design. Based on the study we would like answer that the main challenge in collaborative design might be finding ways to actually realize collaborative design effectively. To be more specific, we found five classes of problems which concern mostly design in general, but particularly collaborative design:

1) Creating understanding
2) Satisfying quality
3) Balancing rigor and relevance
4) Organizing interaction
5) Ensuring ownership

Considering limitations and validity, there is one choice we made regarding the content, which should be at least recognized. We consciously tried to develop a wider perspective to design in practice and theory; we have not limited the analysis to any particular discipline, tradition, design theory or class of artifact. While variation might be relevant to specific disciplines, we feel that at this stage of our study the comparison across approaches will give more insight in collaborative challenges of design. Considering the nature of the literature, many of the projects are IT systems, which could be characterized as mid-sized project, and larger designs may have problems that are not recognized in this specific literature. A key limitation is that we have examined only second hand accounts on the engineering design methodology books included in the dataset. Whether this will affect the conclusion is hard to say, but judging by the review by Finger and Dixon [24, 25] and Joseph [33], the literature is centered around the same problems, particularly representation and theoretical challenges of design. As for other validity considerations, we base our claims of validity on the fact that we did a well structured review of literature, documented it and we also examined and abstracted our textual material in a structured way to extract our conclusions. This way we have grounded our conclusions on a solid body of literature.

The greatest implication for research is to recognize collaborative design as a separate phenomenon from stakeholder view to design. Collaborative design means co-evolution of a solution including the stakeholders in the process, not just negotiating requirements in the front-end phases. While challenges of collaboration do not seem to be unknown to designers, they have not been explicit. Collaboration literature suggests that when confronted with complex and knowledge intensive tasks, groups are often not able to detect and overcome such challenges without support or training [41, 48]. Even if groups are able to accomplish their goals by other means, they can often collaborate more efficiently and effectively using collaboration support [26, 48].

The implication for support tools are that creating shared understanding is a fundamentally important task in collaborative design. Beside that, the special challenges in supporting collaborative design are supporting manageable, organized and rigorous design process over time and different locations. Collaboration support can exist of tools, processes and services that support groups in their joint effort. In the interest of offering support to alleviate these challenges, we can start drafting requirements or, if not requirements, at least implications for support tool development. Especially in complex projects, the need to create understanding of the problem and ability to evaluate the artifact demand a lot from the representations, which can include text, modeling languages and visualizations. Nowadays, the weak link is perhaps the ability to use representations flexibly in different applications and transfer them between users. Regarding quality of the design, a shared understanding not only through representations, but also through social processes and facilitating interactions between the actors is important. In balancing rigor and relevance, the challenge of supporting the activity is how to translate the specification to an artifact so that all parties are satisfied. As always, time is of the essence in design, on multiple levels. The problem may evolve over time, and the design activity has to adapt to the situation in an organized manner, which needs active communication and feedback. The other time-related factor in design is organization of the activity over time, how to manage the process and the collaborators. Perhaps the last challenge of ensuring ownership is the hardest to support, although we argue that by facilitating the process as a whole instead of looking it as a series of separate tasks will help reduce the "throw it over the wall"-mentality and induce greater commitment from problem owners, stakeholders as well as designers. Although, collaboration support can be offered in different shapes, such as facilitation, training and tools or technology [19], these challenges and the ensuing implications suggest that there is need for more flexible tools that can support the design process longitudinally instead of chopping the process to individual activities. Existing research already offers powerful means in form of design patterns to overcome this challenge. Design patterns are already considered a "lingua franca" for design [23]. In form of a pattern language, design patterns capture best practices for a specific domain, as e.g. computer-mediated interaction [47] or group facilitation tools and techniques [55]. In practical terms, it seems that a great deal would be achieved by integrating the existing tools to work together to support both conceptual and social dimensions of group work.

As for practical implications, we expect that if we can find methods and best practices, e.g. in form of design patterns, to involve stakeholders without thwarting rigor, collaborative design should result in better artifact, more user satisfaction, more consensus about design choices and more sustainable results.

Acknowledgements

Kalle Piirainen gratefully acknowledges that this research has been conducted partly under a personal grant from the Academy of Finland.

References

1. Ackoff, R.L.: The Art of Problem Solving. John Wiley & Sons, Chichester (1978)
2. Acosta, C.E., Guerrero, L.A.: Supporting the Collaborative Collection of User's Requirements. In: The International Conference on Group Decision and Negotiation, Karlsruhe (2006)
3. van Aken, J.E.: Management Research as a Design Science: Articulating the Research Products of Mode 2 Knowledge Production in Management. British J. of Management 16, 19–36 (2005)
4. van Aken, J.E.: Design Science and Organization Development Interventions: Aligning Business and Humanistic Values. The J. of Applied Behavioral Science. 43(1), 67–88 (2007)
5. Benbasat, I., Zmud, R.W.: Empirical Research in Information Systems: The Practice of Relevance. MIS Quarterly 23(1), 3–16 (1999)
6. Bensabat, I., Zmud, R.W.: The Identity Crisis Within the IS Discipline: Defining and Communicating the Discipline's Core Properties. MIS Quarterly 27(2), 183–194 (2003)
7. Boehm, B.W.: A Spiral Model of Software Development and Enhancement. IEEE Computer 21(5), 61–72 (1988)
8. Boehm, B., Bose, P.: A collaborative spiral software process model based on Theory W. In: The Third International Conference on the Software Process, Reston VA, pp. 59–68 (1994)
9. Boehm, B., Gruenbacher, P., Briggs, R.O.: Developing Groupware for Requirements Negotiation: Lessons Learned. IEEE Software 18(3) (2001)
10. Boland Jr., R.J.: The Process and Product of System Design. Management Science 24(9), 887–898 (1978)
11. Brady, R.H.: Computers in Top-Level Decision Making. Harvard Business Review, 67–76 (1967)

12. Briggs, R.O., Kolfschoten, G.L., de Vreede, G.J.: Toward a Theoretical Model of Consensus Building. In: The Americas Conference on Information Systems, Omaha NE (2005)
13. Briggs, R.O., Kolfschoten, G.L., de Vreede, G.J., Dean, D.L.: Defining Key Concepts for Collaboration Engineering. In: The Americas Conference on Information Systems (2006)
14. Checkland, P.B.: Systems Thinking, Systems Practice. John Wiley & Sons, Chichester (1981)
15. Churchman, C.W.: The Design of Inquiring Systems: Basic Concepts of Systems and Organization. Basic Books Inc., New York (1971)
16. Couger, J.D.: Creative Problem Solving and Opportunity Finding. Boyd And Fraser, Danvers (1995)
17. Cross, N.: Science and Design Methodology: A Review. Res. in Engineering Design 5, 63–69 (1993)
18. Dennis, A.R., Garfield, M.J.: The Adoption and Use of GSS in Project Teams: Toward More Participative Process and Outcomes. MIS Quarterly 27(2), 289–323 (2003)
19. Dennis, A.R., Wixom, B.H., Vandenberg, R.J.: Understanding Fit and Appropriation Effects in Group Support Systems Via Meta-Analysis. MIS Quarterly 25(2), 167–183 (2001)
20. Dorst, K.: Viewpoint – Design research: a revolution-waiting-to-happen. Design Studies 29, 4–11 (2008)
21. Dorst, K., Cross, N.: Creativity in the design process: co-evolution of problem-solution. Design Studies 22(5), 425–437 (2001)
22. Drucker, P.F.: The Effective Executive, London (1967)
23. Erickson, T.: Lingua Francas for design: sacred places and pattern languages. In: The Conference on Designing Interactive Systems, pp. 357–368. ACM Press, New York (2000)
24. Finger, S., Dixon, J.R.: A Review of Research in Mechanical Engineering Design. Part I: Descriptive, Prescriptive, and Computer-Based Models of Design Processes. Res. in Engineering Design 1, 51–67 (1989)
25. Finger, S., Dixon, J.R.: A Review of Research in Mechanical Engineering Design. Part II: Representations, Analysis, and Design for the Life Cycle. Res. in Engineering Design 1, 121–137 (1989)
26. Fjermestad, J., Hiltz, S.R.: Group Support Systems: A Descriptive Evaluation of Case and Field Studies. J. of Management Information Systems 17(3), 115–159 (2001)
27. Guanaes Machado, R., Borges, M.R.S., Orlando Gomes, J.: Supporting the System Requirements Elicitation though Collaborative Observations. In: Briggs, R.O., Antunes, P., de Vreede, G.-J., Read, A.S. (eds.) CRIWG 2008. LNCS, vol. 5411. Springer, Heidelberg (2008)
28. Hengst, M., den Kar, E., van de Appelman, J.: Designing mobile information services: user requirements elicitation with GSS design and of a repeatable process. In: The Proc. of 37th Hawai'i International Conference on Systems Sciences (2004)
29. Hevner, A.R., Ram, S., March, S.T., Park, J.: Design Science in Information Systems Research. MIS Quarterly 28(1), 75–105 (2004)
30. van den Hoven, J.: Design for Values and Values for Design. Information Age. J. of the Australian Computer Soc. 7(2), 4–7 (2005)
31. ISI Web of Knowledge, http://apps.isiknowledge.com/
32. Janis, I.L.: Victims of Groupthink. Houghton Miffin, Atlanta (1972)
33. Joseph, S.: Design systems and paradigms. Design Studies 17, 227–239 (1996)
34. Kassarjian, H.H.: Content Analysis in Consumer research. The J. of Consumer Res. 4(1), 8–18 (1977)
35. Khurana, A., Rosenthal, S.R.: Integrating the Fuzzy Front End of New Product Development. Sloan Management Review 1997, 103–120 (Winter 1997)

36. Koen, P., Ajamian, G., Burkhart, R., Clamen, A., Davison, J., D'Amore, R., Elkins, C., Herald, K., Incorvia, M., Johnson, A., Karol, R., Seibert, R., Slavejkov, A., Wagner, K.: Providing Clarity and a Common Language to the "Fuzzy Front End". Research-Technology Management 44(2), 46–55 (2001)
37. Maher, M.L.: Process Models for Design Synthesis. AI Magazine 11(4), 49–58 (1990)
38. March, S.T., Smith, G.: Design and Natural Science Research on Information Technology. Decision Support Systems 15(4), 251–266 (1995)
39. Markus, M.L., Majchrzak, A., Gasser, L.: A Design Theory for Systems That Support Emergent Knowledge Processes. MIS Quarterly 26(3), 179–212 (2002)
40. Mitroff, I.I., Betz, F., Pondly, L.R., Sagasty, F.: On Managing Science In The Systems Age: Two Schemas For The Study Of Science As A Whole Systems Phenomenon. TIMS Interfaces 4(3), 46–58 (1974)
41. Nunamaker Jr., J.F., Briggs, R.O., Mittleman, D.D., Vogel, D., Balthazard, P.A.: Lessons from a Dozen Years of Group Support Systems Research: A Discussion of Lab and Field Findings. J. of Management Information Systems 13(3), 163–207 (1997)
42. Purcell, A.T., Gero, J.S.: Drawings and the design process: A review of protocol studies in design and other disciplines and related research in cognitive psychology. Design Studies 19(4), 289–430 (1998)
43. Rittel, H.W.J., Webber, M.M.: Dilemmas in a General Theory of Planning. Policy Science 4, 155–169 (1973)
44. Schildt, H.A., Mattson, J.T.: A dense network sub-grouping algorithm for co-citation analysis and its implementation in the software tool Sitkis. Scientometrics 67(1), 143–163 (2006)
45. Schildt, H.A.: SITKIS: Software for Bibliometric Data Management and Analysis v2.0. Helsinki University of Technology, Institute of Strategy and International Business (2002), http://www.hut.fi/~hschildt/sitkis
46. Schümmer, T., Lukosch, S., Slagter, R.: Using Patterns to empower End-users - The Oregon Software Development Process for Groupware. International J. of Cooperative Information Systems. Special Issue on 11th International Workshop on Groupware (CRIWG 2005) 15, 259–288 (2005)
47. Schümmer, T., Lukosch, S.: Patterns for Computer-Mediated Interaction. John Wiley & Sons, Chichester (2007)
48. Schwarz, R.M.: The Skilled Facilitator. Jossey-Bass Publishers, San Francisco (1994)
49. Schön, D.A.: The reflective practitioner. Temple-Smith, London (1983)
50. Simon, H.A.: The New Science of Management Decision. Prentice Hall, New York (1960)
51. Simon, H.A.: The Structure of Ill Structured Problems. Artificial Intelligence 4, 181–201 (1973)
52. Simon, H.A.: The Sciences of the Artificial, 3rd edn. MIT Press, Cambridge (1996)
53. Starkey, K., Madan, P.: Bridging the Relevance Gap: Aligning Stakeholders in the Future of Management Research. British J. of Management 12, 3–26 (2001)
54. Takeda, H., Veerkamp, P., Tomiyama, T., Yoshikawa, H.: Modeling design processes. AI Magazine 11(4), 37–48 (1990)
55. de Vreede, G.J., Davison, R., Briggs, R.O.: How a Silver Bullet Lose its Shine - Learning from Failures with Group Support Systems. Communications of the ACM 46(8), 96–101 (2003)
56. Walls, J.G., Widmeyer, G.R., El Sawy, O.A.: Building an Information Systems Design Theory for Vigilant EIS. Information Systems Research 3(1), 36–59 (1992)
57. Webber, R.P.: Basic Content Analysis, Quantitative Applications Social Sciences Series. Sage, Thousand Oaks (1990)

The Application of Design Patterns
for the Adaptation of a Modeling Tool
in Collaborative Engineering

Michael Klebl[1], Monika Hackel[1], and Stephan Lukosch[2]

[1] FernUniversität in Hagen
Department for Cultural and Social Sciences
58084 Hagen, Germany
{michael.klebl,monika.hackel}@fernuni-hagen.de
[2] Delft University of Technology
Faculty of Technology Policy and Management, Systems Engineering Department
2600 GA Delft, The Netherlands
s.g.lukosch@tudelft.nl

Abstract. In order to improve design processes in mechatronical engineering, concurrent processes and interdisciplinary phases replace the hitherto existing sequential processes. This change places high demands on the interdisciplinary collaboration in teams of engineers from different disciplines. Collaborative functions in engineering tools support this collaboration. The aim of this contribution is to demonstrate the application of design patterns for the design of adapted collaborative features within an adequate engineering tool. This adaptation addresses the requirements and needs of process organization in different engineering departments.

Keywords: CSCW, Design Patterns, Collaborative Engineering, Socio-technical systems, Social Scientific Research, Mechatronics.

1 Introduction

Design patterns are the documentation of a proven technical solution regardless of a specific implementation – formalized as well as generally understandable. They are directed at supporting the development of solutions in specific cases in relation to specified classes of problems and associated contextual conditions. After the idea of design patterns was transferred from architecture [1] to software engineering, the use of design patterns for the development of information systems was limited to software development and software architecture, thus being purely focused on technology [5]. More recently, the notion of design patterns is also applied to other areas of designing socio-technical systems. Design patterns addressing the organization of communication and collaboration expand the scope far into the field of social interaction. Schümmer & Lukosch [9] offer a catalogue of design patterns, placing themselves between a purely technical orientation and a purely social basis, and thus aim at systematically creating a connection here. However: in the design of socio-technical

L. Carriço, N. Baloian, and B. Fonseca (Eds.): CRIWG 2009, LNCS 5784, pp. 262–269, 2009.

systems one question regularly arises – whether the technical design patterns are capable of resolving difficulties in social interaction.

In this contribution we address this question in a specific case study from a joint research project including partners from the industry field. In the second part of this contribution we outline the project AQUIMO that frames a methodology for interdisciplinary mechatronical engineering. The implementation of this process model is accompanied by the development of an adaptable modeling tool for mechatronic engineering and an associated qualification program. A specific intention within the project AQUIMO is the demand for the adaptability of the process model, the modeling tool and the qualification program. Using exemplary difficulties in interdisciplinary collaboration, we discuss the use of design patterns for the design of adapted collaborative features within the modeling tool in the third part. In the fourth part of this contribution we review whether, how and to what extent design patterns can be applied to the design and adaptation of collaborative systems and hence support solution search, solution analysis and solution decision.

2 Aims and Foundations

Despite the intention of mechatronic engineering to combine mechanical, electronic and information systems engineering in an interdisciplinary, i.e. mechatronical, design process [3], many companies still develop their products step by step. In order to foster the implementation of an interdisciplinary design process, the project AQUIMO devises an adaptable process model, which is supported by the development of an adaptable modeling tool for mechatronic engineering and an associated qualification program. Three medium-sized companies, leading providers for their specific market in mechanical and plant engineering, a software company and two university-level institutions are participating in the project. The combination of process model, engineering tool and qualification program will enable to replace subsequent processes in engineering, which are concentrating on discipline specific aspects in engineering, through interdisciplinary or parallel processes. This interdisciplinary collaboration, taking place in teams of engineers from different disciplines, allows the reduction of failure rates and saves expenses based on an early design matching.

Accompanying social scientific research investigates changes in work organization that result from the implementation of the process model and its adaptation to the specific requirements in the participating companies. Here, a formative approach towards analysis and evaluation was selected. This approach is based on the approach of "developmental work research" as set forth by Yrjö Engeström [4]. Hence, this approach refers to activity theory and connects to the tradition of the action research. This meets the requirements on social research in the project AQUIMO for three reasons: At first, engineers are considered as experts for their field of action. As participating practitioners they contribute directly to the research process. At second, developmental work research attains a multi-perspective view of the iterative development cycles within the project. At third, modeling and understanding activity systems using the typical triangular structure of developmental work research allows the integration of an individual perspective to a social perspective by referring to tools, signs and, artifacts as well as to rules and division of labor.

The purpose of the investigation in the early stage of the project is to gain a differentiated picture of the difficulties in the interdisciplinary collaboration. For this purpose a methodic procedure has been devised consisting of individual guided interviews, the structuring of statements by means of concept maps and a group discussion [7]. The results of these surveys in the participating companies serve as the source for the deepened requirements analysis for both the engineering tool and the qualification program in the following iterative development cycles. In the next part, we will select some significant results from these surveys that are related to the support of collaboration by means of computer mediated interaction.

3   Design Patterns Applied: Two Examples

In the following, we present two examples which highlight the challenges in interdisciplinary collaboration. These examples report on our survey in engineering teams at the companies involved in the AQUIMO project.

3.1   Workspace Awareness

Group awareness is an essential concept for coordinated collaboration which is based on shared artifacts. Gutwin & Greenberg [6] have shaped the term "workspace awareness" to sum up the need for awareness that arises in small teams working in a shared workspace. For the companies in our study, this is the central challenge for interdisciplinary collaboration. Especially the questions on what someone else is working as well as what has been changed recently occurred to be the most important ones. As a reason for that, we consider the fact that all three companies are not geographically distributed and that there are only short ways in between the engineers' workplaces. Due to this, in all three companies a lot of informal teamwork occurs, resulting in "local knowledge" within the work processes [8]. Therefore, informal teamwork is very much appreciated in all involved companies. As a result, many appointments are made spontaneously and independently from the overall work organization of the companies. This also means that a lot of information about revisions on the shared artifacts is exchanged informally. There is no structured record of changes and changes cannot be tracked systematically. This explains that other challenges of group awareness like presence awareness, group structural awareness or emotional awareness are not of central interest. However, the challenges focusing on workspace awareness vary differently in the three companies:

(A) One of the companies directs their demand to be informed about changes in design documents to the intended workflow, i.e. a notification about changes should happen accordingly to the intended workflow and thereby trigger responsibilities and dependencies.
(B) A second company already uses automatic change notifications. However, this functionality is used so broadly that engineers either ignore the notifications or even turn them off.
(C) For the third company change notifications are not an urgent problem that needs to be addressed by a technical solution. Important changes are solely communicated informally in accordance with everyday work requirements due to the spatial and personal proximity of the employees.

When considering these cases, we already identified three design patterns from the cluster on how to maintain asynchronous group awareness and one additional pattern from the cluster on how to protect users.

(1) ACTIVITY LOG [9]: Provide access to a log that remembers all activities that users perform on shared artifacts – modifications as well as read accesses.
(2) PERIODIC REPORT [9]: Inform users periodically about the changes that took place between the time of the current report and the previous one.
(3) CHANGE INDICATOR [9]: Indicate whenever an artifact has been changed by an actor other than the local user.
(4) ATTENTION SCREEN [9]: Allow users to filter the information that reaches them. Use meta-information such as sender details, or content information such as important keywords, to distinguish between important and unimportant information.

When considering the application of these patterns with respect to the affordances of each company, the following adaptations have to be performed:

(A) When the knowledge about changes in shared documents is understood as a way to organize work, an ATTENTION SCREEN (4) and a PERIODIC REPORT (2) can effectively be used to organize work with respect to roles or persons. The compilation of information from time to time as well as the selection of information according to the necessities of the recipients will reduce the information load and thus lead to efficient workspace awareness. In order to address the requirement to prompt responsibilities and dependencies within the intended workflow, in this company the configuration of an ATTENTION SCREEN as well as of a PERIODIC REPORT should be restricted to the manager of a group, a department, or a project.
(B) For handling the large amount of information, a PERIODIC REPORT (2) is the suitable tool to address the need for workspace awareness in this company as well. Therefore, the basic functionality of change notifications is equivalent to company (A). However, since change notifications have already been introduced and rejected by the engineers, they have to be offered as an optional feature, not as an obligation, in order to avoid definite refusal of acceptance. The engineers themselves should be able to adapt the configuration. In addition a CHANGE INDICATOR (3) is also a suitable method to notify the engineers that are reluctant to notifications by e-mail. Both implementations appear to be adequate to the notion of giving more control to the team of engineers than to the management.
(C) In the current situation, engineers in this company communicate changes by means of direct communication. Here, an ACTIVITY LOG (1) can improve the traceability of changes. Report functionalities allow for a comprehensive review on changes whenever users have to recall the activities in the shared workspace. In addition, a CHANGE INDICATOR (3) as implemented for the second company (B) helps to avoid mistakes in case the direct communication of changes has been forgotten.

While seeking for a solution, the chosen design patterns immediately help to design useful collaborative functionality. However, it has to be noted that in case (A) and (B) different actors have the right to adapt the notification functionality and thereby different actors can steer the flow of information as well as the work process.

3.2 Division of Work and Perspective Taking

Mechatronical engineering results in a division of work by type: Engineering activities are split up in tasks that are performed by different engineers according to their knowledge background (e.g. mechanics, electrical engineering, or software development). This division of work by type results in further specialization of the collaborating engineers. Therefore, engineers in different disciplines have discipline-specific knowledge and attitudes. However, a knowledge-based collaboration is only feasible, if team members are capable to make assumptions on the knowledge and the motivation of their team members and thus base their own work on this background. Hence, perspective taking, i.e. the capability to adopt other persons' point of views, is an essential foundation for human communication and interaction [2]. In the companies within the AQUIMO project, specific problems arise from missing or insufficient capabilities in perspective taking:

(A) In one of the companies the engineers in each discipline have only a vague idea of the importance of their work for the other disciplines to continue with their tasks at hand. Due to this, necessary information is often communicated too late.
(B) Due to tight mode of collaboration and the importance of informal teamwork, all employees in the second company are aware of the information demands of the different disciplines in the everyday work processes. However, we still identified a need to exchange domain-specific knowledge between the disciplines.
(C) In order to improve the possibilities of perspective taking, the third company completely changed their way of working by establishing an interdisciplinary workgroup. Thereby, the company achieved first positive results. Nevertheless, the question remained how knowledge that is specific to one discipline can be obtained and utilized for several engineers of one discipline across different interdisciplinary workgroups.

We identified the following patterns for computer-mediated interaction to address the issues in relation to interdisciplinary knowledge transfer as well as the communication of knowledge that is specific to one discipline:

(1) SHARED ANNOTATION [9]: Provide means for entering comments on specific messages or parts of a document. Display all users' comments with the content.
(2) FLAG [9]: Provide users with a way to flag important artifacts. the flags so that flagged artifacts can be easily found later.
(3) FEEDBACK LOOP [9]: Provide an easy way for readers to contact the author. Create a user interface element close to the content questions and feedback.
(4) FAQ [9]: Maintain a list of frequently asked questions that have been answered already. Treat this list as the knowledge repository of the group.
(5) SEED FOR THOUGHT [10]: Provide a set of facts that trigger new thoughts. Place these facts in the communication space.

When considering the different development departments within the evaluated companies, we suggest based on the above patterns the following adaptations to improve the issues in relation to perspective taking:

(A)   A FEEDBACK LOOP (3) for the information that is already exchanged between the different disciplines can increase the awareness for information as well as interaction quality. However, since the engineers often communicate directly and thus messages are not recorded, also feedback would be provided not in written communication. Hence, a FEEDBACK LOOP is a social issue rather than a technical solution. In addition to the FEEDBACK LOOP, we therefore recommend the use of SHARED ANNOTATIONS (1) to annotate the shared design documents in the workspace provided by the collaborative engineering tool. Functionalities to browse, search, and view these annotations help to build basic understanding between the different disciplines as well as to foster a feedback culture.

(B) In this company, SEED FOR THOUGHT (5) can be used to trigger the interdisciplinary exchange of information in the case of problems that require innovative solutions. In order to facilitate the dissemination of information that prompts a discussion, a common communication space is needed where engineers can publish information on innovative solutions. The functionalities of a weblog would be a helpful for this communication space. In addition, a FEEDBACK LOOP (3) is a suitable tool to offer more possibilities to adopt and understand the viewpoint of other engineers. Here, the FEEDBACK LOOP connects to written communication in the common communication space. Thus, technical features are apt to foster the feedback culture.

(C) The central challenge within this company is to capture knowledge specific to one discipline. SHARED ANNOTATIONS (1), FLAGS (2), as well as FAQs (4) can be used to capture this knowledge. As in the first company (A), SHARED ANNOTATIONS are used to annotate the shared design documents as well as discipline-specific documents with information on recent developments, standards or specifications. However, editing, browsing, and searching these annotations require the distinction between the different disciplines. This way, engineers direct their annotations to a target group with a similar background of knowledge, beyond the own engineering team. FLAGS that differentiate between the target groups foster orientation, as well as FAQs on topic specific for each discipline help to elaborate the shared knowledge base. In this company, a prerequisite for these solutions is to identify the different target groups and allow these groups to document and classify their knowledge.

The above patterns help to improve the identified problems in relation with the adoption of views. However, the patterns itself appear unspecific. Neither SEED FOR THOUGHT nor FEEDBACK LOOP requires a specific tool support. The successful application of the patterns is thus not based on technical requirements. Instead it is important to differentiate between interdisciplinary knowledge and the knowledge within one domain. For this, the different target groups have to be identified and supported with a suitable role model. Additionally, the personal commitment to provide information and feedback has to be improved. Both prerequisites can only be achieved by changing the individual attitudes of the engineers.

4 Conclusions and Future Work

In the previous section, we showed how the interdisciplinary collaboration within a mechatronical engineering process can be changed by using patterns for computer-mediated interaction. We used the patterns to search, analyze and identify solutions for problems which we identified together with engineers that are experts in their respective domain. Our results allow us to assess whether and how design patterns for computer-mediated interaction can be used in practice:

- In general, design patterns for computer-mediated interaction can be used for designing the collaborative functionalities of a software tool. This is especially the case when specific functions need to be tailored to the needs of a collaborating team [11]. We have shown this for workspace awareness as well as for perspective taking. For these issues a number of well-established design patterns as well as familiar and accepted tool solutions exist.
- In addition to the identified problem classes and the corresponding context, the application of design patterns is dependent on the intended goals. The definition of rights for managing automatic notification in our examples makes this especially explicit. When the notification mechanism is used for work organization, the adaption takes place on the management level and not on the individual level. For the application of design patterns it is thus important to consider the interests and personal goals of the involved stakeholders when using technical systems. A problem alone does not specify the goals for working on the problem and a design pattern should not mislead to anticipate the goal in the suggested solution.
- The application range of design patterns seems to be limited when social as well as psychological factors heavily influence the effect of an adaption. In such a case, it is important to find a balance between technical and non-technical solutions. For example if collaboration is based on direct communication, a technical solution can be considered as obsolete, impersonal, and disturbing. For the collaboration domain in question within this contribution, changing the way of work by organizational development, offering personal development opportunities by means of the related qualification program or simply clarifying processes as well as roles can possibly also lead to improvement.

With this article we contribute to the systematic contemplation of how to apply design patterns according to problems as well as requirements of real world application scenarios. In future, we intend to implement and critically evaluate the implementation and the use of the modeling tool that is adapted according to the work organization in the companies engaged in the project. As a result of the problem analysis presented here, we state that design patterns can be applied to the design and adaptation of collaborative systems and hence support solution search, solution analysis and solution decision. However, our initial question on whether and to what extend the technical design patterns are capable of resolving difficulties in social interaction is always present in this design and implementation process. It is decisive to permanently balance between technical and non-technical solutions.

Acknowledgments

Financial support for this research and development project is provided by the German Federal Ministry for Education and Research (Bundesministerium für Bildung und Forschung, BMBF) as part of the framework program "Forschung für die Produktion von morgen" ("Research for tomorrow's production"). Its administration is handled by the Project Management Agency for Production and Manufacturing Technologies at the Karlsruhe Research Centre (Projektträger Forschungszentrum Karlsruhe Produktion und Fertigungstechnologien, PTKA-PFT).

The authors also wish to thank the three mechanical and plant engineering companies where the research activities are taking place, and especially the engineers involved for their contribution to the study that has been presented here.

References

1. Alexander, C., Ishikawa, S., Silverstein, M.: A Pattern Language. In: Towns, Buildings, Construction. Oxford University Press, New York (1977)
2. Boland Jr., R.J., Tenkasi, R.V.: Perspective Making and Perspective Taking in Commuities of Knowing. In: DeSanctis, G., Fulk, J. (eds.) Shaping Organization Form: Communication, Connection, and Community, pp. 327–368. SAGE, Thousand Oaks (1999)
3. Bradley, D.A., Loader, A.J., Burd, N.C., Dawson, D.: Mechatronics: Electronics in Products and Processes. Chapman and Hall, London (1991)
4. Engeström, Y.: Learning by Expanding: An Activity-theoretical Approach to Developmental Research. Orienta-Konsultit Oy, Helsinki (1987)
5. Gamma, E., Helm, R., Johnson, R., Vlissides, J.: Design Patterns. In: Elements of Reusable Object-Oriented Software. Addison-Wesley, Reading (1995)
6. Gutwin, C., Greenberg, S.: A Descriptive Framework of Workspace Awareness for Real-Time Groupware. Computer Supported Cooperative Work (CSCW) 11(3), 411–446 (2002)
7. Hackel, M., Klebl, M.: Qualitative Methodentriangulation bei der arbeitswissenschaftlichen Exploration von Tätigkeitssystemen [Triangulation of Qualitative Methods for the Exploration of Activity Systems in Ergonomics]. Forum Qualitative Sozialforschung / Forum: Qualitative Social Research 9(3) (2008)
8. Randall, D., Rouncefield, M., Hughes, J.A.: Chalk and cheese: BPR and Ethnomethodologically Informed Ethnography in CSCW. In: Marmolin, H., Sundblad, Y., Schmidt, K. (eds.) ECSCW 1995: Proceedings of the Fourth European Conference on Computer-Supported Cooperative Work, pp. 325–340. Kluwer Academic Publishers, Stockholm (1995)
9. Schümmer, T., Lukosch, S.: Patterns for Computer-mediated Interaction. Wiley, Chichester (2007)
10. Schümmer, T., Lukosch, S.: READ.ME – Talking about computer-mediated communication. In: Zdun, U., Hvatum, L. (eds.) EuroPLoP 2006. Proceedings of the 11th European Conference on Pattern Languages of Programs, 2006, pp. 317–342. UVK Universitätsverlag Konstanz GmbH, Konstanz (2007)
11. Schümmer, T., Lukosch, S., Slagter, R.: Using Patterns to Empower End-Users – The Oregon Software Development Process for Groupware. International Journal of Cooperative Information Systems (IJCIS) 15(2), 259–288 (2005)

Communication Patterns to Support Mobile Collaboration

Andrés Neyem[1], Sergio F. Ochoa[2], and José A. Pino[2]

[1] Department of Computer Science, Pontificia Universidad Católica de Chile
aneyem@ing.puc.cl
[2] Department of Computer Science, Universidad de Chile
{sochoa,jpino}@dcc.uchile.cl

Abstract. The mobility of the collaborators, the diverse technologies available to support them and the continuous change in the collaboration scenarios bring new challenges to design, implement, and reuse communication software for these complex systems. This article presents a design patterns system to help modeling the communication services required to support mobile collaboration. These patterns serve as educational media for developers, students or researchers on how to design communication services for mobile collaborative applications. The patterns also foster the reuse of proven solutions.

Keywords: Communication patterns, mobile collaboration.

1 Introduction

Advances in mobile computing and wireless technologies are opening new possibilities to use them to support various collaborative activities. However, efforts to understand the implications that mobile work and mobile collaboration have on groupware design are still a research subject [5]. Mobile groups are highly varied in the ways they organize work, in the physical dispersion of mobile workers, and in the styles of collaboration [1]. Several researchers have described and classified these mobility variants [6] and levels of coupling among mobile collaborators [10].

These research contributions show there is a wide variety of work scenarios where mobile collaboration needs to be supported. The design and implementation of the groupware solutions for those scenarios imply several requirements and challenges derived from the type of work to be supported and the features of the work and activity contexts [5]. Neyem et al. [8] presented a list of requirements to design the coordination services; however, that proposal does not include the communication support. Modeling the communication support is always an important part of the groupware application design, because the coordination and collaboration mechanisms will depend on it.

In order to help design this application concern, this paper proposes a design patterns system. Next section describes those patterns. Section 3 discusses related work. Section 4 presents the conclusions and future work.

L. Carriço, N. Baloian, and B. Fonseca (Eds.): CRIWG 2009, LNCS 5784, pp. 270–277, 2009.
© Springer-Verlag Berlin Heidelberg 2009

2 Communication Patterns System

A patterns system is a network of interconnected patterns which may help to define sequences of structure-preserving transformations. This section presents a patterns system to deal with communication services for mobile groupware. The particular patterns definition and the relations among them can ease the designers' work during the design process [12].

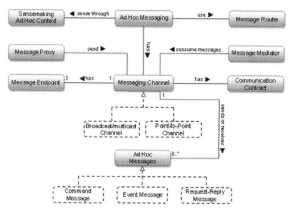

The communication patterns system shown in Fig. 1 focuses on the provision of services for message exchange among mobile workers' applications. Based on that, a mobile collaborative application can send messages (notifications, commands, or events) to other users in the wireless network. Next sections briefly describe these patterns following the nomenclature proposed in [12]. An extended version of the system can be found in [9].

Fig. 1. Communication Patterns System for Mobile Groupware Applications

2.1 Ad Hoc Messaging

Context. A mobile groupware environment consists of several applications connected via a wireless network. These applications need to interact with each other, exchange data or access each other's services in order to coordinate the work performed by a group of mobile users (collaborators).

Problem. Mobile nodes of the wireless network have frequent disconnections and variable transmission rates. Therefore, a communication infrastructure enabling applications to communicate and coordinate on-demand (loosely-coupled) is needed.

Solution. The solution is based on an *ad hoc messaging* system. A messaging system move messages from one mobile node to another one, since nodes and the wireless networks connecting them are inherently unreliable. A messaging system overcomes these limitations by repeatedly trying to transmit the message until it succeeds. Under ideal circumstances, the message is successfully transmitted on the first try, but circumstances are often not ideal. On the other hand, it is important to consider the messaging system interconnects several existing mobile groupware systems, and thus it must be able to communicate, share data and operate in a unified manner in response to a set of common requests.

2.2 Messaging Channel

Context. A mobile groupware environment keeps separate the groups of mobile applications needing to communicate among them by using an ad hoc messaging system. Thus, the ad hoc messaging system keeps a set of connections enabling applications to communicate by transmitting information in predictable ways.

Problem. The ad hoc messaging system might proactively deliver the message, or it might hold the message until the recipient is able to retrieve it. Therefore, this system must provide a means to connect a set of clients and service providers, which communicate by sending and receiving messages.

Solution. Structuring the solution based on *messaging channels*. Messaging Channels are logical pathways to transport messages through message-based communication. Each channel endpoint has a sender/receiver application. A sender application is a program sending a message by writing the message to a channel, and a receiver retrieves a message from a channel. Two typical messaging channels need to be supported: a) *Point-to-Point Channel*, ensures only one consumer consumes any sent message; and b) *Broadcast/multicast Channel*, describes a single input channel splitting into multiple output channels - one for each receiver.

On the other hand, when an application is writing a custom transport, it must decide which message exchange patterns are required for each message delivery. The application can use three message exchange patterns when it sends a message: a) *One-way:* a client sends a message using a fire and forget exchange; b) *Request-Response*: a message is sent, and a reply is received; and c) *Dual*: the sender and receiver send messages to each other, using one-way or request-response messaging.

2.3 Ad Hoc Messages

Context. Mobile applications use a messaging channel to communicate among them. A message channel is the pipe transporting messages between a sender and a receiver.

Problem. Mobile groupware applications need to communicate with each other in a unified and on-demand manner. Therefore, some way enabling transmission of data units from one application to another one in a loosely coupled fashioned is needed.

Solution. Any data transmitted via an ad hoc messaging system, must be converted into one or more messages to be sent through messaging channels. The message itself is simply a data structure (such as a record or an object) consisting of two basic parts: a) *Header*: information describing the data being transmitted and to be used by the ad hoc messaging system; and b) *Body*: the data being transmitted.

The sender application can send three types of messages: a) *Command Message*: to control another application or a series of other applications; b) *Event Message:* for reliable, asynchronous event notification between applications; and c) *Request-Response Message*: it is used when a client needs a reply from a provider.

2.4 Message Router

Context. Ad hoc mobile networks are not supported by infrastructure (i.e. access points). The nodes in the network use wireless communication and a dynamic topology for data dissemination and gathering. These features let ad hoc networks be useful to support on-demand collaboration between mobile workers.

Problem. The current wireless communication norms supporting mobility have a limited communication threshold. For example, the Wi-Fi (IEEE 802.11x) threshold is about 200 meters in open areas and 20 meters in built areas. Most groupware solutions need to extend the communication threshold as much as possible to increase the interaction scope; hence, routing mechanisms are required to allow it.

Solution. The solution involves the provision of a distributed message router (it is present in each mobile unit) which consumes messages from various message channels, it splits them in packets and then it routes them to the receiver. On the other side, the recipient router rebuilds the message based on the received packets. The packet delivery service should offer an intermediate solution between the routing and flooding techniques in order to achieve high reliability with moderate performance degradation. The routing algorithm can follow a gossip-based approach, where each node forwards a message with some probability, reducing the overhead of the routing protocols. This algorithm exhibits bimodal behavior in sufficiently large networks: in some executions, the gossip dies out quickly and hardly any node gets the message; in the remaining executions, a substantial percentage of the nodes get the message.

2.5 Message Mediator

Context. Any mobile collaborative application consists of a number of services that provide functionalities supporting the collaboration process (e.g. data synchronization, peers discovery or context sensing). These services are divided among several computing devices exposing their logic and computation.

Problem. The services used as support of the collaboration process are designed to carry out a particular task. However, if additional functionality is included in these services, they become complex, difficult to coordinate, maintain and evolve.

Solution. The solution proposes a separation of concerns of the services supporting the collaboration process. The idea is to use a message mediator as an intermediary between the messaging channel and the service providers. The mediator is a component consuming messages from a channel and coordinating and distributing them to the corresponding service providers, and vice versa. The Service Provider is a component receiving service requests from the mediator, processing them, and returning the corresponding results. Therefore, when a message mediator receives a message, it identifies the corresponding service provider and sends the message to that component. The service provider could be dynamically created by the mediator or could be selected from a pool of available components. Each provider can run in its own thread in order to process messages concurrently.

2.6 Message Proxy

Context. Mobile collaborative applications need to interact with each other, exchanging data or accessing remote services in order to allow collaboration among mobile workers. When developing the client-side applications, clients need to access remote services using remote invocation methods.

Problem. Accessing the services of a remote component requires using a specific data format and networking protocol. Hard-coding the format and protocol directly into the client application makes it difficult to maintain, evolve and reuse.

Solution. The solution proposes to use a message proxy in the client's address space, which is in charge of managing the invocations to remote services. The message proxy provides a service interface allowing groupware applications to treat the remote services as local ones. The proxy maps the client invocations to specific message formats and protocols, which are used to send these invocations across the network. Thus, the proxy transforms the concrete service invocation and its parameters, into a particular message which is understandable by the network. The proxy uses the regular mechanism to send the message to the remote service provider. Then, the message proxy transforms the results returned by the remote provider, in order to deliver a piece of data that is understandable for the client. The proxy is a no-blocking component; therefore, this resource will not easily become a bottleneck.

2.7 Message Endpoint

Context. The mobile groupware application and the messaging system are two separate software pieces. The application provides functionality for the users, whereas the messaging system manages the messaging policies implementing communication processes. Whenever delivering a service request, the messaging channel, the message proxy and mediator must identify the next node in the communication chain.

Problem. Univocally identifying the service provider and consumer in a mobile ad hoc network is not enough to support messages delivery. Identification of a number interim components participating in the communication chain (e.g., the IP address of the mobile devices, and the user/service/mobile application ID) is also needed.

Solution. The interacting services must connect to the messaging channel using a specialized component named message endpoints. These components represent post addresses where messages can be delivered. The information of both endpoints (sender and receiver) is part of the message. Depending on the communication stage, the component processing the message will be able to extract the information related to the endpoint to determine the next one in the communication chain. When a client has data to communicate, it passes this data to its associated message endpoint, which first converts the data into a message understandable by the messaging system, and it adds the information related to the corresponding endpoints. Then, it submits the message to the messaging system, which analyzes the endpoint information stored in the message in order to identify the destination node. Then, the message mediator of the destination node analyzes the endpoint information to identify the corresponding service provider. Something similar occurs when the results are sent back to the service requester.

2.8 Communication Contract

Context. Mobile groupware applications should be able to communicate with each other, even if they run in heterogeneous remote computing devices. Therefore, the messaging between those components must be compatible, or at least understandable by their respective runtime environments.

Problem. It is inconvenient to pre-establish the rules governing the interactions among mobile nodes in mobile ad hoc collaboration. This is so because the interaction scenario is highly dynamic, and the users are not sure about the features of the next interaction process. Interoperability, flexibility and scalability of the solution depend on the way this issue is addressed.

Solution. The solution is to use contracts to regulate the communication between a service provider and a consumer. Contracts are agreements made by two parts and related to a particular issue. These agreements allow the applications to have a common understanding on the communication protocols, operations, and message structures exchanged between two nodes. A contract is typically defined on the fly and it is the result of a negotiation process. Thus, applications become flexible and interoperable when using contracts. Four contract types have been identified: a) *Protocol Contract* specifies transport protocols, message encodings, security options, and transactional capabilities; b) *Service Contract* describes the operations a mobile application can perform when interacting with other applications; c) *Data Contract* describes the structure of the data that will be exchanged between the applications; and d) *Message Contract* describes the structure of the messages to be exchanged.

2.9   Sensemaking Ad Hoc Context

Context. By context we mean the variables which can influence the behavior of mobile applications; it includes internal resources (e.g. memory, CPU speed or screen size) and external resources (e.g. bandwidth, and signal stability) of mobile devices. Both types of variables are relevant to provide support for coordination processes.

Problem. Contextual information is changing all the time while doing mobile collaborative work. Mobile collaborative applications have to sense it, store it and appropriately use it. Since this information is used by the groupware system to dynamically adapt its behavior, such information has to be available all the time and it has to be as complete as possible.

Solution. The solution to this problem is the creation of an ad hoc context manager. This component has to be fully distributed and it must store, update and monitor status of the context variables. Such information should be used by the coordination layer and then stored in the public (shared) space, because it could be shared with other mobile nodes, and the updates of that information should be incremental in order to avoid losing other contextual information.

Mobile groupware applications will adapt their functionality based on that information to cope with the changes in the work scenario (e.g., a mobile worker gets isolated or networking support is not available anymore). It must be noted the context manager has to be carefully engineered in order to reduce the use of limited resources, such as battery, CPU, memory or network bandwidth. A service-oriented approach can be useful to design and implement this component, because it deals with the heterogeneity of computing devices and resources shortage.

2.10   Patterns System Summary

A study presented by Neyem et al. [8] shows a list of requirements to consider during the design of mobile collaborative applications: *autonomy, interoperability, use of hardware resources, low coordination cost, awareness of users' reachability and deployment ease.* Thus, it is possible to draw a correspondence matrix considering the proposed patterns and these requirements (Fig. 2). The matrix allows developers to select one or more design patterns to deal with a particular requirement.

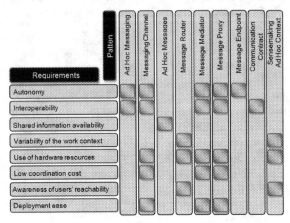

The proposed patterns system has been implemented on a middleware platform and a variety of applications are currently using these coordination services [9]. These applications include mobile collaborative software to support disaster relief operations, to conduct inspections in construction sites and to manage exams in computer science courses [9].

Fig. 2. Correspondence matrix

3   Related Work

There are several experiences reporting the design and use of collaborative mobile applications [3], [5], [12], however they do not describe the strategies used to deal with the typical communication services in mobile groupware. A similar situation occurs with the middleware and platforms supporting distributed work [7], [11]. Therefore these potential design solutions cannot be reused in future applications.

Some researchers have proposed communication for fixed networks [2], [4]. However, the contextual variables influencing the collaboration scenario (e.g. communication instability and low feasibility to use servers) and the mobile work (e.g. use of context-aware services and support for ad-hoc coordination processes) make such solutions unsuitable to support mobile collaboration.

4   Conclusions and Further Work

Developing mobile groupware applications to support collaborative work is a challenging task. Since this is a recent research area, software must overcome challenges in technical implementation, collaboration support and users' adoption. This paper presents a patterns system to support the design of communication services required by mobile collaborative applications. These patterns deal with most of the stated requirements. They also serve as educational and communicative media for developers, students or researchers on how to design communication mechanisms for mobile collaborative applications. They also foster the reuse of proven solutions.

At the moment, these patterns have shown to be useful to design both, mobile groupware applications and a middleware to support collaborative systems [9]. The reuse of these designs has been quite simple for the authors and development team members. However, persons outside the development team have not experienced the system yet. Therefore, future work is required to carry out evaluations with external groupware developers in order to determine the real contribution of this proposal.

Acknowledgments. This work was partially supported by Fondecyt (Chile), grants N°: 11060467 and 1080352 and LACCIR grant N° R0308LAC004.

References

1. Andriessen, J.H.E., Vartiainen, M.: Mobile virtual work: A new paradigm? Springer, Heidelberg (2006)
2. Avgeriou, P., Tandler, P.: Architectural Patterns for Collaborative Applications. Int. J. of Computer Applications in Technology 25(2/3), 86–101 (2006)
3. Baloian, N., Zurita, G., Antunes, P., Baytelman, F.: A Flexible, Lightweight Middleware Supporting the Development of Distributed Applications across Platforms. In: Proc. of CSCWD 2007, pp. 92–97. IEEE CS Press, Australia (2007)
4. Guerrero, L.A., Fuller, D.: A Pattern System for the Development of Collaborative Applications. J. of Information and Software Technology 43(7), 457–467 (2001)
5. Herskovic, V., Ochoa, S.F., Pino, J.A.: Modeling Groupware for Mobile Collaborative Work. In: 13th CSCWD Int. Conf., pp. 384–389. IEEE CS Press, Chile (2009)
6. Kristoffersen, S., Ljungberg, F.: Mobility: From Stationary to Mobile Work. In: Braa, K., Sorensen, C., Dahlbom, B. (eds.) Planet Internet, pp. 137–156. Planet Internet, Lund (2000)
7. Mascolo, C., Capra, L., Zachariadis, S., Emmerich, W.: XMIDDLE: A Data-Sharing Middleware for Mobile Computing. J. on Pers. & Wireless Comm. 21(1), 77–103 (2002)
8. Neyem, A., Ochoa, S.F., Pino, J.A.: Coordination Patterns to Support Mobile Collaboration. In: Briggs, R.O., Antunes, P., de Vreede, G.-J., Read, A.S. (eds.) CRIWG 2008. LNCS, vol. 5411, pp. 248–265. Springer, Heidelberg (2008)
9. Neyem, A.: A Framework for Supporting Development of Groupware Systems for Mobile Communication Infrastructures. Ph.D. Thesis, Universidad de Chile (November 2008)
10. Pinelle, D., Gutwin, C.: A Groupware Design Framework for Loosely Coupled Workgroups. In: 9th ECSCW Conference, pp. 65–82 (2005)
11. Russello, G., Chaudron, M., van Steen, M.: An experimental evaluation of self-managing availability in shared data spaces. Sc. of Comp. Programming 64(2), 246–262 (2007)
12. Schümmer, T., Lukosch, S.: Patterns for Computer-Mediated Interaction. John Wiley & Sons, West Sussex (2007)

A Model for Designing Geocollaborative Artifacts and Applications

Pedro Antunes[1], Gustavo Zurita[2], and Nelson Baloian[3]

[1] University of Lisbon,
Department of Informatics of the Faculty of Sciences, Campo Grande, Lisbon, Portugal
paa@di.fc.ul.pt
[2] Universidad de Chile,
Department of Information System and Management of the Economy and Businesses
School, Diagonal Paraguay 257, Santiago de Chile, Chile
gzurita@ing.puc.cl
[3] Universidad de Chile,
Department of Computer Science of the Engineering School, Blanco Encalada 2120,
Santiago de Chile, Chile
nbaloian@gmail.com

Abstract. There are many human activities for which information about the geographical location where they take place is of paramount importance. In the last years there has been increasing interest in the combination of Computer Supported Collaborative Work (CSCW) and geographical information. In this paper we analyze the concepts and elements of CSCW that are most relevant to geocollaboration. We define a model facilitating the design of shared artifacts capable to build shared awareness of the geographical context. The paper also describes two case studies using the model to design geocollaborative applications.

Keywords: Geocollaboration, sensemaking, collaborative capacity.

1 Introduction

Since thousands of years mankind has used maps printed in stone, textile, papyrus and paper to support various tasks involving navigation. Nowadays, we use a great variety of electronic devices like handhelds and Tablet-PCs to accomplish the same purpose. Smartboards and other interactive large multi-touch displays allow virtual navigation on 3D maps. And people locate physical landmarks using mobile devices and GPSs. Furthermore, the widespread availability of mobile and wireless technology, combined with advances made in human-computer interaction, user-interfaces and visualization, turn possible the computer support to multifaceted activities requiring geospatial information and collaboration, also known as geocollaboration.

According to [1-3], geocollaboration is a complex computer supported collaborative working situation where people execute diverse tasks using geospacial information. These tasks may involve exploring [2] and/or interpreting geographically-related data [4], mapping data into meaningful representations [5], and taking geospatial

L. Carriço, N. Baloian, and B. Fonseca (Eds.): CRIWG 2009, LNCS 5784, pp. 278–294, 2009.

decisions in various kinds of situations, like crisis management [6, 7], building planning, knowledge creation and management [8, 9], and strategy making [10]. Thus geocolaboration may be defined as the study of collaborative tasks where the information concerning location plays a fundamental role, as well as the development of methods, tools and frameworks facilitating these activities.

A central issue in geocolaboration concerns modeling the collaborative tasks performed by a group of people and involving geospacial information. This activity requires conceptualizing work scenarios around different modalities of time, place, space and context.

There have been some conceptual framework proposals addressing some of the mentioned modalities in specific application scenarios like crisis management, policy creation on-the-field, urban planning [3], military strategy, and mining exploration [1]. The role of these frameworks is to facilitate the application design according to a comprehensive set of technological, social and cognitive requirements. In this paper we organize all these elements in a generic framework which main purpose is also to facilitate application design and development.

2 CSCW Concepts Used in Geocolaboration

The notion of place has been considered fundamental to understand CSCW. The time/place map proposed by Johansen et al [11] has been one of the most prevalent CSCW taxonomies in the research literature (see, e.g. [12]). The distinction between same-place and different-place has a focus on accessibility rather than geographical nature, determining the overall architecture and functionality of the system. In particular, the time/place map is based upon the discussion of DeSanctis and Gallupe [13] about the different support to remote and local groups.

Some subsequent developments of the time/place map continue to emphasize the accessibility constraints. For instance, the expansion of the place dimension in three categories – co-located, virtual co-located and remote –, addresses the infrastructure capabilities to access each other in a team [14-16].

Going beyond the accessibility constraints imposed by the technology, place has also been regarded by social theorists as a fundamental constraint to communication. Studies of media richness [17] and media naturalness [18] show that communication mediated by technology looses several important features such as nonverbal cues, rapid feedback and arousal. In this line of reasoning, the notion of place is fundamental to adapt the medium to the group and task, and conversely adapt the group and task to the medium.

The conceptual change from place to space introduces a more broad concern with geographical relationships such as location, distance and orientation [19, 20]. Places exist in spaces [20, 21]. Dix et al [19] proposed a taxonomy of space considering physical and virtual places, and Cartesian and topological locations.

Rodden [22] analyzed the relationships between context, places and spaces. He proposed a conceptual model of virtual spaces using focus and nimbus. Focus and nimbus are subspaces that, respectively, map the attention and presence of elements in spaces. Also related with context, we find the distinction between private and public spaces, the former pertaining to things and actions belonging to one single individual and the later shared among a group [8, 23].

The notion of virtual space is fundamental in Collaborative Virtual Environments [24]. Virtual spaces are interactive, shared, malleable, populated and may be navigated. According to MacEachren and Brewer [12], interaction involves the aggregation of participants, topology of connections and dissemination of information. The navigation is not necessarily spatial but may also be logical. For instance, the rooms-metaphor defines navigation in virtual spaces like discussion forums [25] that are not spatially organized but rather organized according to the associated set of activities. Virtual spaces may assume complex structures, such as clusters, stacks, lists, tables, rooms, etc. [26]. Users should then be able to navigate these structures. Collaborative visualization, as an enabler of interaction and collaboration, is naturally another major challenge to consider in virtual spaces [8, 27]. Collaborative visualization involves at least data exchange, shared control and dynamic interaction [12].

Concerning the relationship between physical space and navigation, we also find in the literature the distinction between wandering, visiting and traveling [28]. In the same line of reasoning, Dix et al [19] proposed different levels of mobility: fixed, mobile, autonomous, free, embedded and pervasive. Cheverst et al [29] studied the relationships between mobility, location awareness and location services to derive important requirements such as flexibility, visibility and context-sensitivity. Davis [30] analyzed the challenges posed by the relationship between mobility and information access, including the removal of time/space constraints to communication and knowledge work, improved access to decision makers and increased ability to receive and process information.

We should also analyze the notion of workspace. According to Snowdon et al [24], a place has inherent a set of activities that occur there, while a workspace is just a container of places with ongoing activities. We may distinguish two categories of workspaces: structured and georeferenced workspaces.

The structured workspace organizes (logically or physically) several activities in coherent sets, which are nevertheless independent from the place itself. A group editor is a good example of this type of workspace, since the workspace serves to organize different activities, like writing and revising, while maintaining a coherent view of the whole [31].

Liechti [32] studied the relationship between context and workspace, defining peripheral awareness as the understanding of the activities being carried out by others nearby one's place. Peripheral awareness is naturally related with the notions of focus and nimbus, but also with notification and attention. Gutwin and Greenberg [33] expanded this view to account for the whole space, defining workspace awareness as the understanding of another person interactions in a shared workspace using a basic set of questions: who, what, where, when, and how.

A georeferenced workspace organizes activities dependent on the geographical place where they are carried out. We find in the literature innumerous examples of georeferenced workspaces. For instance, Collaborative Spatial Decision-Making (CSDM) tools and Spatial Decision Support Systems (SDSS) fundamentally rely on geographical places to support decision-making [12, 34, 35]. Less attached to the physical property of workspaces, we find synthetic collaborative environments for geo-visualization [26, 36, 37]. And we also find proposals combining physical with virtual georeferenced workspaces, like the Geo-Spatial Hypermedia system proposed by Grønbæk et al [26].

Table 1. Major geocollaboration concerns

Space					
	Place		Accessibility	Same-place, different-place	
				Co-located, virtually co-located remote	
			Mediation	Nonverbal cues, rapid feedback, arousal	
	Geographical relationships		Location	Cartesian, topological	
			Distance		
			Orientation		
	Awareness		Location awareness		
	Physical space		Mobility	Wandering, visiting, traveling	
				Fixed, mobile, autonomous, free, embedded, pervasive	
				Flexibility, visibility, context-sensitivity	
				Information access	
	Virtual space		Context	Focus, nimbus	
				Private, public	
			Interaction	Aggregation, topology, dissemination	
			Navigation	Spatial, logical	
			Collaborative visualization	Data exchange, shared control, dynamic interaction	
			Structure	Clusters, stacks, lists, tables	
	Workspace		Collaborative capacity	Individual, collective, coordinated, concerted, negotiated	
		Structured workspace	Workspace awareness	Who, what, where, when, how	
			Peripheral awareness	Focus, nimbus, notification, attention	
		Georeferenced workspace	Geographical relationships	Location	Geographical places, synthetic places, combined places
		Social space	Social awareness	Cultural meaning, history	
			Sensemaking	Perception, interpretation and anticipation	
			Embodied interaction		

Antunes et al [2] proposed adopting the notion of collaborative capacity to characterize geocollaboration. Collaborative capacity is a measure of the organizational ability to respond to problems and challenges. The theory was developed by Nunamaker et al [38] and has been tested by other researchers [39-41]. It identifies four levels of increasing ability for successful collaboration, ranging from the individual, collective and coordinated to the concerted level. The theory is that organizations will increase their potential to create value by increasing their collaboration levels. To the four categories we add one more extending the collaborative capacity beyond the concerted level: the negotiated level. The reason to propose this additional level is the observation that the capacity to negotiate conflicting views in concerted work increases the quality of the outcomes [42].

Dourish [21] and Brewer and Dourish [43] make the distinction between spatial (structured or georeferenced) and social workspaces, the former more focused on the physical context and the later more adequate to understand broader issues related to social practice beyond the physical reality. In this context, social workplaces combine physical affordances with social interaction, cultural meaning, experience and knowledge.

Weick [44] developed the notion of sensemaking to better understand what occurs in social spaces. Sensemaking is an ongoing process aiming to create order and make retrospective sense of what occurs through the articulation of several cognitive functions like perception, interpretation and anticipation [44]. It has also been associated to collaboration [45] and preliminary decision-making activities like "understanding the situation" or "getting the picture" [46].

Sensemaking is a cognitive function necessary to build awareness of the different elements occurring in the workplace: the team members, their activities and the available physical and virtual artifacts. Sensemaking also serves to build awareness of the relationships between workplaces within workspaces, including geographical relationships and mobility. And finally, sensemaking also contributes to build awareness of the relationships between action and environmental response [47].

Dourish [48] proposed the notion of embodied interaction to account for the embedded relationships between social and spatial spaces. These relationships seem quite common in our everyday experience. Dourish exemplifies with metaphorical expressions like "his position is indefensible" [48], revealing how embedded spatial concepts are in our communication.

From the discussion above we realize that geocollaboration results from a complex interaction between various concepts. Firstly, we shall consider the relationships between place and space, physical and virtual, spatial and social, place and work. Secondly, context and awareness seem fundamental to characterize what may be occurring in spaces, weather physical or virtual. And finally, both mobility and navigation, the former more related with physical spaces and the later more associated to virtual spaces, are also fundamental factors to ponder when analyzing geocollaboration. In Table 1 summarizes the above discussion.

3   Related Work

Table 2 summarizes the description of what we regard as the most relevant works in the field of geocolaboration published in the last two years. Most of these works are aimed to support crisis management. Some of them correspond to case studies (noted as T in Table 2) [7], while others propose applications or prototypes (noted as A/P) including various geocollaboration characteristics like geographical relationships, awareness, mediation and accessibility.

In the table, the R column shows the work's reference number. M/F indicates the field, defined as Knowledge Construction (KC), Decision Making (DM) and Crisis Management (CM). An asterisk in the A/P column indicates the work corresponds to an application or prototype, while an asterisk in the T column indicates it is an ethnographical study or case analysis identifying application requirements. An asterisk in the MD column indicates the work is about mobile devices. The I column indicates the study deals with interaction. N means the study deals with navigation and V with visualization. Regarding geographic issues, SP indicates users are in the same place and DF in different places. D means the distance between two points is calculated at some stage of the work. An asterisk in column O means the work deals with orientation issues. Regarding awareness issues, L means the work deals with location, and S with social awareness. W and P indicate the work deal with workspace and peripheral awareness, respectively. Regarding mediation and accessibility, CM means the work proposes or implements co-located mediation, VM virtually co-located mediation, SI synchronous interaction and AI asynchronous interaction.

Table 2. Characterization of selected published research works

R	M/F	A/P	T	SPACES												PLACES			
				Structure				Geographical relationships				Awareness				Mediation and accessibility			
				MD	I	N	V	SP	DP	D	O	L	S	W	P	CM	VM	SI	AI
[8]	KC	*	?	*	*	*			*	*	*	*	*	*	*		*	*	
[4]	DM	*			*	*			*								*	*	
[9]	CM		*		*				*					*	*			*	*
[1]	DM	*			*		*	*		*	*	*		*		*		*	
[50]	CM		*	*	*	*	*		*			*			*			*	
[49]	CM		*	*		*			*			*							*
[51]	CM	*			*	*	*		*						*			*	
[6]	CM		*	*	*	*	*		*			*		*				*	*
#		4	4	4	7	6	4	1	7	2	2	5	1	4	4	1	2	7	3

Half of the works referenced in Table 1 adopt mobile devices to capture data in the field, while a central server processes and aggregates this information. On the other hand, half of the works describe a proposal rather than a concrete application [3, 6, 49, 50]. Only one of the works describes an application where users work in the same place, thus supporting co-located mediation. The works described in [6, 50, 51], propose theoretical models where data is gathered in the field and synchronized in a central server. Most of the works are meant for people working in different places and only two of them use GPS to mark physical locations. Only one work considers social awareness. Regarding mediation and accessibility, two works consider virtual mediation, most of them synchronous.

Most works stress the importance of designing simple user-interfaces and some of them suggest including speech recognition [52, 53]. Most works adopt gestures to control the system functionality, including marking locations on maps, identifying and associating the users' comments and building awareness [7].

Other surveyed works, not shown on Table 2 for conciseness, describe the advantages of using synchronously connected mobile devices with other physically or virtually distributed systems. eMapBoard [5], is a geo-collaboration tool for disaster management offering real-time analysis components. It implements a client/server architecture and is intended to be used in control centers. Hence, it lacks the support for mobile devices. GeoMAC (Geospatial Multi-Agency Coordination Group) is a web-based tool originally designed for fire managers accessing online maps of current fire locations and perimeters [54]. Detailed real-time information cannot be provided and it does not allow distributed collaboration. Toucan Navigate (http:www.infopatterns.com)

is a P2P-based collaborative geographic information system allowing whole teams to concurrently interact with a map regardless of the physical location of its members. The annotations over the map are shared and updated automatically. However, it does not support mobile on-field operators. GeoConference allows exploiting geographic data using standard services. In a geo-conference, participants share information in a synchronized geo-referenced workspace [55]. The GeoConference system includes tools to manage users, workgroups and geo-data access. It is mainly used in control centers and no real support for mobility is provided. ArcPAD is a mobile client/server GIS product developed by ESRI4. ArcPAD uses handheld and mobile devices, and provides field operators with the ability to capture, analyze, and display geographic information [56]. It cannot be used for ad-hoc and on-field collaboration.

4   Introducing a Geocollaboration Model

The proposed geocollaboration model is organized in two ladders (see Figure 1). The first ladder defines places. In general terms, places deal with accessibility and mediation. We define places as a combination of three model elements: teams, tasks and artifacts. Teams of co-workers manipulate artifacts to accomplish tasks.

The artifacts may be physical or virtual. We regard the manipulation of artifacts in a working context not an end in itself but a mean to manage the knowledge necessary to accomplish tasks. A place is therefore where the work is being done: a team accomplishes a particular task using some specific artifacts in a place. In this ladder we do not relate the notion of place with the physical location of the team.

In the second ladder we consider space. Spaces contain multiple places and the corresponding teams, artifacts and tasks. Spaces bring additional context to the above elements. We organize these contextual elements in four major categories: virtual, physical, social and awareness elements. Concerning the virtual properties, they fundamentally define artifacts according to structure, interaction, navigation and visualization.

The physical properties such as location, distance, orientation and mobility may characterize teams and tasks. If all team members are in the same physical place, then the place is co-located. If they are in different physical places and work in different tasks, then the places are remote. And if the team is dispersed across different physical locations but working in the same task, then the place is virtually co-located. This combination of elements is sufficiently abstract and flexible to afford the most common logical and physical arrangements that we find in CSCW.

Regarding mobility, we consider the team members may either be fixed or wandering around the space. The artifacts may be fixed in one place, mobile or pervasive in the whole space.

Spaces also deal with the social dimension, including collaborative capacity, sensemaking and embodied interaction. We realize that collaborative capacity has a significant impact in teams, tasks and artifacts. Teams with more collaborative capacity need shared artifacts to organize their work places according to increasing levels of communication, interaction, coordination, collaboration and negotiation support. The impact of this view is mostly associated with tasks and artifacts: designing more collaborative artifacts will support more collaborative tasks, and should consequently lead to an increased collaborative capacity [2].

Still regarding the social dimension, we realize that sensemaking concerns the articulation of teams, tasks and artifacts. Sensemakers will explore different work arrangements in spaces and places, moving around artifacts in dynamic and exploratory ways. Weick [47] illustrates this type of behavior with the example of bringing an airplane back to an aircraft carrier, where several people working in different places must contribute to complete successfully the operation.

We regard the sensemaking function in conjunction with collaborative capacity: sensemaking may be individual or collective, but we posit the collective construction of sensemaking will lead to increased organizational ability to make sense of the ongoing situations. And we finally emphasize that artifacts are instrumental to support increasing levels of sensemaking, from individual to collective.

In our model, the notion of embodied interaction is mostly attached to physical artifacts and the technology they provide to support communication, collaboration, mobility, and obtaining information from and about the environment.

Finally, we consider the relationship between space and awareness. In the proposed model, awareness concerns understanding who are the team members, where and what tasks are being performed, what artifacts are being manipulated, and what relationships are established between places and geographical locations.

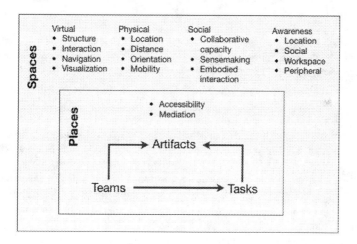

Fig. 1. Geocollaboration model

The two ladders previously described allow us to define a design process in two complementary steps. The first ladder assumes a descriptive view that we consider instrumental to analyze the "as-is" work situation:

- Identifying the work elements: Who are the team members? Which tasks they accomplish? Which are the relevant work places and spaces? Which artifacts are used?

- Identifying their relationships: What tasks the team members accomplish? What artifacts they manipulate to accomplish specific tasks? In what workplaces and workspaces reside the artifacts? How the team members, artifacts and tasks move around spaces?

Then, the second ladder departs from the "as-is" to analyze the "to-be" situation. This step leads the designer towards analyzing how artifacts may develop:

- Collaborative capacity: What is the current level of collaborative capacity? How to develop artifacts with increased collaborative capacity?
- Sensemaking: What is the current level of sensemaking? How to increase sensemaking?
- Awareness: Do artifacts support the diversity of location, social, workspace and peripheral awareness? How can awareness be improved?

In Figure 1 we summarize the fundamental constructs of the proposed geocollaboration model. We note artifacts emerge as the central model element, not only because they are responsible for articulating teams, tasks, places and spaces, but because they became responsible for increasing awareness and collaboration support.

5 Applications

In this section we describe two application developments using the geocolaboration model presented in the previous section.

5.1 Redesigning a Geological Inventory Process

This case concerns work redesign at a public agency responsible for inventorying and valuing the Portuguese geological resources. The core activities of this agency include studying and mapping the existing resources, developing risk maps, and producing geographical information systems. The case study was focused on the geological inventory process.

The geocollaboration model organizes the analysis in two ladders. This structured approach was followed by the case study. The first ladder was defined during interviews with several experts from the agency. We identified two workplaces: the office and the field. In general, the inventory process requires multiple visits to the field to elicit various types of data, intertwined with consolidation activities done in the office. The visits to the field tend to be done by one single person, while the office activities combine individual and collaborative work.

The work arrangement was therefore structured around two different spaces (office and the field) and two different places (visit and consolidation) having one-to-one relationships. Indeed, the inventory process seemed highly dependent on the relationship between place and space: many activities, such as determining the land structure, are mostly done in the physical space, as the experts often need to move around to analyze different cues to determine the exact land structure. But the activities are also highly dependent on the notion of place, especially in what regards confronting the opinions from experts in different fields, such as paleontology, petrology or sedimentology, which are done when consolidating work in the office.

We then continued the study by observing and inquiring people working in the field. We analyzed the artifacts used by the experts. Work in the field is centered on two artifacts: the field book and the combination of a map with a transparent overlay. The map/overlay allows representing the inventory data, while the field book serves to annotate supplementary information, including doubts and concerns.

Finally, in the second ladder, we analyzed the sensemaking and collaborative capacity of the artifacts used by the target organization. We realized several critical issues were hampering sensemaking:

- Whenever doubts occur, workers have to switch places, either because they lack physical context (e.g., to triangulate different physical evidence) or social context (to triangulate with different experts).
- It was often difficult to use the book outside the field, because it would loose context. While consolidating in the office, workers need to reconstitute the visit to put back in context the data recorded in the field book.
- Information was scattered between the field book and map/overlay, which were difficult to co-relate.

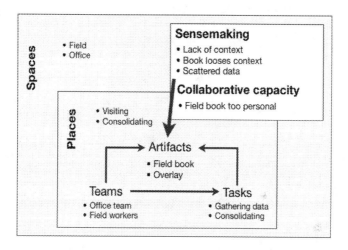

Fig. 2. Summary analysis of the geological inventory process

Regarding collaborative capacity, we also realized the field book is inherently a personal artifact, which looses value when working as a team in the office. The analysis of the geological inventory process is presented in Figure 2. Based on this analysis, we then defined our major technology requirements for work redesign:

- Sharing the field book with the purpose to increase collaborative capacity.
- Integrating the field book with the map/overlay, aiming to increase sensemaking.
- Integrating the two places, visit and consolidation, bringing all relevant stakeholders together to resolve problems as they appear in the field and in the office, aiming again to increase collaborative capacity. This required integrating communication mechanisms (audio and instant text messaging) with the field book.

These requirements lead us to develop a prototype, running on tablet and common PCs, integrating the field book and map/overlay (see Figure 3). This prototype also merged the visit and consolidation activities into one single place distributed across two different spaces: field and office. This allowed the field workers, using tablet PCs, to get in contact with the office workers and immediately exchanging comments on any occurring problems or doubts.

The prototype supports data exchange, shared control and dynamic interaction. It also integrates GPS with instant text messaging. The exchanged instant text messages are preserved in the field book with automatic associations to the geographical position of the field workers, thus keeping the doubts, comments or opinions in their proper context. Because many doubts are resolved in the field, there is less chance to swing back and forth between the office and the field.

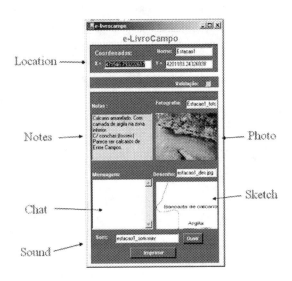

Fig. 3. Developed prototype for application 1

The prototype was evaluated with a field test and contextual interviews with several experts from the agency. The obtained results indicate that the system increased sensemaking and collaborative capability. Related to sensemaking, the participants regarded very positively the expeditious way to locate points and associate them in the field book. Related with collaborative capability, the participants were extremely favorable to the communication between field and office workers, effectively resolving problems occurring in the field and thus simplifying the whole inventory process. More details about this case study can be found in [2].

5.2 Supporting the Evacuation of Crowded Places

This case concerns supporting the police evacuating people from a stadium or any other facility with capacity for hosting thousands of people. The major problems to consider are finding adequate evacuation routes, spreading out people in congested places such as bus and/or metro stations and parking places, and dealing with high-density and fluid crowds. These are frequent problems faced by the police in Santiago de Chile, where sports fields with capacity for 80.000 people were built in surroundings close to the city limits and now surrounded by busy streets and dense inhabited city quarters. In these events, the police will place agents in strategic places to patrol people coming out from the sports field, showing them the planned evacuation ways.

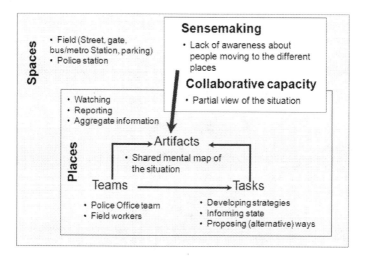

Fig. 4. Summary analysis of the evacuation process

   Normally, each agent will have a radio device through which he/she will report the situation context to the central police station and colleagues. The agents in the central police station try to make a picture of the whole situation based on the scattered information provided verbally by the agent. The central police station will give commands to the agents in the field, managing any exceptional situations that may occur.
   This is a typical geocollaboration situation where location and collaboration are of critical importance. Analyzing the situation, in the first ladder we defined two spaces: police station and field. The workplaces are the police station and some strategic points in the field: stadium gates, streets, parking places, bus and metro stations. In the field, agents have to watch the emergent situations, report them to the central station and give instructions to the mob.
   The main sensemaking task here is building a shared mental map of the situation, understanding the whole picture and anticipating events. This mental map is also the artifact used to communicate with each other. We detect here a main problem hampering this task: sometimes the exchange of voice information does not allow everyone to communicate accurately, timely and within context. Also, the whole context may only be assembled in the central station, often with many delays. The summary analysis of the evacuation scenario is presented in Figure 4. As we see, the central shared artifact is the area map. This map is situated at the central police station and is annotated according to the information people on the field submit by voice.

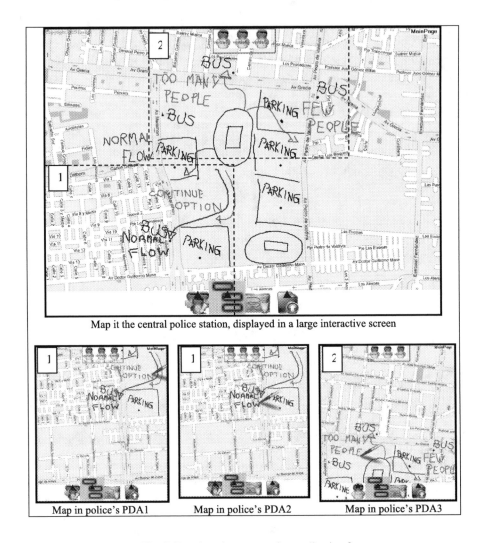

Fig. 5. Developed prototype for application 2

We designed a distributed application to support this activity, allowing agents in the police station to display the map on a big touch-sensitive screen showing the sports field and surrounding areas. The map may be annotated with freehand writing and sketching. Each agent in the field has a portable device, also touch sensitive, showing the portion of the map where she/he is located, which they can also annotate. The map is automatically adjusted according to the agent's position thanks to an incorporated GPS. All applications (the one running in the police station and those running in mobile devices) are synchronized, so that any annotations done in one application will be immediately shown in the others. In Figure 5 we may see some screenshots showing the map available in the police station (upper section) and the mobile devices of three agents in the field (lower section). Another design element focusing on awareness concerns the areas marked and labeled with a number, indicating what areas the agents in the field are controlling. In this way the agents in the station may perceive the whole control of the situation. Furthermore, the agents may synchronize their portion of the map with another agent. This supports the collaboration of two (or more) agents in charge of the same geographical area. In Figure 5 we see that policeman 1 and policeman 2 are working synchronized in the area labeled as 1. This prototype is currently being developed and there is still no data from formal evaluation available.

6   Discussion

This paper proposes a model for designing geocollaborative applications. The model considers the main design goal is supporting a team in the collaborative construction of a shared vision about the relevant conditions in a certain geographical area, in situations where this vision is critical for team members to accomplish their task. The model is based on a set of foundational concepts, including teams, tasks, places, spaces, and artifacts. Artifacts emerge as the central model element, not only because they are responsible for articulating teams, tasks, places and spaces, but also because they are responsible for increasing sensemaking and collaborative capacity.

In order to test the model flexibility and completeness, we used it to design two quite different applications. The model provided guidelines that helped analyzing and articulating the geocollaborative requirements of those applications.

Moreover, the proposed model also allows describing various geocollaboration applications in a uniform and standardized way, and therefore facilitates their benchmarking and classification. In addition, the proposed model allows identifying the major software components and libraries necessary to integrate the geocollaborative elements proposed in the model. Some of these software components have already been developed to implement the applications described in the paper and constitute the basis of a future geocollaborative software platform.

The proposed model may also serve to lay out the conceptual framework necessary to characterize two emerging types of tools: e-planning and e-participation. The aim of e-planning and e-participation is to engage people living in a region or urban area to participate in government decisions and urban planning.

Acknowledgements. This paper was supported by the Portuguese Foundation for Science and Technology (PTDC/EIA/67589/2006) and Fondecyt 1085010.

References

1. MacEachren, A., Guoray, C., Brewer, I., Chen, J.: Supporting Map-Based geocollaboration through Natural Interfaces to Large-Screen Displays. Cartographic Perspectives 54, 4–22 (2006)
2. Omitted to allow blind review: A Conceptual Framework for the Design of Geo-Collaborative Systems. Group Decision and Negotiation 15, 273–295 (2006)

3. Cai, G.: Extending Distributed GIS to Support Geo-Collaborative Crisis Management. Progress in Human Geography 25 (2001)
4. Rinner, C.: Argumentation Mapping in Collaborative Spatial Decision Making. In: Collaborative GIS, pp. 85–102. Idea Group Publishing, USA (2006)
5. MacEachren, A., Guoray, C., Brewer, I., Chen, J.: Visually Enabled geocollaboration to Support Data Exploration & Decision Making. In: Procs. of the 21st International Cartographic Conference, Durban, South Africa, pp. 10–16 (2003)
6. Capata, A., Marella, A., Russo, R.: A Geo-Based Application for the Managemnt of Mobile Actors During Crisis Situations. In: Proceedings of the 5th International ISCRAM Conference, Washington DC, US (2008)
7. Schafer, W., Ganoe, C., Caroll, C.: Supporting Community Emergency Management Planning through a geocollaboration Software Architecture. Computer Supported Cooperative Work 16, 501–537 (2007)
8. Convertino, G., Ganoe, C., Schafer, W., Yost, B., Carroll, J.: A Multiple View Approach to Support Common Ground in Distributed and Synchronous Geo-Collaboration. In: Proceedings of Third International Conference on Coordinated and Multiple Views in Exploratory Visualization, pp. 121–132 (2005)
9. Convertino, G., Zhao, D., Ganoe, C., Carroll, J., Rosson, M.: A Role-Based Multiple View Approach to Distributed Geo-Collaboration. Human-Computer Interaction 4553, 561–570 (2007)
10. MacEachren, A., Cai, G., Sharma, R., Rauschert, I., Brewer, I., Bolelli, L., Shaparenko, B., Fuhrmann, S., Wang, H.: Enabling Collaborative Geoinformation Access and Decision Making through a Natural, Mulimodal Interface. International Journal of Geographical Information Science 19, 293–317 (2005)
11. Johansen, R., Sibbet, D., Benson, S., Martin, A., Mittman, R., Saffo, P.: Leading Business Teams. Addison-Wesley, Reading (1991)
12. MacEachren, A., Brewer, I.: Developing a Conceptual Framework for Visually-Enabled geocollaboration. International Journal of Geographical Information Science 18, 1–34 (2004)
13. DeSanctis, G., Gallupe, R.: A Foundation for the Study of Group Decision Support Systems. Management Science 33, 589–609 (1987)
14. Sharifi, S., Pawar, K.: Virtually Co-Located Product Design Teams. International Journal of Operations & Production Management 22, 656–679 (2002)
15. Rodden, T., Blair, G.: CSCW and Distributed Systems: The Problem of Control. In: Proceedings of the Second Conference on European Conference on Computer-Supported Cooperative Work, pp. 49–64. Kluwer Academic Publishers, Amsterdam (1991)
16. Kim, T., Chang, A., Holland, L., Pentland, A.: Meeting Mediator: Enhancing Group Collaborationusing Sociometric Feedback. In: Proceedings of the ACM 2008 Conference on Computer Supported Cooperative Work, pp. 457–466. ACM Press, San Diego (2008)
17. Daft, R., Lengel, R.: Organizational Information Requirements, Media Richness and Structural Design. Management Science 32 (1986)
18. Kock, N.: Media Richness or Media Naturalness? The Evolution of Our Biological Communication Apparatus and Its Influence on Our Behavior toward E-Communication Tools. IEEE Transactions on Professional Communications 48, 117–130 (2005)
19. Dix, A., Rodden, T., Davies, N., Trevor, J., Friday, A., Palfreyman, K.: Exploiting Space and Location as a Design Framework for Interactive Mobile Systems. ACM Transactions on CHI 7 (2000)

20. Harrison, S., Dourish, P.: The Roles of Place and Space in Collaborative Systems. In: Proceedings of the 1996 ACM conference on Computer supported cooperative work, pp. 67–76. ACM Press, Boston (1996)
21. Dourish, P.: Re-Space-Ing Place: "Place" And "Space" Ten Years on. In: Proceedings of the 2006 20th Anniversary Conference on Computer Supported Cooperative Work, pp. 299–308. ACM Press, Alberta (2006)
22. Rodden, T.: Populating the Application: A Model of Awareness for Cooperative Applications. In: Proceedings of the 1996 ACM Conference on Computer Supported Cooperative Work, pp. 87–96. ACM Press, Boston (1996)
23. Greenberg, S., Boyle, M., Laberge, J.: PDAs and Shared Public Displays: Making Personal Information Public, and Public Information Personal. Personal Technologies 3, 54–64 (1999)
24. Snowdon, D., Munro, A. (eds.): Collaborative Virtual Environments: Digital Places and Spaces for Interaction. Springer, New York (2000)
25. Greenberg, S., Roseman, M.: Using a Room Metaphor to Ease Transitions in Groupware. In: Ackerman, M., Pipek, V., Wulf, V. (eds.) Sharing Expertise. Beyond Knowledge Management, pp. 203–256. The MIT Press, Cambridge (2003)
26. Grønbæk, K., Vestergaard, P., Ørbæk, P.: Towards Geo-Spatial Hypermedia: Concepts and Prototype Implementation. In: Proceedings of the thirteenth ACM conference on Hypertext and hypermedia, pp. 117–126. ACM Press, College Park (2002)
27. Brewer, I., MacEachren, A., Abdo, H., Gundrum, J., Otto, G.: Collaborative Geographic Visualization: Enabling Shared Understanding of Environmental Processes. In: Proceedings of IEEE Symposium on Information Visualization, Washington DC, p. 137 (2000)
28. Kristoffersen, S., Ljungberg, F.: Your Mobile Computer Is a Stationary Computer. In: CSCW 1998 Handheld CSCW Workshop, Seattle (1998)
29. Cheverst, K., Davies, N., Mitchell, K., Friday, A., Efstratiou, C.: Developing a Context-Aware Electronic Tourist Guide: Some Issues and Experiences. In: Proceedings of the SIGCHI Conference on Human Factors in Computing Systems, pp. 17–24. ACM Press, The Hague (2000)
30. Davis, G.: Anytime/Anyplace Computing and the Future of Knowledge Work. Communications of ACM 45, 67–73 (2002)
31. Koch, M., Koch, J.: Application of Frameworks in Groupware—the Iris Group Editor Environment. ACM Computing Surveys (CSUR) 32 (2000)
32. Liechti, O.: Supporting Social Awareness on the World Wide Web with the Handheld Cyberwindow. In: Workshop on Handheld CSCW at CSCW 1998, Seattle, US (1998)
33. Gutwin, C., Greenberg, S.: The Effects of Workspace Awareness Support on the Usability of Real-Time Distributed Groupware. ACM Transactions on Computer-Human Interaction 6, 243–281 (1999)
34. Nyerges, T., Montejano, R., Oshiro, C., Dadswell, M.: Group-Based Geographic Information Systems for Transportation Site Selection. Transportation Research C 5, 349–369 (1997)
35. Armstrong, M.: Requirements for the Development of GIS-Based Group Decision Support Systems. Journal of the American Society for Information Science 45, 669–677 (1994)
36. Manoharan, T., Taylor, H., Gardiner, P.: A Collaborative Analysis Tool for Visualisation and Interaction with Spatial Data. In: Proceeding of the seventh international conference on 3D Web technology, Tempe, Arizona, pp. 75–83 (2002)
37. MacEachren, A., Edsall, R., Haug, D., Baxter, R., Otto, G., Masters, R., Fuhrmann, S., Qian, L.: Virtual Environments for Geographic Visualization: Potential and Challenges. In: Procs. of the 1999 workshop on new paradigms in information visualization and manipulation, pp. 35–40. ACM Press, Kansas City (1999)

38. Nunamaker, J., Romano, N., Briggs, R.: Increasing Intellectual Bandwidth: Generating Value from Intellectual Capital with Information Technology. Group Decision and Negotiation 11, 69–86 (2002)
39. Bach, C., Belardo, S., Faerman, S.: Employing the Intellectual Bandwidth Model to Measure Value Creation in Collaborative Environments. In: Proceeding of the 37th Hawaii International Conference on System Sciences, Hawaii (2004)
40. Qureshi, S., Briggs, R.: Revision the Intellectual Bandwidth Model and Exploring Its Use by a Corporate Management Team. In: Proceeding of the 36th Hawaii International Conference on System Sciences, Hawaii (2003)
41. Qureshi, S., Vaart, A., Kaulingfreeks, G., Vreede, G., Briggs, R., Nunamaker, J.: What Does It Mean for an Organization to Be Intelligent? Measuring Intellectual Bandwidth for Value Creation. In: Proceeding of the 35th Hawaii International Conference on System Sciences, Hawaii (2002)
42. Omitted to allow blind review.: Addressing the Conflicting Dimension of Groupware: A Case Study in Software Requirements Validation. Computing and Informatics 25, 523–546 (2006)
43. Brewer, J., Dourish, P.: Storied Spaces: Cultural Accounts of Mobility, Technology, and Environmental Knowing. International Journal of Human-Computer Studies 66, 963–976 (2008)
44. Weick, K.: The Collapse of Sensemaking in Organizations: The Mann Gulch Disaster. Administrative Science Quarterly 38, 628–652 (1993)
45. Larsson, A.: Making Sense of Collaboration: The Challenge of Thinking Together in Global Design Teams. In: Proceedings of the 2003 international ACM SIGGROUP conference on Supporting group work, pp. 153–160. ACM Press, Sanibel Island (2003)
46. Hasan, H., Gould, E.: Support for the Sense-Making Activity of Managers. Decision Support Systems 31, 71–86 (2001)
47. Weick, K.: Making Sense of the Organization. Blackwell, Oxford (2001)
48. Dourish, P.: Where the Action Is. The MIT Press, Cambridge (2001)
49. Malizia, A., Astorga, F., Onorati, T., Díaz, P., Aedo, I.: Emergency Alerts for All: An Ontology Based Approach to Improve Accessibility in Emergency Alerting Systems. In: Proceedings of the 5th International ISCRAM Conference (2008)
50. Bortenschlager, M., Leitinger, S., Rieser, H., Steinmann, R.: Towards a P2P-Based geocollaboration System for Disaster Management. In: GI-Days 2007 - Young Researchers Forum (2007)
51. Convertino, G., Mentis, H., Bhambare, P., Ferro, C., Carroll, J., Rosson, M.: Comparing Media in Emmergency Planning. In: Proceedings of the 5th International ISCRAM Conference, Washington DC, US (2008)
52. Sharma, R., Yeasin, M., Krahnstoever, N., Rauschert, I., Cai, G.: Speech-Gesture Driven Multimodal Interfaces for Crisis Management. Proceedings of the IEEE special issue on Multimodal Human-Computer Interface 91, 1327–1354 (2003)
53. Fuhrmann, S., MacEachren, A., Dou, J., Wang, K., Cox, A.: Gesture and Speech- Based Maps to Support Use of GIS for Crisis Management: A User Study. In: AutoCarto 2005, Las Vegas, US (2005)
54. Wagtendon, J., Zhu, Z., Lile, E.: Appendix E - White Paper on Pre-Fire Risk Assessment and Fuels Mapping (2004)
55. Siegel, C., Fortin, D., Pellerin, E.: Bringing Geospatial Expertise to Emergency Operations Management with Geoconferencing. In: Proceedings of the 98th Conference of the Canadian Institute of Geomatics, Ottawa, Canada (2005)
56. ESRI: Arcgis 9, Arcpad Reference Guide, USA (2005)

MobMaps: Towards a Shared Environment for Collaborative Social Activism

Luís Gens[1], Hugo Paredes[2], Paulo Martins[2], Benjamim Fonseca[2], Yishay Mor[3], and Leonel Morgado[2]

[1] UTAD, Quinta de Prados, Apartado 1013, 5001-801 Vila Real, Portugal
luisgens@utad.pt
[2] GECAD/UTAD, Quinta de Prados, Apartado 1013, 5001-801 Vila Real, Portugal
{hparedes,pmartins,leonelm,benjaf}@utad.pt
[3] Institute of Education, University of London, London, UK
yishaym@gmail.com

Abstract. Nowadays it is possible to disseminate information to the all world in real time using current communication tools supported mostly by the Internet. The work of several organizations reporting a multitude of problems that our society faces can be sustained by participatory platforms, which stimulate the collaboration of participants all over the world. In this paper we present a technological platform that provides a shared environment for collaborative social activism. We adapted the platform to a particular organization, MachsomWatch that reports human rights abuses in Israelis checkpoints. Finally we present some preliminary results obtained by ethnographic research using the developed platform.

Keywords: Mobile collaboration, georeferenced information, social activism.

1 Introduction

In an era of accelerated evolution of knowledge, the difference between the success and failure of an initiative often depends on the amount and quality of information that one can get from the field and on the field. Technological progress enables the development of solutions to overcome some of the information constraints of field work. The proliferation of mobile devices such as cell phones and PDAs allows field operations to benefit from tools similar to those available for office-based work [10].

The Web as we know it is changing. New wiki and blogging sites are appearing constantly, and have contributed to further develop collaborative work. In our days, people like to participate, to share their opinions, their knowledge, and their culture using computer technologies. By introducing mobile devices into this mix we have mobile social software [1]. At the same time, we see a growing use of geographic information on the Web mostly due to low-cost availability of image and geographic data from map servers, either from public or private organizations. If we can find at our disposal these kinds of technology and people motivated to do collaborative social activism, why don't we take advantage of both of them? Why not put them working together and provide tools to overcome their problems?

L. Carriço, N. Baloian, and B. Fonseca (Eds.): CRIWG 2009, LNCS 5784, pp. 295–302, 2009.

On this paper we start by presenting some social activism scenarios that represent situations that range from everyday life to disaster events, as well as some issues that people face trying to deal with them. In this context we present our concept of social activism. A comprehensive analysis of existing technologies is given afterwards, followed by a technical presentation and an overview of a technological platform proposal. Afterwards an existing prototype implementation is presented associated with a case study. Finally, we conclude with some final remarks.

2 Social Activism Scenarios

In present days, our society faces a multitude of problems. These can vary from human rights violations, natural disasters such as floods, forest fires, earthquakes, etc., and even neglect for people's special needs (sight, hearing or motion limitations, for instance). Some of these are known worldwide, while others remain silent.

Reporting human rights abuses: Every year Amnesty International reports several situation of abuses of human rights around the world. A particular case of these situations occurs in Israel where every year several Non Governmental Organizations (NGO) report repeated human rights abuses performed by army and border police on Palestinians crossing checkpoints. Included in this kind of organizations is MachsomWatch, an Israeli women organization established in February 2001, with more than 500 members (all volunteers) of different backgrounds, personal and social characteristics. The organization monitors military checkpoints trying to resolve the situations and conflicts at the moment and in the local where they occur and after reporting the situations to the global community by making their reports available on the web. As a result of their efforts, the MachsomWatch members expect to improve the conditions in which people that have to use the checkpoints live. In spite of all their effort, there are some issues that can be very difficult to solve, like coordination between different agents on the field, the ability to record incidents, the need to provide rapid real-time response, and to let people all over the world know what is happening.

Urban barriers to accessibility: Many people suffer from limitations such as blindness, hearing loss, motion limitations, etc. For them, moving from place to place can be very challenging. Private and public buildings were built (and in many cases, still are) thinking only of people that do not have such limitations. If one compounds this with the typical problems of a city (street holes, cars parked on sidewalks, etc.), one ends up with an adverse environment. To help solve these problems, several organizations have sprung around the world, to act as a middle man between people and the entities responsible for solving this kind of problems.

Spatial Data Infrastructure applied to disaster management: Spatial data has won a special place in our day to day life. Today we find infrastructures based on spatial data that help on the decision-making level when applied in disaster management. These infrastructures are named Spatial Data Infrastructures (SDI). SDI has proven to be very important and essential in collaborative decision-making in disaster management [9]. Normally, these kinds of infrastructures are composed by different kinds of organisms and/or organizations, collecting and maintaining the database for responding to disasters collaboratively. The aim of the SID is to create an environment where users can retrieve, access, update and disseminate spatial data in an easy and secure way. By allowing the sharing of data, the SDI enables users to save time, efforts and resources.

The technological emergence of the Web into the Web 2.0 gave the people the power to construct their Web. This fact empowered their ability to report some of the above mentioned facts that occur all over the world, enjoying their free speech rights and their moral responsibilities as members of the society to reveal such situations.

The use of collaborative systems to solve situations that affect many people can have a very good impact on a community's ability to uphold rules and care for the wellbeing of community members. By supporting passive citizens to change into a more interactive role, collaborative systems promote a higher sense of community, making people more concerned about their surroundings [6]. Seeing how other people enter information to help someone else is a major incentive, and provides encouragement for others to do the same. Also, because people can now have a direct impact on the situation, they also have an increased concern to see it through, putting extra pressure on the authorities to solve problems more quickly.

3 Technological Background

Some of the challenges of the previous scenarios can be addressed by a system that allows information to be taken to those who need it and collected from those who have it, allowing people to participate, sharing information and knowledge, taking advantage of collaborative work. If we allow people to collect, validate and correct information, we can have quick access to important information that can be used to improve one's actions under some situations [4].

With the massive proliferation of wikis on the Web and the easy access maps provided by free tools like Google Maps, the integration between the two technologies became natural. One of the best-known examples is Wikimapia, a Web site based on a wiki that provides a map interface, where anybody can search, add, and modify information on a certain location. Many other applications implement the same idea (using maps in connection with collaboration tools), e.g. OpenStreetMap and HealthMap. A particularly important example of how these tools can work together to achieve a very important and relevant goal is Scipionus (www.scipionus.com), which was used during the Katrina disaster for people to provide information about problems they were aware of in certain areas, and to allow others to collect news about their relatives, friends or colleagues[3]. This kind of systems, like any other based on large-scale collaboration efforts, is not without flaws, and always faces the same problem: validation of the information inserted in the platform [11].

The ability to use geospatial data retrieved from the field is very relevant to increase the quality of the information gathered [2]. Current tools have several limitations, specially regarding the ability to manipulate spatial data [10]. For greater use of this specific function, new mobile software tools are needed.

The use of mobile devices to allow access to spatial information (e.g., Google Maps Mobile) creates new services and situations for the users [12]. Some examples of integration between maps and mobile devices are PDPal, a spatial annotation application for the Palm PDA; and Yellow Arrow, an application to combine physical marks with information about them [1].

Mobile devices can be used, and are in fact being used, to retrieve information from the field and to add data to maps, allowing for a better understanding of the local and global situation. Tools like ProMed-mail, a tool to help find and distribute information about diseases regarding animals, plants or humans all over the world in a very quick manner [8], are changing the way many organizations respond to unexpected situations with their ability to gather and distribute data easily.

A new and important feature of cell phones is the digital camera. This device has considerably increased the range of uses of the cell phone as a collaborative tool, by providing a way to take pictures in the spur of the moment and to upload them in a single, continuous act. Specific software services, known as photo blogging or moblogging, allow users to use their cell phone cameras to take pictures and upload them to a site or blog where, later, commentaries can be added by any user. There's a wide variety of application in this segment, e.g. MobShare, Kodak Mobile, Buzznet, and Nokia Lifelog [5]. As another example, albeit without the ability to upload photos from the cell phone,is Flickr that (at this time) is the largest site for image sharing [5].

Mobile communication infrastructures are sufficiently diversified to allow at least a minimum level of access in almost every place populated by humans (e.g., satellite communication). They are the ideal platform to provide access to information and contribute with information. A platform based on three basic principles, mobility (providing generalized access and greater freedom of movements, since one does not need to be connected to a physical network to have access to the information), collaborative work, and location-based information, could be used to improve many social activism situations.

4 MobMaps Platform

Currently, there are very few tools that allow maps to be used on mobile devices and, at the same time, offering a way to do collaborative work. Our technological platform proposal tries to fill this gap. Our main objective was to develop a platform to serve as layer between mobile devices and geospatial data services. This platform needed to be device-independent: able to run in any kind of mobile device, be it a cell phone or a PDA.

Moreover, the MobMaps[1] platform intends to be a generic platform for collaborative social activism that can be adapted to the several situations fulfilling specific requirements of each application context.

The MobMaps technological platform has five main components: Mobile Devices; Application Layer; Input/Output Modules; Maps Server; and Information Server.

The Mobile Device application is composed of a module for presenting maps to the user, providing means for navigation, visualization of information and the necessary tools to allow data entry by the user.

The front-end mobile devices perform as little data processing as necessary for the user to access the application. The application can adapt itself to the features of each mobile device, in order to run on devices with different levels of hardware resources.

---

[1] MobMaps derives from the words "Mob" plus "Maps" empathizing the geo-referenciation collaborating people.

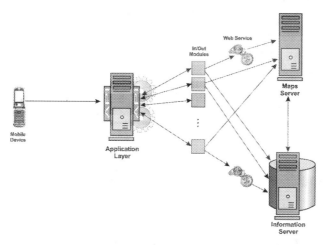

Fig. 1. MobMaps technological platform

Centralized content storage and access control are done at the Information Server and at the Maps Server (Figure 1). These can be developed specifically for use in a scenario, or readily available Web services, linked to the Application Layer through Input/Output modules.

The Application Layer can be called the heart of the entire system behaving as the functional core of the MobMaps platform. It is responsible for the coordination between all the different modules and for most of the needed information processing activities. Included in the major functions of this module are: translation coordination – convert values provided from the mobile device to accepted GIS coordinates; image resizing – adjust image sizes to the mobile devices' characteristics.

In order to provide some of these functionalities, the platform relies on a set of In/Out Modules, which allows the communication and data structure translation between the Appplication Layer and the backend layers composed of Information Server and Maps Server. These modules adapt the request provided from the Application Layer interface and translate that data into information that can be understood by the servers (the map and information server). Then, when the server returns the requested information, the reverse operation occurs. The selection by the Application Layer of the correct In/Out module to be used depends on the original request provided by the user. This middleware layer represents an important architectural implementation as it allows the system to work with all, or at least most, maps servers and information management systems available.

The maps server that can be used with the platform can vary from the use of a complex GIS system to a simple implementation using of a still image. In the specification stages, our main concerns were the single images scenario and the GIS map server scenario. This choice was the result of a particular concern to work on solutions primarily based on largely accepted software or software related standards such as OGC, which guarantees the interoperability of the platform and a large amount of available maps servers systems.

Since MobMaps aimed to provide a way to record, search and access information in a collaborative environment, an Information System needs to record that information.

The complexity of the Information System can vary from a simple text file to a very complicated and elaborated blog or wiki system. Despite the fact that nowadays the use of database based systems has reached almost every corner of the computer world, there are still some systems that, because of size limitation or due to their highly specific nature, have other information structures.

The Information System can have a full, partial or zero cooperation with the map server, depending on the nature of the specific service that is being used. Any and all of these different situations can be implemented with residual effort, thanks to our platform architecture.

5   MachsomWatch: A Case Study

MachsomWatch is, as stated before, an organization of Israeli women from different age, educational, social and economical ranges, that monitors military checkpoints and attempts to resolve and defuse the situations by reporting events on the Web. MachsomWatch volunteers need to be in movement as they act in groups of 2 or 3 volunteers in more than 30 checkpoints in shifts of 3 to 5 hours each day. When on action in the field, the volunteers do not have traditional Internet access to communicate with the head quarters of the organization and report the situations. In this context the most adequate and available communication tool are mobile phones.

Communication is needed in the resolution of certain situations, where local diplomacy with the observers and the militaries cannot solve the violations that occur. In that kind of situations, volunteers need to contact the organization headquarters in order to report and solve the problem. Despite the resolution of the problems, every situation that occurs in a checkpoint is reported and can be accessed in the organization's Web site (http://www.machsomwatch.org/en).

The adaptations of MobMaps platform to the specific case of MachsomWatch has as requirements: the need to be used by volunteers in the checkpoints, therefore a mobile application; must be user friendly, as volunteers are from different ages, social, economical and educational ranges; must enable the access at all time to registers about each checkpoint and the current status of all violations; and have the possibility to add new information in real time to the system as the situation occurs.

Based on these requirements the MobMaps platform was simplified in three main layers: Mobile Devices; Application Layer; and Information Server. As stated in the requirements, in this case study there are only a fixed area and a pre-defined number and locations of checkpoints where situations occurs, so the Maps Server is not needed. In this case a fixed map image is located in the Application Layer server and will be transformed according to the mobile characteristics that came in the requests from the mobile device. As far as the In/Out modules are concerned, the simplification of the platform and the adaptation to a specific case has as consequence the removal of architectural generalizations. In the case study the implementation of the mobile device makes 4 requests to the application layer (Figure 2): (1) map request; (2) checkpoint position request; (3) checkpoint information request; and (4) request to send new information about a checkpoint.

The developed prototype was intended to be tested on the field, so a member of our research group traveled to Israel for 8 days where, using an ethnographic research methodology, he observed, reported and evaluated the working methods of the

Fig. 2. Mobile application: (a) services menu, (b) map and options (c) checkpoint information and (d) upload screen

MachsomWatch. The trip was planned in order to fulfill three major objectives: present the platform to the organization; educate the future users of the platform; and follow some teams in their work at checkpoints.

The field study revealed some important conclusions about the working methodologies of the organization. The first is that the MachsomWatch neither have fixed headquarters nor a permanent team to assist the resolution of situations in the checkpoints. The second is the usage of mobile Internet in Israel, which is very expensive, so most of the volunteers do not have contracts with that functionality. The last conclusion was about the average age of the members of MachsomWatch: more than 85% of the members are more than 55 years old.

In what concerns research on the field four meetings with members of the organization were performed and the work of 5 volunteer teams on the field was observed, reported and analyzed, covering more than 11 chekpoints visits. The work was reported in the blog http://aventurasdeisrael.blogspot.com/ (in Portuguese).

6 Conclusions and Final Remarks

Large amounts of information are hard to process. Usually graphic methods of delivery are used to allow people to filter meaningful information. By seeing it on a map, people can understand how the global status evolved. Collaborative platforms provide a way for all people to share their knowledge creating possibilities for new, innovative approaches to the problem's resolution and can be seen as optimal informative platform. By using mobile devices, these people can access and deliver information in places that classic means of communication cannot reach.

In this paper we have presented some scenarios from real life situations, their problems and how a combination between spatial data, the use of mobile devices and collaborative work can help minimizing them. To allow this combination to work we have presented a technical proposal for porting the collaborative work to the field through the use of mobile devices, maximizing the response to the situations presented on those scenarios. This system addresses the problem of the lack of solutions that exists when people leave the office and need to work, either on the move or in open country. In the process, we have identified issues regarding the communications between mobile applications, map servers and information storing systems.

References

1. Bleecker, J.: Design Patterns for Mobile Social Software. Paper presented at the meeting WWW 2006 Workshop MobEA IV — Empowering the Mobile Web, Edinburg, Scotland (May 2006)
2. Christensen, U.: Conventions and articulation work in a mobile workplace. SIGGROUP Bulletin 22(3), 16–21 (2001)
3. Church, R.L.: Geographical information systems and location science. Computers & Operations Research 29, 541–562 (2002)
4. Gens, L., Alves, H., Paredes, H., Martins, P., Fonseca, B., Bariso, E., Ramondt, L., Mor, Y., Morgado, L.: Mobile software for gathering and managing geo-referenced information from crisis areas. Paper presented at the meeting Proceedings of the Second International Conference on Internet Technologies and Applications (ITA 2007), Newi, Wrexham, UK (September 2007)
5. Jacucci, G., Oulasvirta, A., Salovaara, A.: Supporting the Shared Experience of Spectators through Mobile Group Media. Paper presented at the meeting Proceedings of Group 2005, Sanibel Island, Florida, USA (November 2005)
6. Kalibo, H.W., Medley, K.E.: Participatory resource mapping for adaptive collaborative management at Mt. Kasigau, Kenya. Landscape and Urban Planning 82, 145–158 (2007)
7. Kolbitsh, J., Maurer, H.: The transformation of the Web: How Emerging Communities Shape the Information we Consume. Journal of Universal Computer Science 12(2), 187–213 (2006)
8. Madoff, L.C., Woodall, J.P.: The Internet and the Global Monitoring of Emerging Diseases: Lessons from the First 10 Years of ProMED-mail. Archives of Medical Research 36, 724–730 (2005)
9. Mansourian, A., Rajabifard, A., Zoej, M.J.V., Williamson, I.: Using SDI and Web-based system to facilitate disaster management. Computers & Geosciencies 32, 303–315 (2006)
10. Nusser, S., Miller, L., Clarke, K., Goodchild, M.: Geospatial IT for mobile field data collection. Communications of the ACM 46(1), 63–64 (2003b)
11. Palen, L., Hiltz, S.R., Liu, S.B.: Online forums supporting grassroots participation. Communications of the ACM 50(3), 54–58 (2007)
12. Tezuka, T., Kurashima, T., Tanaka, K.: Toward Tighter Integration of Web Search with a Geographic Information System. Paper presented at the meeting WWW 2006, Edinburg, Scotland (May 2006)

Spatial Operators for Collaborative Map Handling

Renato Rodrigues and Armanda Rodrigues

Departamento de Informática, Faculdade de Ciências e Tecnologia,
Universidade Nova de Lisboa, Quinta da Torre, 2829 -516 Caparica, Portugal
renatolmsrodrigues@gmail.com, arodrigues@di.fct.unl.pt

Abstract. In this paper, we describe an online spatial decision-support system developed to support asynchronous spatial collaboration between physically distributed users, with a focus on defining spatial operators for supporting these activities in a public participation context.

The developed system takes advantage of recent development on maps APIs and Web 2.0 technologies, to provide generic features that can improve spatial decision-making. It implements a web-based public participation GIS (PPGIS) while allowing geographic information based collaboration. The system, is based on a customizable platform, targets different types of spatial collaboration and user expertise, while encouraging debate between participants with similar spatial interests.

The system's features, architecture and interface are presented in detail as well as usability testing procedures realized for different users' roles and expertise.

Keywords: GeoCollaboration, Online Mapping, Geographic Information Systems, Spatial Decision-Making, Public Participation.

1 Introduction

Geographic information plays a vital role in different situations including natural crisis managements, urban planning and public participation. In these complex decision-making problems, individual knowledge is frequently insufficient and a group approach is welcomed. However, existing Geographic Information Systems (GIS) were developed for individual use, not directly supporting group work [1].

Typical activities where the spatial decision-making process can be enriched with group work are those involving the public's participation. These activities are traditionally led in on-site meetings but can also be supported by systems known as PPGIS (Public Participation GIS). The participation of the public in a spatial decision activity may involve difficulties: problems of trust, atmosphere of confrontation, restrictions on time and location of public meetings, just to name a few [2].

Freedom in terms of time and location of participation and the flexibility of anonymous participation can improve the public's willingness to participate and consequently, facilitate the decision-making process.

The aim of this work is to define tools which can be useful for people involved in a spatial decision-making process, specifically one that includes the participation of the public, and needing, at some point, to collaborate with each other. Collaboration, in

L. Carriço, N. Baloian, and B. Fonseca (Eds.): CRIWG 2009, LNCS 5784, pp. 303–310, 2009.

this context, may involve the sharing of information, ideas and events, all with a geographic footprint.

Using recently made available online mapping application development tools [3], a collaborative system was implemented, facilitating geographic data handling on a publicly available interface. The aim of the system is to improve the spatial decision-making process in situations where users are physically distributed and cannot be collaborating in real-time. A configurable platform is featured, enabling various scenarios and user expertise.

The basic element for interaction and collaboration is a map window enabling browsing, annotating, querying and visualizing of geographical data. Users define spatial interests by delimiting relevant geographic regions, on which will be noted, to take part in relevant discussions.

The system was developed to generically support online spatial collaboration and specifically public participation activities and was later incorporated into the Local Agenda21 public participation process.

The rest of the paper is structured as follows. Section 2 presents the background of our work. Section 3 presents the technologies involved, followed by the description of the user interface and the interaction options and Section 4 presents the results from the usability tests. The paper concludes with some conclusions and describes future work.

2 Background

As stated above, although GIS applications are often well suited to support human collaboration towards spatial decision, the available tools do not, by definition, support it [4]. However, various projects have now been developed to allow group work with geospatial information (e.g. [5], [6], [7]).

Asynchronous spatial collaborative systems enable members of a collaborative workgroup to think thoroughly on a task at hand and take a pondered decision, freeing participants in terms of time and place. Relevant research occurred within the context of public participation GIS (e.g. [5]), where two web-based public participation GIS are described.

Online Spatial decision support systems allow physically distributed users to collaborate over the Web, on geographically relevant issues [4]. In PPGIS, early work focused on examining the potential of the Internet to support public participation in environmental decision making, [2, 9]. More recently, web-based PPGIS are being used in local policy decision-making [6], with users spatially reporting local environmental and social issues on a website.

All these experiences regarding public participation lacked a generic approach for supporting discussion and flexibility toward users' expertise. Moreover, web-based mapping systems are inherently complex in comparison with other web applications, leading to less usable systems [8].

Until recently, the development of collaborative systems did not involve associating a geographic footprint with collaborative activities, while online mapping tools lacked flexibility in collaborative issues. Recent development in maps APIs [3] like the Google Maps API, provides developers the capacity for improving existing online spatial decision-making systems, facilitating the availability of spatially collaborative activities. Moreover, Web 2.0 technologies, where users can directly generate content to share with other users can greatly enhance current PPGIS [9]. The creation and sharing of spatial information and different types of media have provided the public new means of expressing their opinion.

The presented work takes advantage of these developments, which provided new means to support online spatial decision-making, and enabled the focus on the development of new tools to support spatial decision-making.

3 Case Study

The Agenda21 is an initiative, proposed by the United Nations (UN), which aims to encourage sustainable development[1] through the participation of concerned parties in the decision making process, and it is being applied at several levels and contexts, by the Portuguese Local Government. The process enforces public participation and Local governments aim to encourage this behavior, providing support to it through a gathering spatial platform – a map - where users can locate their opinions, thus enriching the decision-making process. The limitations involved in common practice are related to compulsory planning meetings, which often take place in an atmosphere of confrontation at specific times and locations, which limits access and debate.

Agenda21 is currently being implemented locally in municipalities and is generally supported by academic research teams, like CIVITAS[2], who supported the definition of requirements for this work. They helped understand the general types of collaboration activities which support public participation in sustainable development as well as specifically in the context of Agenda21.

The aim of the developed system is not to replace public participation meetings but rather to complement them, by enabling people who cannot attend these meetings to participate in the decision-making process.

Meetings between the authors and the CIVITAS expert team resulted on the definition of a set of tools, to be made available by a collaborative system for public participation in a sustainable development context. This result constituted a model of the requirements needed for the design of the generic architecture, described below.

4 Generic Architecture

The system architecture used is based on the one proposed by Chow (2008), as the conceptual architecture of a Web application that uses Maps API. Fig. 1 shows the implemented architecture, which consists on a web application hosted in a web server that will return information upon the request of a web browser. Javascript was used to connect to the Maps API. The system loads and exports Keyhole Markup Language[3] (KML) files and the PHP language is used to connect to a MySQL database, storing

---

[1] http://www.un.org/esa/sustdev/documents/agenda21/english/agenda21toc.htm
[2] CIVITAS (Centro de Estudos sobre Cidades e Vilas Sustentáveis) is a research centre of FCT/UNL, whose mission is to support local government in pursuing local sustainable development (http://civitas.dcea.fct.unl.pt/).
[3] An Open Geospatial Consortium standard language to visualize/present geographic information.

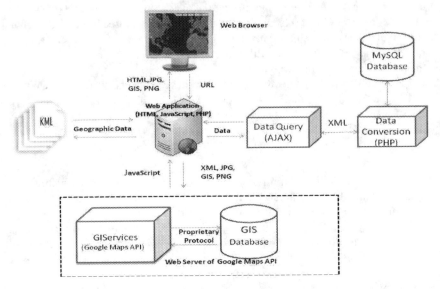

Fig. 1. The architecture of the system (based on [3])

all information. Geographic data is stored through the MySQL spatial extension which provides generation, storage and analysis of geographic features.

The set of geometry types used in the system are based on the OpenGIS Geometry Model (e.g. Points, LineStrings and Polygons, etc) and the chosen Map API was the Google Maps API

Communication between client and server sides, regarding spatial information, is provided through Asynchronous Javascript and XML (AJAX) technology [10].

5 Features

The implemented set of spatial operators resulted from the evaluation of public participation activities led in collaboration with the CIVITAS expert team. They constituted a whole in terms of giving support for setting up a public participation GIS, providing means for the public to actively participate in the decision-making process, involving geographically-referenced information.

Two different roles are identified in the public participation context: participants and moderators. Moderators are responsible for configuring and monitoring the decision-making process, while Participants are contributors to the collaborative process. Different features are available to each role.

5.1 Moderator Features

A system moderator is responsible for configuring the system available to the public, by choosing the appropriate set of tools to support a specific public participation process. In this context, the moderator defines the relevant geographical area for public participation, with can be accomplished by choosing from a list of Portuguese counties (Fig. 2) or by drawing a polygon (Fig. 3).

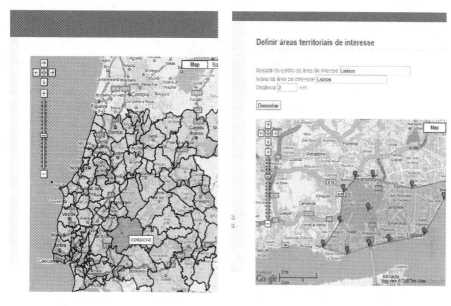

Fig. 2. Define area (by choosing a county)   **Fig. 3.** Define area

The moderator is also responsible for defining categories of information, for adding geographically-referenced news to the site, and for configuring geo-referenced polls (e.g. different proposals for a road's route).

5.2 Participant Features

Participant features include the set of tools found relevant for the participation of the public.

Participants may submit a geographically referenced opinion (Fig. 4). Users choose the topic that is relevant to them and may add a detailed description of his/her opinion, which may include other types of media like images, links and videos. The system cross-references opinions with related geographic footprints (within 1km radius) discouraging the creation of similar topics in the same geographic area, and thus promoting debate between users with similar interests.

Users can also comment each other's opinions. Opinion givers will be noted each time their topics are commented. E-mailing is also available.

The system offers users the possibility of mapping their spatial interests (e.g. an area around his/her house or his/her children's school). This feature works in the same manner as the one in Fig. 2. Users are noted when someone submits an opinion on his/her area of interest. This enables a user to keep track of relevant subjects being discussed and contribute when needed.

As said earlier, users are able to read and comment the news inserted by the moderator, as well as to vote on the polls available on the website. A GeoRSS feed is available to allow users to keep up to date with the public participation process. To take part on this collaborative process, users must register on the site.

Fig. 4. Opinion submission feature. Relevant information includes the author's name, topic location (acquired by reversed GeoCoding), description, associated categories, URL, image and comments.

6   Usability Testing

The aim of the performed usability tests was to assess facility of use by the public and by the relevant technical staff. We also aimed at verifying the usefulness of the implemented features.

The realized usability tests consisted on four parts:

1. Filling in an introductory questionnaire: This included questions on gender, age, area of expertise, experience with computers, experience using digital maps and previous knowledge of public participation processes;
2. Reading a briefing document giving context and explaining the system's goals and features;
3. Evaluation of the users' interaction with the system: they would try to execute a small number of tasks (5-8), involving the features described in section 5. Difficulties and successes were registered;
4. Filling in an evaluation questionnaire to assess the usefulness and the potential of the system. It also included some open-ended questions, to allow users to express their likes and dislikes about the application and to make suggestions.

Different usability tests were made to evaluate the moderator features and the features available to the public.

6.1   Evaluating the Moderator's Features

The moderator's features testers were from the CIVITAS research team. They provided an efficient assessment of the usefulness of this application. They also evaluated the usability of the application for creating and managing a spatial decision-making process, in this context.

The results of the test were quite satisfactory (Mean = 2.66, Standard deviation =0.50), with the moderators accomplishing most (24 in 32) tasks without any help, which is an indication that the system is easy to use. In the open-ended questions, the most praised aspect was the addition of a spatial component in public participation processes and its potential to improve current processes.

6.2   Evaluating the Participant's Features

Two different testing sessions, for different user profiles, were conducted to test the features available to the public. Testing with groups from different backgrounds is important since collaboration in public participation processes usually involves people with different levels of knowledge and expertise.

The first session involved non experts in the subject, thus providing a thorough assessment of the developed features [11].The goal of non-expert testing was to evaluate the usability of the system for the general public. Testers were people from different areas, with some experience with computers although unrelated to public participation or spatial collaboration. Once again, the results were quite satisfactory (Mean = 2.78, Standard deviation =0.58).

The second session involved the technical board of the Oeiras county, professionals acquainted with spatial-decision making problems. The aim here was to evaluate the participant's features with involved professionals. The results were once again,very good (Mean = 2.88, Standard deviation =0.33), with most of the tasks being accomplished easily without any help (22 out of 25). Good usability and increased receptivity for public participation issues were the two main conclusions from the post-testing questionnaire.

These results were important in proving that the system can effectively support spatial collaboration in public participation processes through a set of easy to use spatial features.

7   Conclusions and Future Work

This paper presented a set of tools, implemented in a case-study system, to support online spatial decision-making for physically distributed users, involved in a public participation process. The time-consuming, elevated cost of traditional public participation practices reiterates the need for a generic and low cost approach to public participation. The developed system uses recent developments in maps APIs and Web 2.0 technologies in this context.

Overall, the results from the tests were encouraging, as to the potential of the application to support spatial decision-making processes, where the public is involved. The application's ability to provide easy to use and relevant spatial features to participate in a geo-referenced discussion were also praised.

Future work involves using the system as a parallel medium to a public participation meeting, organized by local government in the Agenda21 process. Another possibility for future development, already under way, is the adding of synchronous tools, enabling users to participate in online meetings with a common geographic interest.

References

1. Muntz, R.R., Barclay, T., Dozier, J., Faloutsos, C., Maceachren, A.M., Martin, J.L., et al. (eds.): IT Roadmap to Geospatial Future, report of the Commitee on Intersections Between Geospatial Information and Information Technology. National Academy of Sciencs Press, Washington (2003)
2. Kingston, R., Carver, S., Evans, A., Turton, I.: A GIS for the public: enhancing participation in local decision making. In: GISRUK 1999, Southampton (1999)
3. Chow, T.E.: The Potential of Maps APIs for Internet GIS aplications. Transactions in GIS, 179–191 (2008)
4. MacEachren, M.A.: Cartography and GIS: extending collaborative tools to support virtual teams. Progress in Human Geography 25, 431–444 (2001)
5. Carver, S., Evans, A., Kingston, A.: Public participation, GIS, and cyberdemocracy: evaluating on-line spatial decision support systems. Environment and Planning B: Planning and Design 28, 907–921 (2001)
6. Kingston, R.: Public Participation in Local Policy Decision-making: The Role of Web-based Mapping. The Cartographic journal 44(2), 138–144 (2007)
7. Cai, G., MacEachren, A.M., Sharma, R., Brewer, I., Fuhrmann, S., McNeese, M.: Enabling GeoCollaborative crisis management through advanced geoinformation technologies. In: Proceedings of the 2005 national conference on Digital government research, pp. 227–288. Digital Government Society of North America, Atlanta (2005)
8. Warren, D., Bonaguro, J.: Usability Testing of Community Data and Mapping Systems (2003), http://www.gnocdc.org/usability/usabilitytesting.html
9. Park, S., Lee, J., Choi, Y., Nam, J.Y.: To improve a public participation decision support system based on web 2.0 technology. In: Proceedings of the Twenty-Eighth Annual ESRI User Conference. ESRI, San Diego (2008)
10. Garret, J.J.: Ajax: A New Approach to Web Applications. Technical report (2005), http://www.adaptivepath.com/publications/essays/archives/000385.php/
11. Alsop, R.: 'What's in your backyard?' A usability study by persona. In: Proceedings of 17th GISRUK Conference, Durham, UK (2009)

Cooperative Model Reconstruction for Cryptographic Protocols Using Visual Languages

Benjamin Weyers[1], Wolfram Luther[1], and Nelson Baloian[2]

[1] University of Duisburg-Essen, INKO, Lotharstr. 65, Duisburg, Germany
[2] U. de Chile, Department of Computer Science, Blanco Encalada 2120, Santiago, Chile
nbaloian@dcc.uchile.cl, {luther,weyers}@inf.uni-due.de

Abstract. Cooperative work in learning environments has been shown to be a successful extension to traditional learning systems due to the great impact of cooperation on students' motivation and learning success. In this paper we describe a new approach to cooperative construction of cryptographic protocols. Using an appropriate visual language (VL), students describe a protocol step by step, modeling subsequent situations and alternating this with the creation of a concept keyboard (CK) describing the operations in the protocol. The system automatically generates a colored Petri subnet that is matched against an existing action logic specifying the protocol. Finally, the learners implement role-dependent CKs in a cooperative workflow and perform a role-play simulation.

Keywords: Cooperative construction, cryptographic algorithm, learning environment, dialog and interaction logic modeling.

1 Introduction

Teaching algorithms is a central part of computer science studies. Cryptographic algorithms and protocols are of special interest for distributed learning environments because of their strong formal requirements as well as the communication partners involved and their widely divergent roles. The protocols deal mainly with the private exchange of information between two or more parties; that is, they enable two or more parties to share specific information without other parties who use the environment being able to access that information. Earlier research and development produced a system called CoBo [1], implementing a distributed learning environment for teaching cryptographic algorithms supporting a component for algorithm visualization. Evaluation of students' success demonstrated the efficacy of this distributed learning system [2], resulting in a collaborative extension described in [3].

In the latter paper, we introduced a workflow for the cooperative creation of concept keyboards (CK) in distributed learning environments. The touch-sensitive surface of the CK detects a finger tip on the sheet of paper lying on it. The sheet of paper is totally flexible and can be redesigned in any way. The workflow presented is motivated by the fact that, if students are involved in the keyboard creation, they are motivated to analyze the algorithm and to build a mental model of the underlying action logic [1]. They have to understand the algorithm and then find an individual solution for a usable keyboard mirroring their own parts in the cooperative human-machine dialog and matching the computational and mental process model.

L. Carriço, N. Baloian, and B. Fonseca (Eds.): CRIWG 2009, LNCS 5784, pp. 311–318, 2009.
© Springer-Verlag Berlin Heidelberg 2009

However, the development of a mental model of the process flow was addressed only implicitly by creating input interfaces. CK creation fails to represent the context in which the operations on the keyboard must be accomplished, which motivates the addition of a further cooperative introduction phase to the protocol. It will start with an initial theoretical instruction by the teacher, followed by a description of the subsequent situations and actions of the protocol which is created by the students step by step, using predefined expressive icons and textual help or an interactive helpdesk system, like the one implemented in CoBo. By combining this creation of a situation in the protocol paired with the creation of a CK with buttons representing actions appropriate to the current situation, the student can incrementally build a complete formal and cognitive model of the entire protocol. Contribution of the presented work is therefore that this creative task completely involves students in the construction process of the protocol, which results in deeper understanding.

The cooperative reconstruction takes place in two phases. The first —on the students' side (described above)—consists of characterizing a certain situation using a visual language (VL) and adding an appropriate operation to a keyboard to launch it. The second phase—on the computer's side—matches the logic just created to an existing action logic specifying the protocol. The students receive direct feedback letting them know whether their created solution is correct or not. If needed, they can consult the helpdesk system to find the correct solution or parts of it. This process is comparable to a model checking approach based on Petri nets (PN).

In an extended scenario [4], an intruder is involved in a communication intended to take place between only two communication partners. The exercise for the students is to modify the algorithm in such a way that the intruder is completely locked out. Because of the flexibility of a VL in describing the protocol situation, it is possible to introduce new steps into the protocol, extending its basic action logic. This can be done with rule-based transformations of PNs [5] created by the process described above. Automatically checking those dynamically created extensions of the former algorithm against a formal description is more complex than the simple model checking approach presented in this paper. Nevertheless, we are convinced that the approach presented in this paper can serve as a basis for the complete generation or dynamic extension of cryptographic protocols and even more complex processes through rule-based transformations of PNs.

2 Recent Work

Collaborative learning has shown group work to have an important impact on achieving learning success [6, 7]. Small group work presents opportunities for students to share insights [8] and observe the strategies of others [9], which is helpful in the context of algorithm learning. Since 1988 many algorithm visualization (AV) systems have been developed (see Stasko et al. [10] and Diehl et al. [11]). Various repositories exist all over the Web for AV, such as [12]. The AV of cryptographic protocols is addressed in Cattaneo et al. [13]. Kerren, Müldner and Shakshuki describe solutions for AV and explanation systems [14] as well as for Web environments based on hypertext languages [15]. Basic research has also been conducted by Eisenberg [16] and others on algorithm visualization.

Archer [17] describes how human beings use visual artifacts and languages to materialize their cognitive thoughts and worlds. This is particularly important for learning environments that focus on the reflection of learned knowledge and therefore the materialization of abstract concepts and mental models [18].

Several tools are implemented for modeling PNs interactively (see [19]). Formal modeling of dynamic systems and controlled model adaptation via Petri net transformations are described by Ehrig et al. [5, 20]. This is done with a stepwise development or replacement of place/transition nets in a well-defined context following established rules.

Earlier research [1, 21] and evaluation [2] was conducted in our group to investigate problems with cryptographic algorithms in distributed learning environments. This research found that students need and want an initiation phase in the learning process where they are introduced to the protocol and become involved in its execution before they begin a free exploration phase. In [3] we introduce a new collaborative design of CKs in the learning process to meet this requirement.

3 Construction of Cryptographic Protocols

This section will describe how protocol reengineering can be implemented using two different formal approaches. The first subsection will provide a brief introduction to the formal definition of VLs followed by a concrete example of a cryptographic protocol. The second subsection will give a short description of the PN type used and describe a special graph transformation for PNs. The last subsection introduces the transformation from a VL to PNs.

Visual Languages: Chang [22] defines a VL as a set of elementary icons and a set of operations that creates complex icons using the elementary ones. Every icon has a physical part, the image, and a logical part, the meaning. The physical part is described as a scalable vector graphic (SVG) and the meaning can be described formally as ontology (such as concept graphs or the Web Ontology Language). An iconic system defines a hierarchical structure of icons with its head icon as root—a whole sentence or a complex icon. An operation on the set of icons defines a physical operation (e.g., the direction-dependent combination of physical parts of icons) and a logical operation describing how to create the new meaning of the resulting complex icon.

A well-known cryptographic protocol for key sharing (NSP) was introduced by R. M. Needham and M. D. Schroeder in [23]. The aim of this protocol is the secure sharing of a session key between two participants A and B using an unsafe communication channel. The characteristic property is the use of a Trusted Third Party (TTP) that has already previously shared keys with A and B. Icons for visually describing the protocol situations are shown in Figure 1. A specific characteristic of this algorithm is the problem of a man-in-the-middle attack, which leads to a cooperative extension scenario for reconstructing cryptographic protocols.

Petri Nets: A Petri Net [24] is a directed graph with two different types of nodes: transitions and places that are mutually connected by (possibly weighted) edges. Every place can hold tokens. These tokens can be consumed or generated by connected transitions depending on incoming (consuming) or outgoing (creating) edges. A transition is enabled if enough tokens are on its incoming places (depending on the weight of the edges) and if there is enough free space for generated tokens on the outgoing places. A net has an initial marking, which is a vector of natural numbers in which every element describes the number of tokens held by a certain place as indicated by the position of the number in the vector.

Jensen [25] introduced colored PNs with additional features. Every token is allocated an individual color which represent simple or complex data types. Special types of places and transitions are introduced, namely, fusion places allowing for synchronization of places and transitions containing programming code, guard functions and time delays. Edges are extended by attributing complex conditions.

Formal Conversion of Visual Sentences to PN-Based Action Logic: De Rosis et al. [26] describe the transformation of a user's mental model of problem solving to a usable interface employing PN transformation based on formal definition of physical and logical relations. Combining de Rosis's concepts and Chang's formal VLs [22] we have developed a new approach to describe cryptographic protocols in a cooperative environment. By converting visual designed context paired with a designed CK to a Petri subnet and matching it to the existing action logic of the protocol using a model tracer, this approach is embedded in a cooperative learning workflow, which is described in the next section.

Figure 1 shows the cooperative interface for reconstruction of the NSP. The left side shows the situation where Alice has received an encrypted message from the TTP and now wants to send it to Bob. The students' task is first (1.) to select an appropriate operation that matches the current situation (in the context of the NSP) and then (2.) to model a changed situation that results from the appliance of the operation.

Thus, they must first add the "send" button to the CK and then move the message icon from the Alice icon to the Bob icon, which results in the situation shown on the right in Figure 1. The second step consists in an icon operation supported by the cooperative environment and the VL.

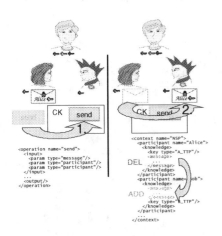

Fig. 1. Reconstruction of a cryptographic protocol

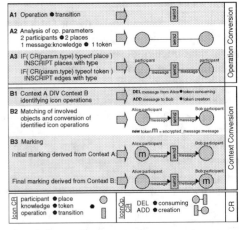

Fig. 2. Conversion Algorithm to convert a VL to a PN-based representation

To realize this reconstruction, we have developed the following three-part system (shown in Figure 3). The Icon Interpreter automates syntactic and semantic analysis of an iconic system by generating XML descriptions from the visually designed situation of the protocol as shown in Figure 1. This generation is based on the implemented XML-based parser and interpreter engine in CoBo. (How to replace this simple way of interpreting a VL that certainly does not meet the requirements of a classical compiler is the subject of further research.)

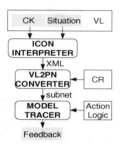

Fig. 3. Architecture

The second part is the VL2PN-converter, which converts the XML description resulting from the Icon Interpreter to a Petri subnet, using the algorithm shown in Figure 2. The algorithm employs a set of conversion rules (CR) stored in the CR database (shown in the lower part of Figure 2) and is split into two phases. The CR has to be implemented by experts or is automatically generated concerning specifications of the VL.

During the first phase, the XML description of the selected operation (resulting from (1.) in Figure 1) is interpreted. In step A1, the operation icon (physically displayed as a button) is converted to a transition using a rule in the set of icon conversion rules (Icon CR). In step A2, the logical part described in XML is analyzed to produce the number of involved items, like places and tokens, again using the CR. Resolved places are connected by undirected edges to the transition. The direction is resolved by the context analysis in the second phase. The information about tokens (a message token in the example) is stored for a later conversion step. The last step, A3, resolves the inscriptions for the edges and places in the basic topology of the subnet. The edge inscription defines the token types that can be consumed or generated by the connected transition. The place inscription is semantic and will be instantiated later with the names of the participants in the protocol.

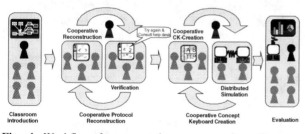

Fig. 4. Workflow for cooperative reconstruction of cryptographic protocols

The second phase of the algorithm (indicated by a capital B) uses the XML description of the context A and the resulting situation (context B). The first step, B1, calculates the difference between context A and context B. The result is a DEL and an ADD icon operation. These operations are converted using the Icon Operation Conversion Rules (Icon Operation CR) in the CR database. The deletion operation (DEL) is interpreted as the consuming of a token, and the addition operation (ADD) as the creating of a token. Therefore, in step B2, the participants are first instantiated, which means the place inscriptions are extended by the concrete participants in the situation (Alice and Bob). Secondly, the edges can be changed to direct ones, creating the time context for the DEL operation on Alice's knowledge (consuming) and the ADD operation in Bob's knowledge (creating). Also, the token found in step A2 can be instantiated by the message, which is initially part of Alice's knowledge. In the last step, B3, the marking is resolved from context A and context B by converting the parts of the knowledge extracted from the XML description to the PN.

The last system component (the model tracer) matches the subnet resulting from the conversion algorithm to the original action logic, modeled as PN. Executing the matching in a step-by-step reconstruction of the protocol, the PN acquires a certain marking, depending on the situation represented. In the next step, the operation is mapped to a transition in the existing PN, and the new marking has to be calculated before the marking of the subnet is matched. If the new marking of the given action logic matches this one of the subnet, the reconstruction step was correct.

4 Cooperative Construction

The workflow for the cooperative construction of cryptographic algorithms is shown in Figure 4. After a short introduction to the protocol in the classroom, students are split into several groups, each working on one instance of the system using one physical interface. The following reconstruction is solved cooperatively in this interface. The students receive a set of operations in the form of buttons as well as a set of icons to describe the situation in which the operation should be selected. First, the students create the initial protocol situation showing the three communication partners. Next, they choose an operation adapted to this situation. The system checks this constellation by applying the conversion algorithm described above and matching it against the existing action logic, which is followed by a discussion-and-decision phase on what to do next. A solution check can yield the following outcomes:

1. The subsequent marking of the action logic cannot be calculated: The applied operation is incorrect, and the students must choose another one.
2. If the converted subnet does not match the subsequent marking of the action logic, the situation created by the students is incorrect.
3. If the marking of the subnet matches the subsequent marking of the action logic, the solution is correct, and the students can continue with the cooperative reconstruction.

This process will be iterated until the final situation is reached (that is, a session key is shared).

The second phase of the workflow is subsequent cooperative creation of role-dependent keyboards using the cooperative system detailed in our previous paper [3] and summarized above. Finally, the group simulates the protocol in a distributed manner, in which each student embodies one role in the protocol using his/her individual keyboard. Possible errors in the creation phase can then be fixed in the next step, reconfiguration of the individual keyboard, which is followed by a new simulation phase.

Through the combination of both approaches in one cooperative workflow, students learning algorithms at differing proficiency levels are optimally supported by the system, Thus, to become familiar with the protocol, beginners proceed step-by-step with the keys organized in the right order on the keyboard, whereas advanced students use a self-configured keyboard that allows them to explore an algorithm and learn from their mistakes. We are confident that this approach represents a solution to the problems diagnosed in the evaluation of CoBo [2]. A new evaluation will be conducted in June 2009 to confirm this claim.

5 Conclusion and Future Work

We have described an approach to a cooperative reconstruction of a step-by-step simulation of cryptographic protocols. This is still a work in progress; nevertheless, what we have accomplished so far represents a significant contribution to our further work. The CoBo system implements all basic components, including algorithm visualization for the VL and a configurator for the CKs. The action logic is implemented for four well-known protocols using colored PNs and can be used for the matching process described above.

What remains to be completed is the implementation of the system providing the visual interface and data structures described in Section 3. The entire workflow also has to be evaluated to identify shortcomings and future work.

There are many possible extensions of this basic approach. The use of a completely flexible icon interpreter as described in [27] offers the opportunity to avoid hardcoding any syntactic or semantic interpretation of a VL. Such interpreters can be generated using new approaches for the interactive compiler-compiler techniques proposed in [28] and [29], enabling extended learning scenarios, like the NSP with intruder. The idea is to completely generate the action logic by using only the described approach based on VLs in combination with CKs. The action logic will not be matched against the existing action logic but be analyzed by more complex methods described in recent papers on graph transformation and rewriting.

Another workflow including a VL compiler and a completely flexible generation of action logic can be used to describe the role-dependent parts of a protocol and then, in a second cooperative step, to bring those partial solutions together to form one coherent PN using graph transformation operations and tools [20]. Once the solution is automatically checked for correctness, the distributed simulation can follow.

This approach and suitable extensions can be used for other algorithms resulting in more complex scenarios. Tools like ConKAV developed in our group [30] were created for simulating complex data structures like AVL trees. Those applications are not as linear and deterministic as cryptographic protocols are. Considerations during the simulation phase are more complex because of the variety of solution methods, which can be better or worse for certain initial values than the standard solution.

References

1. Baloian, N., Breuer, H., Luther, W.: Concept keyboards in the animation of standard algorithms. Journal of Visual Languages and Computing 6, 652–674 (2008)
2. Baloian, N., Luther, W.: Cooperative visualization of cryptographic protocols using concept keyboards. To appear in International Journal of Engineering Education (2009)
3. Weyers, B., Baloian, N., Luther, W.: Cooperative Creation of Concept Keyboards in Distributed Learning Environments. In: Proc. of 13th Int. Conf. on CSCW in Design, pp. 534–539. IEEE, Santiago de Chile (2009)
4. Kovácová, A.: Implementierung des Needham-Schroeder Protokolls in einer verteilten Simulationsumgebung für kryptografische Standardverfahren, Duisburg (2008)

5. Ehrig, H., Hoffmann, K., Padberg, J.: Transformations of Petri Nets. Electronic Notes in Theoretical Computer Science 148, 151–172 (2006)
6. Nickerson, R.R.: On the distribution of cognition: some reflections. Distributed cognitions, 229–259 (1997)
7. Webb, N.M.: Peer interaction and learning in small groups. International Journal of Education Research 13, 21–39 (1986)
8. Bos, M.C.: Experimental study of productive collaboration. Acta Psych. 3, 315–426 (1937)
9. Azmitia, M.: Peer Interaction and Problem Solving: When Are Two Heads Better Than One? Child Development 1, 87–96 (1988)
10. Stasko, J.: Software visualization. MIT Press, Cambridge (1998)
11. Diehl, S.: Software visualization. In: Proceedings of the 27th international conference on Software engineering, pp. 718–719 (2005)
12. Crescenzi, P., Faltin, N., Fleischer, R., Hundhausen, C., Näher, S., Rössling, G., Stasko, J., Sutinen, E.: The Algorithm Animation Repository. In: Proceedings of the 2nd International Program Visualization Workshop, vol. 1, pp. 14–16 (2002)
13. Cattaneo, G., Santis, A., de Ferraro Petrillo, U.: Visualization of cryptographic protocols with GRACE. Journal of Visual Languages & Computing 19, 258–290 (2008)
14. Kerren, A., Müldner, T., Shakshuki, E.: Novel algorithm explanation techniques for improving algorithm teaching. In: Proc. of the 2006 ACM Symposium on Soft. Vis., pp. 175–176 (2006)
15. Shakshuki, E., Kerren, A., Muldner, T.: Web-based Structured Hypermedia Algorithm Explanation system. International Journal of Web Information Systems 3, 179–197 (2007)
16. Eisenberg, M.: The Thin Glass Line: Designing Interfaces to Algorithms. In: Proceedings of the SIGCHI Conference on Human factors in computing systems, pp. 181–188 (1996)
17. Archer, B.: Design as a discipline. Design Studies 1, 17–20 (1979)
18. Meyer, H.: Unterrichts-Methoden. Cornelsen Scriptor, Frankfurt am Main (2008)
19. Petri Nets World,
    http://www.informatik.uni-hamburg.de/TGI/PetriNets/tools/quick.html
20. Ehrig, H., Hoffmann, K., Padberg, J., Ermel, C., Prange, U., Biermann, E., Modica, T.: Petri Net Transformations. Petri Net, Theory and Applications, 534–550 (2008)
21. Baloian, N., Luther, W.: Algorithm Explanation Using Multimodal Interfaces. In: Proceedings of the XXV International Conference on the Chilean Computer Science Society, vol. 21 (2005)
22. Chang, S.-K.: Iconic Visual Languages (1997),
    http://www.cs.pitt.edu/~chang/365/sk1.html
23. Needham, R.M., Schroeder, M.D.: Using encryption for authentication in large networks of computers. Communications of the ACM 21, 993–999 (1978)
24. Baumgarten, B.: Petri-Netze. Spektrum Akademischer Verlag, Heidelberg (1996)
25. Jensen, K.: Coloured petri nets. Springer, Berlin (1997)
26. de Rosis, F., Pizzutilo, S., De Carolis, B.: Formal description and evaluation of user-adapted interfaces. International Journal of Human-Computer Studies 2, 95–120 (1998)
27. Chang, S.-K.: Syntactic Analysis of Visual Sentences,
    http://www.cs.pitt.edu/~chang/365/sk2.html
28. Costagliola, G., Deufemia, V., Polese, G.: Visual language implementation through standard compiler–compiler techniques. Journal of VLs & Computing 18, 165–226 (2006)
29. Pereira, M.J.V., Mernik, M., da Cruz, D., Henriques, P.R.: VisualLISA: A Visual Interface for an Attribute Grammar–based Compiler-Compiler. In: CoRTA 2008 (2008)
30. Baloian, N., Breuer, H., Luther, W.: Algorithm visualization using concept keyboards. In: Proceedings of ACM SoftVis 2005, pp. 7–16 (2005)

Enacting Collaboration via Storytelling in Second Life

Andréia Pereira[1], Katia Cánepa[1], Viviane David[3], Denise Filippo[2],
Alberto Raposo[1], and Hugo Fuks[1]

[1] Department of Informatics – Pontifical Catholic University of Rio de Janeiro (PUC-Rio)
R. Marquês de São Vicente, 225, Rio de Janeiro, RJ, 22453-900, Brasil
[2] Superior School of Industrial Design – State University of Rio de Janeiro
Rua Evaristo da Veiga, 95, Rio de Janeiro, RJ, 22031-040, Brasil
[3] Groupware@LES
R. Marquês de São Vicente, 225, Rio de Janeiro, RJ, 22453-900, Brasil
{asoares,kvega,hugo,abraposo}@inf.puc-rio.br,
vivifelipe@yahoo.com.br, dfilippo@esdi.uerj.br

Abstract. This work presents a collaborative educational game, Time2Play, developed in Second Life, which allows the creation of stories in a collaborative fashion, offering a new form of expression in education. This game is projected for children from 7 to 12 years old, enabling them to express their creativity and imagination by creating and enacting stories of their own. The proposed game can become a resource for implementing collaborative projects in school activities.

Keywords: Collaborative Learning, 3D Collaborative Virtual Environments, Storytelling, Second Life.

1 Introduction

Virtual worlds are interactive multi-user environments where users are represented by avatars to communicate and interact with other users. These environments allow for better immersive interaction than 2D environments regarding simulations creating, experimental learning, scenario modelling, and opportunities for collaboration and co-creation. The learner is immersed in an environment together with other learners and teachers, not necessarily co-located, combining the advantages of distance learning and classroom learning.

Second Life is a 3D virtual environment that simulates some aspects of social real life. This research uses Second Life as a platform for collaborative educational games, featuring Time2Play, a game that allows users to create and enact stories collaboratively.

In their knowledge construction process, children use several languages to express their ideas [1]. This kind of environment enables them to express their thoughts, creativity and imagination and to enjoy learning proactively. Telling stories give learners the possibility to encourage their reading habits and imagination process. They also develop a better knowledge of their own culture by creating important references for their own development. When learners develop a story collaboratively, they learn to collaborate with colleagues and to develop their capacity for interpersonal relationships.

L. Carriço, N. Baloian, and B. Fonseca (Eds.): CRIWG 2009, LNCS 5784, pp. 319–327, 2009.
© Springer-Verlag Berlin Heidelberg 2009

The use of 3D collaborative environments is still poorly exploited in storytelling. This paper investigates the use of 3D collaborative environments such as Second Life in the process of planning and enacting stories. It aims at providing the basis for developers of this type of game regarding 3D objects resources that must be available in these games.

This paper is organized as follows: sections 2 and 3 present a bibliography on storytelling and interactive collaborative virtual environments in collaborative learning. Section 4 discusses Second Life and its use as an educational environment. Section 5 presents the Time2Play game and Sections 6 and 7 present a proof of concept and its evaluation. Section 8 concludes this paper.

2 Storytelling

Interactions with colleagues of the same age, adults and world objects, real or imaginary, are essential to the learning process and to the recognition of the learner as a reference subject. Storytelling communicates, entertains, teaches collaboration and preserves our cultural traditions and memories, playing a key role in the learning development. Telling stories is a natural way of transmitting knowledge among individuals. When a story is told, the author's intention is to transmit knowledge to the others people [2].

People think the world as stories. New events or problems are understood with reference to stories previously understood and these new events are conveyed to other people as stories. The mind can be seen as a collection of stories, of experiences already lived [3].

Through stories learners enrich their learning experience by developing several language forms, expanding their vocabulary and living the imaginary. Stories take learners to another world and make them feel the emotions and feelings it awakens. Narrative allows learners to become immersed in the story and to meditate upon it, favouring the development of artistic sensibility.

The proposal is for learners to release their capacity to create and reinvent the world through play and exploration activities, having their fantasies accepted and exercised and thus, through the magical world of make-believe, being allowed to explore their own limits. With the chance to imagine and create, learners may turn their thoughts to collaborate in solving problems that are important to them.

3 Virtual Environments and Collaborative Learning

Collaboration can be seen as an interplay of communication, coordination and cooperation, Figure 1. Communication is related to conversation, coordination to the management of people, activities and resources and cooperation to the production that takes place in a shared space [4].

Learning to collaborate should be a main goal of education, promoting the division of responsibilities and functions, so that children use this skill in their interactions with society. Collaborative games are effective educational tools that entertain while they motivate, facilitating learning, increasing the capacity of assimilation of what has

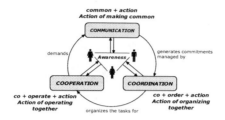

Fig. 1. Diagram of the 3C Collaboration Model

been taught and exercising the player's mental and intellectual abilities. The active exchange of ideas within small groups not only increases interest in what is being learned but also promotes critical thinking and the ability to solve problems.

Technology creates environments and they involve us by providing a set of conditions and objects that we perceive and interact with. 3D Virtual Environments are worlds simulated by a computer that allow three-dimensional exploration by means of dynamic interactions between users and the environment. Objects can be moved or transformed and react to the user; sounds may provide a context for information in the environment. There are different types of 3D virtual environments, but all have features in common, such as [5]:

- Shared space: the environment allows many users to participate concomitantly;
- Graphical Interface: the world shows the space visually, ranging in style from 2D animations to 3D immersive environments;
- Promptness: interactions occur in real time;
- Interactivity: the environment allows the user to modify, develop, construct or send customised content;
- Persistence: the environments continue to exist whether or not individual users are connected;
- Socialisation/Community: the environment allows and encourages the formation of social groups such as teams, corporations, clubs, republics, neighbourhoods, etc.

This latter feature, unlike the others, is social rather than technical, demonstrating that these environments are capable of awakening the socialisation of thought, forming groups of individuals who have the same desires and get together to achieve them, an important characteristic for collaboration. There are many possible applications for 3D virtual environments, from the ability to teach classes, do business and conduct conferences and meetings, to the creation of virtual exhibitions and even the 3D visual recreation of buildings and places that no longer exist.

An important feature is the fact that these systems are dynamic: scenarios change in real time as users interact with the environment. Collaborative 3D Virtual Environments favour discussions through the exchange of information and ideas, group problem-solving and the construction and reconstruction of thinking, motivating participation by awakening respect among partners and a sense of joint responsibility. This results in greater interaction among learners and participants.

4 Second Life and Educational Technology

Second Life is a digital, online 3D world created and maintained by its residents, who shape it according to their needs and imagination. Used as an educational tool it is configured as a step beyond e-learning, as the environment provides a space for the exchange of experiences featuring interaction among students and schools.

Many universities and companies are using Second Life for training. As an example we can mention the Sci-Lands Project, which features a mini-continent focused on scientific and technological education and is formed by important universities and organizations such as NASA, NOAA, the National Physical Laboratory, Elon University, the University of Denver, and the Imperial College London, among others [6]. In Brazil, the Pontifical Catholic University of Rio de Janeiro (PUC-Rio) opted for using Second Life as an environment for experimentation in collaborative activities. The objective is to create an environment of innovation and exchange of knowledge and learning where everything can be created and constructed. There, students produce most of the contents of courses and projects and even establish their own virtual companies. In this environment students are not passive recipients like they would be in a classroom; they participate in the construction of content, creating value together with their teachers and collaborators [7].

Second Life has modelling tools and its own scripting language. This makes Second Life a platform for learner experimentation as they can create their own experiences and learning environments rather than be mere passive consumers of learning.

5 Time2Play – The Game

Time2Play is a game which enables the creation of stories and the recreation of known stories in a 3D environment. In the game each learner has an avatar for enacting stories. Each avatar in Second Life has already an inventory of objects, clothes, textures, animations and sounds, among other elements.

Upon entering Second Life the avatar is teleported to the auditorium. The auditorium features seats, simulating a real theatre and enabling other avatars to watch the story being enacted. The audience avatars can also participate in the story that is being staged, donning new clothes and helping the performance to become even more interesting.

On the backstage of the auditorium, participants find three rooms. The central room has three panels with options for special effects, scenes and objects, as shown in Figure 2.

Fig. 2. The panels of central room

In the Scenarios panel there are several scenarios based on themes such as beach, forest, snowy park, among others. When a learner clicks the forward or backward arrows, the images change displaying all the existing scenarios. Clicking on the image displayed on the panel selects a scenario, which then appears at the centre of the auditorium (Figure 3).

Fig. 3. Auditorium with scenario

The scenarios may be changed with the inclusion of new objects provided by the Objects panel. These changes will take effect only during the use of the scenario, and if another group chooses the same scenario it will appear to them in its original form. Through the Special Effects panel it is possible to include in the scenarios effects such as snowing, floating hearts, butterflies and flying fairies among others.

Some objects in the scenarios have animations. These animations simulate, for instance, fire, wind, water and lighting effects. There are also animations for the avatars, which through the use of spheres available in various places in the scenario or in some of the objects, may sit in different postures. The scenarios also play environmental sounds based on the theme they represent.

There is a room to the left of the closed section of the auditorium with panels that provide clothing, hair, skin, makeup and accessories for the characterization of the avatars according to the theme of the story that they will enact.

This room also features the Characters panel. This panel enables learners to transform their avatars into non-human characters by acquiring an alternative form such as a robot, cat, witch, etc. (Figure 4).

Fig. 4. Characters

Learners can choose among different animations for enriching their avatars during the performance in the room to the right of the backstage. Animations are separated into categories for either male or female avatars, and categories in which the animations can be used by both male and female avatars.

The avatars use voice and text chat as a communication medium to narrate the stories and talk to each other.

6 Time2Play – Proof of Concept

This study evaluated user interaction within Time2Play and the manner in which the process of building stories together in Second Life's shared environment takes place. The evaluation method used is the usability test. The usability test is a process in which representative participants evaluate a product regarding specific usability criteria [8]. This method consists in installing a prototype and evaluating its overall quality. This evaluation is performed through utilisation of the prototype by the user and by observation of his or her navigation, understanding and interaction with the environment. Three sessions were held with the participation of 8 students from age 7 to 12, divided into groups of two or three. Each learner used a pre-created avatar and navigation limitations on account of their age (younger than 18). At the beginning of each session a questionnaire was administered to figure out learner profiles. This questionnaire is presented in Table 1.

Table 1. Profiling Questionnaire

Questionnaire - Time2Play
Identification of user profiles: 1. How old are you? 2. What do you think of storytelling? 3. To what school do you go? In which grade are you? 4. Do you use a computer? If yes, Where? How often? What types of programs do you use? 5. Do you play videogames? If yes, Where? How often? What types of games do you like best? 6. Do you do group work at school? If yes, do you enjoy doing that? 7. Do you know Second Life? If yes, which islands have you visited? Do you like to play in Second Life?

None of the students who participated in this study had any prior knowledge of the usage of the Second Life environment. All participants use the computer at least twice a week, and make use of games and the internet. Only three do not play videogames.

All learners perform group work at school and enjoy this type of activity; only one child reported not liking to tell stories and preferring to listen to them. The tests began with a presentation on Second Life, where the basic actions that can be performed in the environment were demonstrated, such as walking, flying, sitting, using the inventory and communicating through text and voice chat, this being the basic knowledge needed to use the game. This process of learning the environment took about 30 minutes.

Then, a challenge consisting in the collaborative building and enacting of a story using the scenarios, avatar clothing and animations available in the game was proposed. The process of planning, telling and enacting the story took about 40 minutes.

During the tests the behaviour of learners in the process of collaborative construction of stories according to the 3C model of collaboration [5] was observed. The observations to be made are:

- How communication takes place in the environment;
- What communication difficulties are present;
- How coordination takes place during the activity;
- What coordination difficulties are present;
- How cooperation takes place in the environment;
- What cooperation difficulties are present;

The data were collected through direct observation of children performing the task. These observations were made by four people: a pedagogue and three people from the computer science field. An interview with the children was held at the end of the tests giving them the opportunity to express their views on the game and the technology. The script of this interview can be seen in Table 2.

Table 2. Interview

Interview - Time2Play
1. What did you think of creating stories with your friends?
2. What did you think of playing in Second Life?
3. What did you think of Time2Play?
4. How was it like creating stories together with your friends inTime2Play?
5. Would you like to have other objects to include in your stories? If yes, which objects?
6. Would you like to exclude any objects among those that were available? If yes, which ones?

The involvement of learners in the prototyping process makes it possible to understand the nature of their relationship with this type of technology, providing feedback and guidance in the development of games in these environments.

7 Discussion

The sessions show evidences that it is feasible to use the Time2Play game as an environment for collaborative learning. During the sessions it was possible to observe that learners older than 9 were more concentrated than the younger ones, and that the latter encountered difficulties in the operation of the environment and in some occasions lost interest, momentarily changing the focus of their attention to other activities such as going to the bathroom, asking for food and walking around the room.

In all groups one of the learners took the leading role, directing the whole process of construction and enacting of the story. That learner conducted the group in the decisions regarding which topic they would talk about, who would perform each role, who would start the story and at what point who should do what during the enacting of the story.

Communication was held mostly through voice chat, but it was observed that both chat channels were used for the three dimensions of collaboration; for example, the learner who assumed the role of coordinator alternated between text and voice chat to delegate tasks to the other participants. Sometimes learners used text chat to narrate parts of the story, but it was used mostly for private communication among students to convey something they did not want the observers to hear, such as requests for going to the bathroom, when they were too shy to tell the story or to make a malicious comment like "you are boring".

While cooperating, the role of the narrator was seldom used, as all learners told parts of the stories, one of them starting the story to be complemented by the others, giving a direction to the story they were enacting. They all performed the roles of narrators and actors, some being more capable of further developing the role of actors than the others. The learners often physically followed their avatars with their own bodies, dancing and making gestures to imitate them in the real world (Figure 5).

Fig. 5. Children using Time2Play

The need for additional characters-avatars or Non-Player Characters-to perform roles that would steer the story to other directions was observed. Some groups addressed this problem imagining that this additional character was present, without using any object to identify it but giving it life and a voice. Other groups asked some of the observers to join the story to play the role they had imagined. This provides evidence that there is a need for Non-Player Characters to assist in the learners' creation process, thus enabling even a small group of participants to have a story with few participants.

8 Conclusion

The use of technology in education should stimulate creativity and curiosity, and interaction should favour questioning and decision-making and foster collaboration. The creation of learning environments with images, text and multimedia contents enables more dynamic and interactive contexts and immersion enables a better understanding of places, situations and circumstances.

The use of 3D Virtual Environments such as Second Life in educational environments to foster collaboration among learners contributes to the awareness of the actions of participants and to the interaction among themselves and with the environment. In these environments activities are developed in a way that do not require especial I/O devices, sticking to the already well known keyboard, mouse, speaker and microphone.

Used as educational resources these environments bring an opportunity for teachers to propose, use and conduct collaborative educational activities in which students cease to be spectators and become co-creators who observe and participate in their creation.

This study provides evidence that collaborative games in 3D environments for storytelling such as Time2Play may be used by learners between 9 and 12 years old. The demand for additional characters, imagined or requested, brings evidence of the need to create non-player character for this game. The communication channels available were used in the three dimensions of collaboration, being voice chat was the most used one, pointing to the need for microphones, speakers and headphones.

This research has the objective of providing basis for collaborative 3D virtual environment software developers to assess resources that should be made avaialable in educational activities in these environments. Moreover, it also allows the assessment of the potential of environments aimed at supporting the educational development of learners in collaborative activities.

It also contributes to enriching the learning process by enabling, from an early age, contact with technologies which are nowadays increasingly closer to society, especially to children and teenagers, through television, video games, computers and the internet.

References

1. Carvalho, A., Salles, F., Guimarães, M.(orgs): Desenvolvimento e Aprendizagem. Editora UFMG, Belo Horizonte (2002)
2. Carminatti, N., Borges, M.R.S., Gomes, J.O.: Collective Knowledge Recall: Benefits and Drawbacks. In: Fukś, H., Lukosch, S., Salgado, A.C. (eds.) CRIWG 2005. LNCS, vol. 3706, pp. 216–231. Springer, Heidelberg (2005)
3. Schank, R.C.: Tell Me A Story – Narrative and Intelligence. Northwestern University Press, Evanston (1995)
4. Fuks, H., Raposo, A.B., Gerosa, M.A., Lucena, C.J.P.: Applying the 3C Model to Groupware Development. International Journal of Cooperative Information Systems (2005)
5. Virtual Words Review. What is a Virtual Word?
   http://www.virtualworldsreview.com/
6. Scilands. SciLands Virtual Continent, http://www.scilands.org
7. Cunha, M., Raposo, A., Fuks, H.: Educational Technology for Collaborative Virtual Environments. In: The proceedings of CSCWD, 12th International Conference on CSCW in Design, Xi'an, China, April 16-18 (2008)
8. Rubin, J.: Handbook of Usability Testing: How to Plan, Design and Conduct Effective Tests. John Wiley & Sons, Inc., New York (1994)

An Approach for Developing Groupware Product Lines Based on the 3C Collaboration Model

Bruno Gadelha, Ingrid Nunes, Hugo Fuks, and Carlos J.P. de Lucena

Department of Informatics, Pontifical Catholic University of Rio de Janeiro (PUC-Rio)
R.M.S Vicente, 225, Gávea, Rio de Janeiro - RJ, Brazil, 22453-900
{bgadelha,ionunes,hugo,lucena}@inf.puc-rio.br

Abstract. Software Product Lines (SPLs) are a new software engineering technology that aims at promoting reduced time and costs in the development of system families by the exploitation of applications commonalities. Given that different Groupware applications typically share a lot of functionalities, Groupware Product Lines (GPLs) have emerged to incorporate SPL benefits to the Groupware development. In this paper, we propose an approach for developing GPLs, which incorporates SPL techniques to allow the derivation of customized groupware according to specific contexts and the systematic reuse of software assets. Our approach is based on the 3C Collaboration Model that allows identifying collaboration needs and guiding the user to select appropriate features according to their collaboration purpose. A GPL of Learning Object repositories, named FLOCOS GPL, is used to illustrate the proposed approach.

Keywords: Groupware development, software product lines, learning objects.

1 Introduction

Over the past decades, the use of applications to support the collaboration among groups, namely Groupware, is gaining significant popularity. This popularity is due to the fact that sharing the same space and time is becoming a hard task for people with busy schedules. This kind of applications shares a particular set of common requirements [1], resulting in the need of specific software development techniques to support groupware development. One key issue in groupware development is the huge amount of time wasted on implementing infrastructure aspects like protocols, synchronism, session management and others, leaving little time for implementation of innovative solutions [2]. In addition, there are several common functionalities shared by different groupware applications, such as forum and chat, resulting in the development of applications with a common core, but differ in some functionalities, which are according to specific contexts. This scenario calls for approaches that take advantage of these common functionalities, while allowing the customization of applications.

This scenario has been explored by Software Product Lines (SPLs) [3, 4], a new trend in the context of software reuse. SPLs investigate methods and techniques to build and customize families of applications through a systematic method. This approach exploits the commonalities among family members promoting several benefits, such as a reduced time-to-market, lower development costs and quality improvement. A SPL

L. Carriço, N. Baloian, and B. Fonseca (Eds.): CRIWG 2009, LNCS 5784, pp. 328–343, 2009.

[3] refers to a family of systems sharing a common, managed set of features to satisfy the needs of a selected market and that are developed from a common set of core assets in a prescribed way. According to Czarnecki et al. [5], a feature is a system property that is relevant to some stakeholder and is used to capture commonalities or discriminate among products in a SPL.

The issues previously mentioned found on groupware development have also been addressed by several groupware component-based approaches [6, 7]. These approaches have used components [8] mainly for two reasons: product lines and tailorability. Despite accelerating groupware implementation, we have identified some deficiencies in the proposed groupware componentization approaches: (i) they do not cover the domain analysis phase. Domain analysis is essential to capture commonalities and variabilities in a domain, and therefore to identify the features to be supported by the components. In addition, domain analysis defines the scope of the SPL, stating the possible products to be derived from the SPL; (ii) the approaches propose to assemble service components to build customized products; however they do not provide means to deal with fine-grained variabilities allowing customization into the services; and (iii) they do not provide feature traceability. It is essential in SPLs to trace features, because it allows one to know which components implement which features, thus enabling the selection of appropriate artefacts of a SPL in the product derivation process.

In this paper, we present an approach that addresses these identified deficiencies in the development of families of groupware, which we refer to as Groupware Product Lines (GPLs) from now on. The main goal of the approach is to develop GPLs leveraging practices of the SPL development in order to incorporate the benefits provided by SPLs. A GPL is composed of a reusable infrastructure that provides common groupware services and has a production plan based on feature traceability to allow the derivation of applications customized to users' specific needs. The features to be addressed by the GPL are identified and defined in the domain analysis, which is based on the 3C Collaboration Model [9, 10], whose purpose is to classify features into three categories (Cooperation, Communication and Coordination) and to guide the user to select appropriate features according to his needs. Our approach is illustrated with the FLOCOS GPL, which consists of a product line of Learning Object repositories.

The remainder of this paper is organized as follows. Related work is presented in Section 2. Section 3 presents and details the proposed approach for developing GPLs. In Section 4, we describe the FLOCOS GPL development in order to illustrate our approach. Section 5 presents and discusses relevant issues that arose from our experience with the GPLs. Finally, conclusion and directions for future work are presented in Section 6.

2 Related Work

Several component-based approaches have been proposed in the literature in order to minimize the efforts in the groupware development. The main idea of these approaches is to encapsulate many of difficult technical aspects, e.g. session management and communication protocols, which are typically present in the groupware development. Among the motivations of component-based development, these proposals are mainly concerned with tailorability [11, 12] and product lines [7, 13, 14]. The first is related to the idea of assembly by the final user. In the latter, components are developed with the aim of reusing them in various systems, reducing the total investment and maintenance costs.

Examples that illustrate these approaches are: (i) GroupKit [13], which addresses the development of synchronous groupware by providing a toolkit that hides from the programmer complexities intrinsic to this type of application; (ii) FreEvolve [11], which was designed to enable groupware customization through composing components or tailoring existing applications to final users; (iii) DreamTeam [14], which is a component-based platform to support building synchronous groupware; and (iv) Gerosa et al. [6, 7] propose structuring collaborative systems using components that encapsulate the technical difficulties of distributed and multi-user systems and reflect the concepts of collaboration modelled by the 3C model.

Despite the fact that these approaches provide useful techniques to reduce groupware development efforts and to tailor them to specific contexts, they do not cover the whole GPL development process. They lack some essential activities such as the domain analysis, which is responsible for eliciting common and variable requirements of the GPL, and tracing features, which provide means of systematically derive customized groupware from a GPL, based on a feature selection. In summary, the approaches aim at developing reusable software assets or even provide them, but these assets are reused in an *ad hoc* way.

Based on these identified deficiencies, we propose our approach for developing GPLs in next section. The approach incorporates activities of the SPL development, resulting in the evolution of the reuse from an *ad hoc* to a systematic way.

3   An Approach for Developing Groupware Product Lines

In this section, we present and describe our approach for developing GPLs. It incorporates activities and model from both groupware and SPL development. The approach not only provides means of developing reusable assets to build groupware, but also concerns with identifying the GPL features and providing feature traceability, whose main purpose is to allow a systematic derivation of members of the groupware family.

As a motivation for our approach, we present a typical scenario of families of groupware. The family is a set of Learning Management Systems (LMSs). Several LMSs can be seen available on the market or in academic environments. The members of the LMS family have a set of common collaborative services, such as e-mail, chats, discussion forums, links and repositories. However, other services are present in just some or even in only one LMS. These commonalities and variabilities are depicted in Table 1 (adapted from [7]). It shows collaboration services found in the following LMSs: AulaNet (http://www.eduweb.com.br), TelEduc (http://teleduc.nied.unicamp.br), AVA (http://ava.unisinos.br), WebCT (http://www.webct.com) and Moodle (http://www.moodle.org). The use of a GPL approach allows the development of reusable assets, and to derive each one of these LMS configurations in a systematic way. Moreover, any valid combination of the features can be derived from this reusable infrastructure to meet specific user needs.

Table 1. Collaborative Services in LMS [7]

	Communication Services							Coordination Services									Cooperation Services												
	Mail	Discussion List	Forum	Mural	Brainstorming	Chat	Messenger	Agenda	Activities Report	Participation Monitoring	Questionnaire	Tasks	SubGroups	Resource	Guidance	Voting	Repositories	White Board	Search	Glossary	Links	Cooperative Journal	Classifier	Wiki	Contact Manager	Peer Review	FAQ	Notes	RSS
AulaNet	X	X	X			X	X	X	X	X	X	X				X	X				X							X	
TelEduc	X		X	X		X		X	X	X	X	X	X				X		X								X	X	
AVA	X		X	X		X		X	X	X	X	X			X		X		X		X						X	X	
WebCT	X		X			X		X	X		X	X				X	X	X	X	X	X	X						X	
Moodle		X			X	X	X	X	X	X	X	X	X			X	X		X	X	X		X		X		X	X	X

Analyzing groupware families, like LMS for example, as a whole (domain analyses) and its possible variations makes possible to deploy customizable groupware according user's needs.

Next sections describes the two main steps of our approach, which are: (i) Domain Analysis, which analyzes the domain using the 3C Collaboration Model to guide the requirements elicitation and identifies common and variable features (Section 3.1); (ii) Domain Design and Implementation, which aims at defining an architecture that supports the defined variability and the derivation of applications in a systematic way (Section 3.2).

3.1 Domain Analysis

In the domain analysis, the main concepts and activities in a domain are identified and modelled using adequate modelling techniques. The common and variable parts of a family of systems are identified, defining the scope of the GPL indicating which products can be derived from it. The GPL domain analysis differs from the SPL domain analysis because of the need of the collaboration analysis. The analysis of Collaboration in our approach is made in accordance with the 3C Collaboration Model [9, 10] as illustrated by Figure 1.

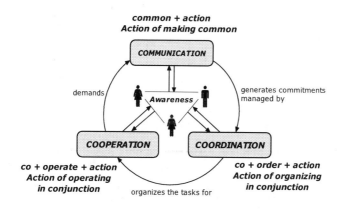

Fig. 1. 3C Collaboration Model

During communication there is an exchange of messages targeting future common action. Coordination deals with the resources and their interdependencies needed for accomplishing the accorded plan of action. Cooperation is the action realized by the joint operation of group members in the shared space. Collaboration, therefore, depends on the interplay of these dimensions. Using this model as a guide to develop groupware means developing tools to support each "C" in the model.

To support communication, [15] says that the designer of communication tools defines the elements that will set the communication channel between the interlocutors, considering the specific groupware user needs like purpose, dynamics, time and space. According to group needs, aspects like privacy and information overload must be analyzed.

To support coordination, the designer must think about what must be coordinated. If the aim is to coordinate people, the focus of the coordination must be the tools that provide communication and the context. Coordinating resources is related to the shared space where actions are made (cooperation). Coordinating tasks consists in managing interdependencies between tasks that are carried out to achieve a goal [15]. So, the designer must keep these different aspects of coordination when designing tools to support this task.

To support cooperation, the designer must think on how the actions on the shared space will take place. A set of tools for storage and maintenance of artefacts (documents, spreadsheets, presentations and others) must be present. Depending on the objective of the group, it is necessary some tools for online synchronous edition of these artefacts.

After identifying GPL requirements in accordance to the 3C Collaboration Model, variabilities inside the domain must be identified. For instance, a GPL of groupware whose purpose is cooperation may need a feature to provide communication. In one product of the GPL, a synchronous communication is need and in another product, an asynchronous one. This results in a feature "Communication" that has two different alternatives for instance chat and e-mail. Feature modelling is the activity of analyzing and capturing the common and variable features of SPLs and their respective interdependencies, which is introduced in the GPL development. It was originally proposed by the FODA method [16], and has been used in several SPL approaches in order to analyze the domain and its variabilities. According to Czarnecki et al. [5], a feature is a system property that is relevant to some stakeholder and is used to capture commonalities or discriminate among products in a SPL. Features are essential abstractions that both customers and developers understand.

The features are organized into a coherent model referred to as a feature model, which models the features of a GPL as a tree, indicating mandatory, optional and alternative features. This tree representation is named feature diagram. Mandatory features are the ones that must be present in any derived product of the GPL, and is part of the identity of the product. For instance, in a Learning Object repository, there must be functionalities that allow storing and retrieving them. Optional features aggregate functionalities to a derived product that are necessary just in certain context, so it can or not be present in a product of the product line. Alternative features in turn are features that vary from one product to another. An example in collaborative systems is that communication can be achieved by exchanging asynchronous messages, e.g. e-mail, or synchronous, such as chat.

A feature model refers to a features diagram accompanied by additional information such as dependencies among features, and it represents the variability within a system family in an abstract and explicit way. The constraints provide information about valid configurations of the feature model. For instance, two different features can be mutually exclusive, or the inclusion of a certain feature requires the inclusion of another.

In order to model GPL features, we propose an extension of the feature model to provide information of both collaboration and variability concerns. We have named this extended version 3C-Feature Model. In collaboration systems, there are several supporting features that are intrinsic to this kind of system. As a consequence, the feature model is split into two layers: functionality layer and infrastructure layer. The first refers to the features that provide functional behaviour to the user; however features are represented by different symbols to indicate the purpose of the feature (Communication, Cooperation and Coordination) according to the 3C Collaboration Model. These notations are complementary to the ones already used in traditional feature models, which are: (i) filled circles to represent mandatory features (present in all products of the GPL); (ii) empty circles to represent to represent optional features (present in some product of the GLP); and (iii) a line connecting alternative features (one of them must be chosen for the product being derived). The latter is composed of features that support some functional features. A feature model with our proposed adaptations is presented in Figure 4.

3.2 Domain Design and Implementation

The design and implementation of a GPL must result in an architecture that supports the identified variability, which characterizes the main difference of GPL and single groupware architectures. The GPL architecture is modelled using typical UML diagrams; however they are adapted to provide explicit variability information. We adopted the stereotypes <<kernel>>, <<optional>> and <<alternative>>, proposed by the PLUS method [17], to indicate which design elements are part of the GPL core, are present in some products and varies among products, respectively.

Supporting variability is typically achieved by taking into account feature modularization and adopting techniques to promote it. The main benefit of having features modularized is that it allows an easy (un)plugging of a feature; therefore it improves the systematic derivation of products and consequently reduces time and costs of the derivation process. Different implementation techniques can be used to modularize features in the code [18], e.g. polymorphism, design patterns, frameworks, conditional compilation and aspect-oriented programming. A particularity of Groupware is that functionalities typically rely on common services. For instance, chat, e-mail and forum are three different options to support communication. Therefore, we propose the idea of having a Layered architecture [19], in which one of the GPL architecture layers provides the business services, which uses services of the next layer, which in turn provides infrastructure services.

In addition to these techniques, component-based SPL development approaches propose the decomposition of the SPL architecture into components with well-defined interfaces, and therefore they can be easily changed. These approaches have the same ideas component-based groupware development approaches. Consequently, this convergence of interests calls for studies that investigate how these approaches can be used together and how they can be used in the GPL development; however this has not been addressed in this paper yet.

Besides modelling and implementation the GPL reusable assets, it is important to keep traceability links between features and the elements that implements them. This information allows choosing the appropriate implementation elements according to the selected features during the product derivation process.

Figure 2 summarizes the idea of our approach. The main result of the domain analysis is the 3C-Feature Model. Later, the GPL is designed considering the identified variability and, in sequel, the reusable assets are implemented. Links among the GPL artefacts ensures the traceability of the feature in order to allow an effective product derivation process. In our approach, these traceability links are expressed by means of tables, which map features to classes that implement them.

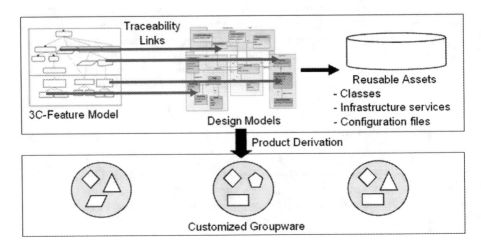

Fig. 2. Approach Overview

4   Illustrating Our Approach with the FLOCOS GPL

This section presents and describes the FLOCOS (Functional Learning Object Collaborative System) GPL, using the approach previously presented. FLOCOS consists of a GPL of Learning Object (LO) repositories based on 3C Model. FLOCOS products are a particular kind of LO repositories because it supports a special kind of LOs, not dealt with by other works: Functional Learning Objects (FLOs) [20]. In order to familiarize the reader with the FLOCOS GPL domain, we first present relevant concepts related with FLOCOS (Section 4.1), and later we describe how the FLOCOS GPL was developed using our approach (Section 4.2).

4.1   FLOCOS Overview

According to SCORM [21], LOs have the following characteristics (the "ilities"): (i) Accessibility: the ability to locate and access instructional components from one remote location and deliver them to many other locations; (ii) Adaptability: the ability to tailor instruction to individual and organizational needs; (iii) Affordability: the ability to increase efficiency and productivity by reducing the time and costs involved in delivering instruction; (iv) Durability: the ability to withstand technology evolution and changes without costly redesign, reconfiguration or recoding; (v) Interoperability: the ability to take instructional components developed in one location with one set of tools or platform and use them in another location with a different set of tools or platform; and (vi) Reusability: the flexibility to incorporate instructional components in multiple applications and contexts.

Considering LOs software artefacts that have the aforementioned "ilities", Java applet applications, web services, web applications, software components and software agents, all fall into this category. Gomes et al. [22] propose the definition of Functional Learning Objects, stating that they are computational artefacts having functionality that enables interaction between entities, whether digital or not, being used/reused in the mediation of the teaching-learning process. They propose a Functional Learning Object Metadata (FLOM) to properly describe the technical characteristics of this class of objects.

Even though FLOs adequately support teaching and learning processes, they tend to require a major implementation effort demanding a skill rarely found in docents. So, in order to successfully meet the increasing demand for these objects, promoting reusability and tailorability is essential.

Therefore, FLOCOS products promote accessibility, reusability and affordability by the use of FLOM for describing FLOs. The products provide a shared space in which users can collaborate by posting, maintaining, searching and discussing FLOs. These functionalities provided by FLOCOS can be customized according to specific user needs, or even whole functionalities can be needed just in some contexts. In next section, we present the development of the FLOCOS GPL.

4.2 Developing FLOCOS GPL

The FLOCOS GPL developing process encompassed mainly two steps: (i) the domain analysis, where the features provided by GPL were identified and classified in two dimensions: first according each C of 3C Model, and after they were classified as mandatory and optional (Section 4.2.1); (ii) the design and implementation, in which the FLOCOS GPL architecture was designed and implemented supporting the variability identified in the previous step and then implementation elements were mapped onto GPL features for traceability purposes (Section 4.2.2).

4.2.1 Domain Analysis

FLOCOS GPL purpose is to give support to the cooperation among members of a group in a shared space. However, even when focusing on a specific "C" of the 3C Collaboration Model, in this case cooperation, communication and coordination should also be treated, given that there is also an intra-relationship between them [23].

In this context, each FLOCOS GPL product is a shared space comprising the FLOs, and the discussion, experience reports, message logs, recommendations, user notifications and action histories related to each one of them. Figure 3 illustrates members cooperating on a Functional Learning Object.

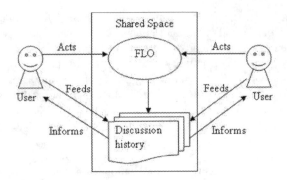

Fig. 3. FLO Cooperation Model

The first step consisted on identifying the features that the GPL must provide to support the FLO Cooperation Model presented in Figure 3, as follows:

- **Communication:** The feature identified to support this "C" is the Discussion. Two kinds of communication are available for FLOCOS products: chat (synchronous) and forum (asynchronous). Therefore, the best communication form can be chosen for a certain context;
- **Coordination:** The features identified to support this "C" are *User registration* and *Notifications*. The *User registration* feature is related to the coordination of the shared space once it controls who will operate in it. The feature *Notifications* is related to the coordination of the tasks once it alerts for events occurred in the shared space and is expected some reaction of the user after the notification;
- **Cooperation:** The features identified to support this "C" are *FLO registration, Experience reports, Actions history* and *Recommendations*. These features are related to the actions taking place in the shared space. Users act over FLOs registered in the system, relate their experience on FLO usage, the system recommends FLOs to users according their preferences and users can view what other users do with the FLOs registered.

Table 2 shows the classification of FLOCOS GPL features on the 3C Model.

Table 2. FLOCOS GPL Features and 3C Model

	Communication	Coordination	Cooperation
User registration		▓	
FLO registration			▓
Discussion	▓		
Experience reports			▓
Actions history			▓
Recommendation			▓
Notifications		▓	

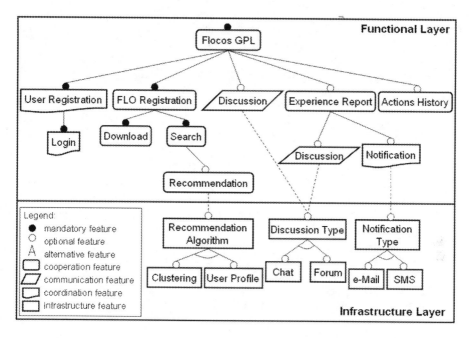

Fig. 4. FLOCOS GPL 3C-Feature Model

Once the features were identified, they were analyzed and classified into three different categories: mandatory, optional and alternative. Mandatory features are essential to any LO repository, optional features are necessary just in some specific contexts, and alternative features varies from one product to another.

The identified functional features were then organized in FLOCOS feature model, which is depicted in Figure 4. The feature model not only shows features with their variability information (mandatory and optional), but also their purpose. The purpose of the products derived from the FLOCOS GPL can be seen be the root of the feature model (Cooperation). Additionally, some constraints were specified in order to indicate dependencies among the features. These are the constraints:

1   Report discussion is available only if the Discussion feature is selected for a product;

2   User notification feature depends of the presence of Experience reports feature on the product;

3   FLO recommendations feature is available only if Actions history feature is selected for a product.

The information provided by the feature model and its constraints indicate the possible products that can be derived of the FLOCOS GPL.

Next, we describe the functional features that comprise the FLOCOS GPL:

• **User registration.** This is a mandatory feature. To access FLOs deployed in FLOCOS, the user must access the system. To do so, he must sign up through this feature. Once registered, the user is able to access other functionalities.

- **FLO registration.** That is the core functionality of LO repositories, and therefore a mandatory feature of FLOCOS GPL. Users, duly registered, can post FLOs, making them available to other system users.
  - o **Recommendation.** FLOCOS enables its users to assign ratings for FLOs evaluation. These ratings provide feedback to the creators and maintainers of the FLO and are used to guide the search for FLOs in the system. The search engine ranks the results based on these ratings plus access and download statistics provided by the Actions history. From the 3C Model viewpoint, it is a functionality with a cooperation purpose. In addition, this feature presents an autonomous behaviour considering that it automatically monitors the user actions to evaluate FLOs. Two options for recommendation algorithms are provided: clustering and based on user profile.
- **Discussion.** A discussion is associated with each FLO registered in the system through the Discussion functionality. FLO users may report their experiences and discuss deficiencies in the development and application of the object in different contexts. This is an optional feature in the GPL, given that a customer may require a simple repository needing no discussion. The Discussion feature has an alternative infrastructure feature associated to it – chat and forum – therefore, a derived product can have the most appropriate discussion type.
- **Experience report.** Similarly to the Discussion, each FLO registered in the system is associated with experience reports. Such reports inform experiences of successes and failures in the use of a specific FLO available in the system. This is another optional feature. For example, in the early stages of a FLO development, a simple repository associated with discussions would suffice the developers' needs. This feature has two sub-features. The Discussion is similar to the one described previously.
  - o **Notifications.** In order to inform users about actions taken on FLOs posted by them, they receive a notification for every message sent to a forum, or to each experience report. This is also an optional feature, considering that some users may consider these notifications irrelevant or feel overloaded by the massive amount of messages generated by the feature. From the 3C Model viewpoint, it is a functionality with a coordination purpose. As the system must be monitored to send messages notifications, this feature also provides an autonomous behaviour.
- **Action history.** FLOCOS keeps a history of the actions of its users in the environment. Through this functionality, it is possible to know what FLOs are more accessed those with the more active discussions, more active users in the environment and so on. This is an optional feature of FLOCOS GPL, given that keeping log register may cause an overhead in the system, which is a cost that some users would not like to pay.

4.2.2 FLOCOS GPL Design and Implementation

In order to provide a flexible architecture to support the variability provided by the FLOCOS GPL, the architecture of the GPL adopts several software engineering practices, such as design patterns and agent technology, to bring low coupling among the different features of the GPL.

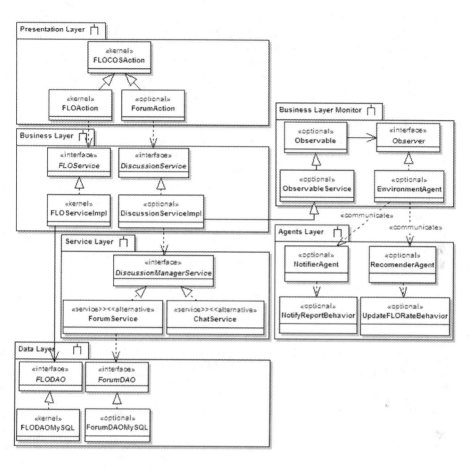

Fig. 5. FLOCOS GPL Architecture (partial)

Giving that FLOCOS products are web-based applications, a pattern widely used to structure this kind of application is the Layer architecture pattern [19]. The layers that comprise the architecture are: (i) Presentation – responsible for processing the web requests submitted by the system users; (ii) Business – responsible for structuring and organizing the business services; (iii) Service – responsible for providing infrastructure services that can be used by the business services; and (iv) Data – aggregates the classes of database access of the system, which was implemented using the Data Access Object (DAO) design pattern. Figure 5 shows a partial view of FLOCOS GPL class diagram illustrating its architecture. The four presented layers are present at the left side of the figure, being composed of classes whose purpose is according to which layer it is part of. As it can be seen, each feature is implemented by a set of classes (Action, Service, and DAO) that is independent of other features. Moreover, the <<kernel>> and <<optional>> stereotypes indicate which classes are part of the GPL core and which are not.

Some features in FLOCOS GPL provide an autonomous behaviour. Software agents are used to implement these features, because they are a powerful abstraction to model this kind of behaviour. A software agent enjoys mainly the following properties: autonomy, reactivity, pro-activeness and social ability. Nunes et al. [24] proposed the Web-MAS architectural pattern, based on the Layer architecture pattern, which allows the extension of web applications to incorporate autonomous behaviour. So, we adopted this pattern, adding the Business Layer Monitor and Agents Layer components to FLOCOS GPL architecture (see Figure 5) to implement the *User Notifications* and *Recommendation* features. These two components are at the right side of Figure 5. The Business Layer Monitor contains the *EnvironmentAgent*, which monitors and propagates the execution of business operations performed in the Business Layer. The Agents layer, in turn, is composed of two agents (*NotifierAgent* and *RecommenderAgent*), which were implemented to provide functionalities needed for the *User Notifications* and *Recommendation* features.

Table 3. FLOCOS GPL Features/Implementation Classes Mapping (partial)

Feature	Classes
User registration	UserAction; UserService; UserServiceImpl; UserDAO; UserDAOMySQL
FLO registration	FLOAction; FLOService; FLOServiceImpl; FLODAO; FLODAOMySQL
Discussion	DiscussionAction; DiscussionService; DiscussionServiceImpl
Experience Reports	ExpReportAction; ExpReportService; ExpReportServiceImpl; ExpReportDAO; ExpReportDAOMySQL
...	...

Finally, the last activity to be performed in this step is to provide traceability links between the FLOCOS GPL features and the implementation elements. Each feature is implemented by a set of classes. As a consequence, we built a table that maps each feature onto the classes that implements it. Table 3 shows a partial view of this mapping between features and their implementation classes. As a result, when one wants to derive a product that contains a certain feature, the related classes must be selected to be part of the product.

5   Discussion

In this section, we present and discuss some lessons learned and challenges that we have identified with our experience with the development of GPLs.

Providing mass customization. The use of the SPL practices and techniques for developing groupware enabled its mass customization, and in the specific case of FLOCOS GPL, in a semi-automatic way. The configuration of a specific product of FLOCOS GPL is made through the edition of a Spring configuration file (XML file) and the selection of classes and other implementation artefacts related to the selected features for the product. The Spring framework offers a model to build applications as a collection of simple components (called beans) that can be connected or customized using dependency injection. Spring container uses the XML configuration file to specify the dependency injection on application components.

Furthermore, there are some model-based tools, such as GenArch [25] and pure::variants, which automate the derivation process, reducing even more the time and effort needed to generate new products.

It is important to notice that a GPL is different from adaptable systems in the sense that its purpose is to accelerate groupware development enabling the product derivation according to user's needs. A common architecture allows to deriving different system versions. On the other hand, adaptable systems are single systems, which can be configured. One direction of future work is to explore how GLP architectures can benefit building adaptable systems.

Easier introduction of new features. Software systems typically evolve by the adding of new features that provide new functionalities to the users or change/improve the existing ones. The adoption of an architecture that has modularized features and low coupling among components helped on the introduction of new features to FLOCOS GPL. The techniques used in SPL engineering, and now incorporated to GPLs, make a system better maintainable as stated in [26]: "The same design techniques that lead to good reuse also lead to extensibility and maintainability over time". In addition, one way of developing Groupware is by a prototyping approach. It is based on the experience with new features, receiving users' feedback and adapting collaborative systems to meet user needs. As a consequence, we believe that SPL techniques, which improve feature modularization and therefore allow (un)plugging features, make sense for prototyping Groupware.

Use of software agents to tasks automation. Several systems have evolved with the addition of functionalities or features that automate tasks that require human intervention. Examples are monitoring the system to check user interaction and reducing searching time by the use recommendation techniques, which can take into account user preferences. Due to the pro-active and autonomous nature of software agents, they are an appropriate technology to implement this kind of feature. This approach was adopted in FLOCOS GPL to notify users about the posting of messages in forums and experience reports and to FLOs recommendation. Moreover, using the Web-MAS architectural pattern brought flexibility in the adding of these features with a low impact.

6 Conclusion

Groupware software development has to deal with common technical issues, such as communication and security. In addition, different groupware applications typically share a lot of functionalities and this fact can be exploited by the development of reusable assets that can be used to build customized applications. Several approaches have been proposed in order to address particularities of this kind of applications. However, they failed to provide complete solutions to the groupware development.

In this paper, we proposed an approach that addresses Groupware Product Lines (GPLs) and aims at solving identified issues in the groupware development. Our approach advances the groupware development by incorporating Software Product Line

(SPL) techniques into the GPL development process, in order to build not single applications, but families of groupware. The groupware family is first analyzed in terms of features based on the 3C Collaboration Model to capture the commonalities and variabilities in the domain, while identifying the features' purpose according to the model. This information is documented in our extended 3C-Feature Model. Variabilities may have different degrees of granularity, such as a whole functionality (e.g. Discussion) or a functionality customization (e.g. Recommendation). Later, the family is designed and implemented taking into account the identified variability. In addition, features are mapped onto the implementation elements in order to allow an effective product derivation process. Our approach is illustrated with the FLOCOS GPL, which allows to systematically deriving customized versions of Learning Object repositories. We also discussed some lessons learned with our experience in the GPL development.

This paper addresses ongoing research on the GPL development. We are currently investigating how an existing component-based approach for developing groupware [6, 7], also based on the 3C Collaboration Model, can be used in a complementary way with our approach. In addition, we aim at adopting product derivation tools in order to automate the GPL derivation process.

Acknowledgments

This research is partially supported by project "Desafios" from CNPq. Bruno Gadelha, Ingrid Nunes, Hugo Fuks and Carlos José Pereira de Lucena receive grants from CNPq. Carlos José Pereira de Lucena and Hugo Fuks also receive grants from FAPERJ project "Cientistas do Nosso Estado."

References

1. Tietze, D.A.: A Framework For Developing Component-Based Co-Operative Applications. Ph.D. Dissertation, Technischen Universität Darmstadt, Germany (2001)
2. Greenberg, S.: Multimedia Tools and Applications 32(2), 1380–7501 (2007) ISBN 1380-7501
3. Clements, P., Northrop, L.: Software Product Lines: Practices and Patterns. Addison-Wesley, Reading (2002)
4. Pohl, K., Böckle, G., van der Linden, F.J.: Software Product Line Engineering: Foundations, Principles and Techniques. Springer, New York (2005)
5. Czarnecki, K., Helsen, S.: Feature-based survey of model transformation approaches. IBM Systems Journal 45(3), 621–645 (2006)
6. Gerosa, M.A., Pimentel, M., Fuks, H., Lucena, C.J.P.: Development of Groupware based on the 3C Collaboration Model and Component Technology. In: Dimitriadis, Y.A., Zigurs, I., Gómez-Sánchez, E. (eds.) CRIWG 2006. LNCS, vol. 4154, pp. 302–309. Springer, Heidelberg (2006)
7. Gerosa, M.A., Raposo, A., Fuks, H., Lucena, C.J.P.: Component-Based Groupware Development Based on the 3C Collaboration Model. In: SBES 2006, Brazil, pp. 129–144 (2006)
8. Szyperski, C.: Component technology – what, where, and how? In: Procedings of the 25th International Conference on Software Engineering (ICSE 2003), pp. 684–693. IEEE, Los Alamitos (2003)

9. Ellis, C.A., Gibbs, S.J., Rein, G.L.: Groupware - Some Issues and Experiences. Communications of the ACM 34(1), 38–58 (1991)
10. Fuks, H., Raposo, A.B., Gerosa, M.A., Lucena, C.J.P.: Applying the 3C Model to Groupware Development. IJCIS 14(2-3), 299–328 (2005)
11. Won, M., Stiemerling, O., Wulf, V.: Component-Based Approaches to Tailorable Systems. In: End User Development, pp. 1–27. Kluwer, Dordrecht (2005)
12. Slagter, R.J., Biemans, M.C.M.: Component Groupware: A Basis for Tailorable Solutions that Can Evolve with the Supported Task. In: ISA 2000, Wollongong, Australia (2000)
13. Roseman, M., Greenberg, S.: Building real time groupware with GroupKit, a groupware toolkit. ACM Transactions on Computer-Human Interaction 3(1), 66–106 (1996)
14. Roth, J., Unger, C.: Developing synchronous collaborative applications with TeamComponents. In: Designing Cooperative Systems: the Use of Theories and Models, COOP 2000, pp. 353–368 (2000)
15. Fuks, H., Raposo, A., Gerosa, M.A., Pimentel, M., Lucena, C.J.P.: The 3C Collaboration Model. In: The Encyclopedia of E-Collaboration, Ned Kock (org), pp. 637–644 (2007)
16. Kang, K., Cohen, S., Hess, J., Novak, W.: Peterson.: Feature-oriented domain analysis (foda) feasibility study. Technical Report CMU/SEI-90-TR-021, SEI, Carnegie-Mellon University (1990)
17. Gomaa, H.: Designing Software Product Lines with UML: From Use Cases to Pattern-Based Software Architectures. Addison-Wesley, Reading (2004)
18. Alves, V.: Implementing Software Product Line Adoption Strategies. PhD thesis, UFPE, Brazil (2007)
19. Fowler, M.: Patterns of Enterprise Application Architecture. Addison-Wesley Professional, Reading (2002)
20. Gomes, S., Gadelha, B., Mendonça, A.P., Amoretti, M., Marc, S.: Functional Learning Objects and Actual Metadata Limitations. In: Proceedings of XVI SBIE – Brasilian Symposium on Informatics on Education. Juiz de Fora- MG, Brazil (2005)
21. SCORM 2004 2nd Edition Overview. Advanced Distributed Learning (July 2004)
22. Gomes, S., Gadelha, B., Mendonça, A.P., Castro Jr., A.N.: A Functional Learning Object Metadata Proposal. In: Proceedings of XVIII SBIE – Brasilian Symposium on Informatics on Education. São Paulo - SP (2007)
23. Fuks, H., Raposo, A., Gerosa, M.A., Pimentel, M., Filippo, D., Lucena, C.J.P.: Inter- e Intra-relações entre Comunicação, Coordenação e Cooperação. In: Proceedings of IV Collaborative Systems Brasilian Symposium, Brazil, pp. 57–68 (2007)
24. Nunes, I.O., Kulesza, U., Nunes, C., Cirilo, E., Lucena, C.: Extending Web-Based Applications to Incorporate Autonomous Behaviour. In: WebMedia, Vila Velha (2008)
25. Cirilo, E., Kulesza, U., Lucena, C.: A Product Derivation Tool Based on Model-Driven Techniques and Annotations. JUCS 14, 1344–1367 (2008)
26. Coplien, J., Hoffmann, D., Weiss, D.: Commonality and Variability in Software Engineering. IEEE Software 15(6), 37–45 (1998)

Negotiation-Collaboration in Formal Technical Reviews

Giovana B.R. Linhares[1], Marcos R.S. Borges[1], and Pedro Antunes[2]

[1] Federal University of Rio de Janeiro, Graduate Program in Informatics (PPGI),
Rio de Janeiro, RJ, Brasil
giovana_linhares@hotmail.com, mborges@nce.ufrj.br
[2] University of Lisbon
Faculty of Sciences, Campo Grande, 1749-016 Lisbon, Portugal
paa@di.fc.ul.pt

Abstract. This paper discusses the negotiation-collaboration process: a binomial process mixing collaboration, negotiation and argumentation. We applied the negotiation-collaboration process to Formal Technical Reviews, commonly adopted to verify the functional specification of software. We developed a groupware tool demonstrating the dynamic of the negotiation-collaboration process in Formal Technical Reviews. And we provide results from an experiment with the tool in a software engineering firm. The obtained results demonstrate the negotiation-collaboration process promotes bigger participation in FTR.

Keywords: Formal Technical Reviews, Negotiation, Software Quality Assurance.

1 Introduction

Formal Technical Reviews (FTR) are recommended software quality assurance activities in software engineering [1]. A FTR is a collaborative endeavor involving designers, developers and testers. It serves fundamentally to verify, at various points in the product development lifecycle, if the product is being engineered with quality and coherency with the specification, i.e. the product supplies the right solution to the requirements specified by the client.

In spite of having a common goal, the participants in a FTR often develop conflicting perspectives, interpretations and positions regarding the product quality. This type of conflict justifies the negotiation-collaboration process: a process aiming to integrate conflict management in collaboration. We thus have, on the one hand, the FTR activity that has to be fulfilled by a group of persons and, on the other hand, the negotiation-collaboration process necessary to accomplish the FTR activity with success.

Groupware may simultaneously support the FTR activity and the negotiation-collaboration process. Unfortunately, resolving conflicts and getting to consensus is a complex problem. One major intricacy is dealing with the main assumptions behind conflict resolution: (1) the interlocutors have diverse profiles, interests, viewpoints and strategies that should be respected and often promoted to reach quality results; (2) in this context, reaching consensus requires a collective cognitive effort, understanding the different positions and seeking creative consensus solutions; (3) while also taking into consideration that the process should be as fast and affective as possible.

L. Carriço, N. Baloian, and B. Fonseca (Eds.): CRIWG 2009, LNCS 5784, pp. 344–356, 2009.
© Springer-Verlag Berlin Heidelberg 2009

Many approaches to FTR emphasize collaboration to the detriment of negotiation, for instance adopting a strict focus on shared information. This strategy naturally seeks to enforce consensus. Such an approach may however fail, either because the conflicts may remain dormant, just to arise later in the product lifecycle; or the conflicts may escalate to unacceptable levels, making it more difficult if not impossible to accomplish the intended goals. It is therefore necessary to develop an integrated and balanced view of collaboration and negotiation.

The problem discussed in this paper concerns the lack of negotiation-collaboration balance observed in the current groupware tools [2]. The research described in this paper tries to supplant this lack of balance by integrating models of collaboration, argumentation and negotiation. This research guided the development of a groupware tool supporting FTR in the software engineering Functional Specification phase.

The paper is organized in six sections. In Section 2 we present the theoretical foundations that guide FTR. In Section 3 we describe how the negotiation-collaboration process occurs during FTR and delineate its major requirements. Section 4 describes the developed prototype. Section 5 describes an experiment carried out with the prototype. And in Section 6 we present some conclusions from this research.

2   Theoretical Foundations

Fagan [3] developed the FTR technique while working for IBM. A FTR is defined as an activity practiced in groups, based on formal procedures and designated roles, aiming to discover defects in documents and code. FTR is an activity belonging to the more general software verification and validation process and, even more broadly, integrated in the software quality assurance process [4].

The International Software Testing Qualifications Board provides a general vision of the FTR process organized in several phases [5]. The several phases are described as follows:

- Planning: select the team, allocate functions, define input and output criteria, and select the products to be revised;
- Kick-off: distribute documents, explain the objectives, processes and documents to the participants, and check the input criteria;
- Individual preparation: work done by each participant before the review meeting, taking notes of the potential defects, questions and comments;
- Review meeting: analysis of the work submitted for review, identification and discussion of defects, and decision about the acceptance or not of the product;
- Re-work: addressing the defects found in the review meeting (carried out by the authors);
- Accompaniment: inquiring if the defects were forwarded, obtaining project metrics and checking the output criteria.

Also in accordance with the ISTQB [5], the roles and responsibilities involved in the FTR process include:

- Manager: who assumes responsibility and takes the final decisions during the review, allocates time and determines if the revision's objectives were achieved;
- Moderator: who leads the review, including planning and accompanying the re-work. If necessary, the moderator mediates conflicts. The moderator is responsible for the success of the review meeting;
- Author: the person who submits the document/product for review;
- Reviewers: the persons, having technical and/or business knowledge, who identify, analyze and describe the defects found in the document/product under review;
- Secretary: the person that documents what happened in the meeting, registering the reviewed items and defects found.

With a small differentiation of roles, Pressman [1] indicates the review must have the participation of the review leader, several reviewers and the producer. The producer and author have identical roles, i.e., they deliver the reviewed product. However, the manager and moderator roles are substituted by the more focused role of review leader; and the secretary task is simply carried out by one of the reviewers designated by the leader.

Antunes et al. [2] adopted a different perspective over the roles and responsibilities involved in the FTR process. They discussed the different attitudes the participants may assume during a FTR, identifying two stereotyped attitudes: highly collaborative and highly conflicting. Of course these extreme attitudes may be detrimental to the quality of the FTR process, one because it may lead to groupthink [6, 7] and the other because it may lead to a failed task. Interestingly, dealing with conflict has been considered a way to avoid groupthink [8] and collaboration is also a viable way to overcome conflict. Therefore, the two extreme attitudes, as well as all the other possible attitudes in between, should be reconciled. Ramires et al. [9] developed a groupware tool implementing the Software Quality Function Deployment (SQFD) [10] technique to study the integration of collaboration and conflict. The tool integrates argumentation and negotiation support with SQFD.

The research described in this paper extends the previous works on collaboration and conflict integration in two different streams. The first stream concerns developing the negotiation-collaboration process as a generalization of the previous SQFD process, integrating aspects of collaboration, negotiation and argumentation. The second stream considers developing a groupware tool supporting FTR.

Regarding the negotiation-collaboration process, its theoretical foundations may be found on the confluence of three different views over group work and decision-making:

- Collaboration - Where the fundamental assumption is that a group of people communicate, coordinate activities, share a workspace and construct shared awareness, including awareness of who are the group members, what are they doing, and where are they located in the workspace [11].
- Negotiation - Considering there are several conflicting parts that must bargain to reach a mutually satisfactory outcome. Bargaining will thus involve communication, offers and counteroffers, the definition of a settlement space, and search for mutually beneficial agreements [12, 13].

- Argumentation-based negotiation - Complementing the negotiation view with the perspective that communication is fundamental to justify the own negotiation stance and persuade the other parts to change their stance [14].

3 The Negotiation-Collaboration Process

A problem statement triggers the negotiation-collaboration process. In our case study, the problem statement concerns the approval or rejection of a specification document or piece of code. After being triggered, the process evolves according to three phases:

- **Presentation of proposals:** Several proposals to resolve the problem may be presented by the participants;
- **Argumentation-based negotiation:** The participants have to reach a consensual position about each proposal. Communication is necessary to confront the individual positions, for and against the proposals, and provide arguments to substantiate the adopted positions;
- **Decision:** Having obtained a consensual set of positions, a decision must be made. This step involves analyzing the implications of the adopted positions and the subsequent steps necessary to put them into practice.

Notice that argumentation and negotiation are entangled. It should also be considered the presentation of proposals may be done during the argumentation-based negotiation.

3.1 Data Model

The data model of the negotiation-collaboration process is organized around the following elements: proposals, scores, positions, arguments, results and decisions. A proposal may be regarded as an action-statement that must be analyzed and discussed by the group. Figure 1 depicts this model.

The participants in the process should individually score every proposal. We currently support three scores: accept, reject and accept with restrictions. The occurrence of different scores for a proposal indicates there are divergent positions about the proposal. The positions are thus inferred automatically from the scores. In order to enforce argumentation, the reject and accept with restrictions scores should necessarily be complemented with arguments.

The goal of the negotiation-collaboration process is not necessarily to obtain consensual scores for every proposal. Several rules may be defined by the organization regarding what results should be drawn from the individual scores. The following rules may be considered: majority voting, where the result corresponds to the score selected by the majority of the participants; consensus, i.e. there is only a result if it corresponds to the score selected by all participants; and moderated, where the moderator should decide the result based on the participants' scores.

After obtaining a result for each proposal, the whole collection of results should be subject to a final verdict and lead to a decision. Again, several organizational rules may be adopted to obtain the final verdict. We adopted the following rules to deal with FTR: (1) general acceptance, only if there are no reject solutions; (2) general reject, if there is at least one reject solution; (3) postpone, if there is at most a predefined number of accepts with restrictions; and (4) general acceptance otherwise.

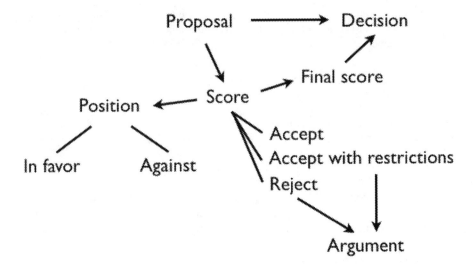

Fig. 1. Data model of the negotiation-collaboration process

3.2 Factors Affecting the Process

Each negotiation-collaboration process, although structured according to the phases previously described, has its own dynamics and depends on a set of factors that interact between themselves, interfering with the process outcomes. We highlight the following contextual factors:

- Level of conflict - As the level of conflict increases, so does the cognitive effort to reach consensus. At the limit, a destructive level of conflict will lead to a failed process. The number of suggested proposals, positions and arguments may serve to measure the level of conflict.
- Number of participants - A large number of participants may also turn it more difficult to reach consensus.
- Status differences - Status differences address the dependence relationships between leaders and subordinates. Groups having significant status differences may be negatively affected by the dependence on people with more power [15]. The balance between the participants' proposals, positions and arguments may serve to measure the effects of status differences.
- Problem involvement - A low involvement with the problem may turn it more difficult to contribute to the process. The number of suggested proposals, positions and arguments may serve to measure the problem involvement.
- Group expertise - The lack of expertise about the problem under discussion may also affect the process outcomes. This factor may be measured by assessing the quality of the presented arguments.

3.3 Quality Criteria for Assessing the Process

It is fundamental to define quality criteria for assessing the negotiation-collaboration process. However, the selection of criteria is quite challenging. Let us consider, for example, a situation where a decision is immediately reached after a small number of proposals, positions and arguments; and contrast it with another situation in which, after a long argumentation, several proposals were discussed.

We may assume the first case has low quality while the second case has high quality. This assumption may however be misleading. For instance, it may well be the case that the first case has low complexity and relevance, and the adopted solution is not only adequate but also efficient. On the contrary, the second case may correspond to a situation where conflicts may have lead to a suboptimal solution, having the additional cost of spending too much time to finish the process.

When considering negotiation processes, quality has been fundamentally associated with efficiency. For instance, the distance between the agreed solution and the best possible solution that could be obtained by continuing the negotiation process, designated value-left-on-the-table, is commonly used to evaluate the quality of negotiation processes [16]. This approach is however more adequate to bargaining than to negotiation-collaboration, since the former is influenced by the zero-sum game while the later is more influenced by "satisfying" trade-offs [17].

When considering collaboration processes, quality tends to be measured according to a diverse set of variables categorized as efficiency, effectiveness, satisfaction and consensus [18]. This suggests the quality of negotiation-collaboration processes should also be measured according to a combination of criteria, for which we suggest:

- Efficiency - Time to complete the argumentation-based negotiation;
- Depth of analysis - Average number of arguments against and in favor;
- Participation - Number of arguments.

3.4 Principles of Negotiation-Collaboration

The negotiation-collaboration process should be founded on sound principles. One such principle is transparency: the process and the adopted rules should be clear to all participants. Equality should also be considered, giving the same rights and privileges to all participants. Confidentiality is also necessary to preserve several aspects of the negotiation. For instance, the positions and arguments, and the individual settlement should be kept confidential to avoid personalizing the discussion.

The process should also be impartial, not guiding the participants towards a particular position or participant's profile. And finally, another important principle to ponder is the Win-Win efficiency. This principle is related with the negotiation and considers that the process should promote the maximization of the gains of all parties.

4 Negotiation-Collaboration in FTR

In this paper we assume the first phase of the process (the presentation of proposals) has already been carried out and a series of individual proposals and comments has been produced. We also consider that, before starting the argumentation-based negotiation, the moderator analyzes the data delivered by the individual reviews and polishes a list of proposals that will trigger the argumentation-based negotiation.

This approach avoids managing duplicates, equivocal language and ensures the consistency of contents and format. The moderator turns doubts, problems, comments, alternatives and solution into proposals for assessment by the reviewers during the argumentation-based negotiation.

The proposals are delivered to the reviewers in the beginning of the negotiation phase, but new proposals may be delivered during the negotiation, if necessary. To ensure confidentiality, the proposals are dissociated from the original authors.

The participants are asked to register their respective scores. Each reviewer associates a score to a proposal, reflecting his/her judgment about the proposal (0 => is not an error, 1 => is a light error or 2 => is a serious error). In case the chosen score is a light error or serious error, the reviewer is requested to complement the score with arguments, consisting of small text sentences. The arguments should be linked to the Functional Specification Document. Examples include: "the item cannot be related with the Functional Specification and should be removed"; "the item does not comply with the specification of function X"; or "the item fails to implement requirement Y".

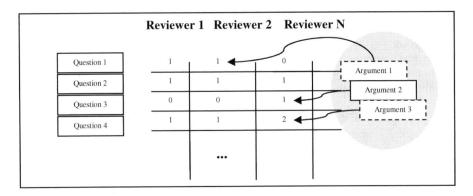

Fig. 2. Argumentation of divergent positions

The positions in favor and against each proposal are automatically calculated. The divergences are displayed to the reviewers, although not exposing the identities of the opponents. In case there are no positions against, the proposal is immediate "closed" and the consensual score is immediately determined. For efficiency reasons, a proposal may not be reopened in the same session.

For efficiency reasons, the negotiation is controlled by a timeout mechanism. The moderator is responsible for setting the timeout. The reviewers are notified when the timeout time is reaching. After the timeout, the process moves to the decision activity.

Figure 2 illustrates the outcomes of a negotiation-collaboration session. Notice that in the illustrated example there are no arguments associated to the proposal 2 (row 2) because the scores were consensual.

After evaluating the arguments associated to one proposal, any reviewer may change his/her own position and add additional arguments. This procedure may be repeated until closing a proposal. All the updates to positions and arguments are immediately visible to the reviewers. The changes in positions imply a corresponding update of the associated arguments.

As previously mentioned, not always a consensus score may be obtained for a proposal. The moderator may handle this situation in three different ways: majority voting,

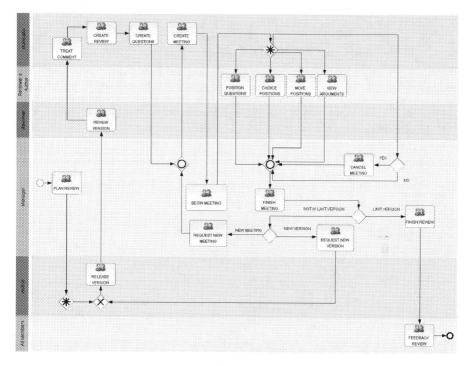

Fig. 3. Illustration of the FTR process

consensus and moderator's decision. The moderator should select one of these rules before starting the session and should make it public to guarantee the transparency principle. Figure 3 summarizes the FTR process.

It is always important to assess the individual contributions to the negotiation-collaboration process. An inspiration for a tool to summarize relevant information given to the moderator is the *participameter* [19]. The type of information delivered to the moderator in our case is shown in Table 1.

Table 1. Individual assessment information

Assessment	
Participation Positioning	Number of proposals from a participant in relation with the total number of proposals.
Participation Arguments	Percentage of arguments from the participant in relation with the total number of registered arguments.
Punctuality	Average of time to complete the task, as a percentage of time assigned to the task.
Contribution of Arguments	Number of arguments from a participant that contributed to the final score in relation with the total number of arguments.
Flexibility to converge	Number of score changes to converge with the majority, in relation with the total number of score changes to converge with the majority.

4.1 The FTR Tool

The tool was built using the .Net framework and the language CSharpe. Being a Web application, it can be used at any time and place. The adopted database manager was the SQL Server. To illustrate the prototype, we present two of its screens. Figure 4 shows the screen where the participants register their positions regarding the proposals specified by the moderator. The different positions are displayed to all participants. Figure 5 shows the functionality to support positions with arguments. It also allows visualizing the arguments from the other participants.

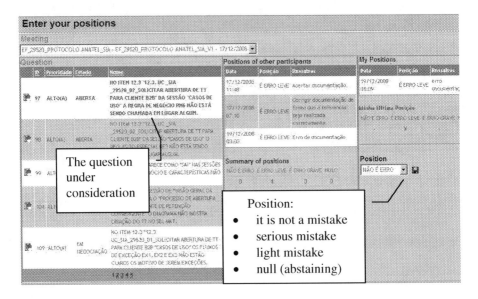

Fig. 4. Registering the participants' positions

   In the screen showed in Figure 4 the desired position for each question must be selected, as a "serious", "light" mistake, " it is not a mistake " or "null" (in case of abstaining). When the user is positioning for the first time, no positions already taken for his colleagues are shown. However, when the user positioned him already in a first time and desires to alter his position, he can know the position of the other participants; this information is available in the "position of others" column.

   In the screen showed in Figure 5, the position on which is desired register one argument must be selected: either positively, with the appearance to "favor"; or negatively, with the appearance "against". After this selection, the user must register his argument and, to confirm the operation, pressing the button "save". In this screen is possible to know the arguments already registered by other participants through the field "arguments of others". The arguments of the currently user are showed in another field "my arguments".

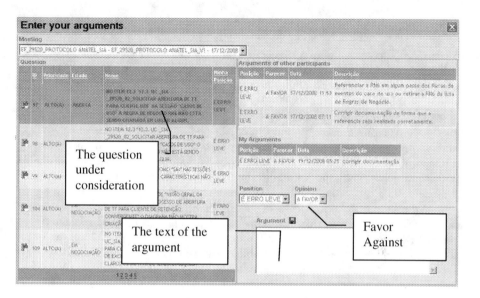

Fig. 5. Registering arguments

5 Evaluation

An evaluation action was carried out in a real-world organization operating in the telecommunications industry. The main purpose of the evaluation action was getting qualitative insights about the negotiation-collaboration process and FTR tool.

The evaluation action was set up to compare two treatment conditions: the control treatment, based on the FTR process currently used by the target organization; and the experimental treatment, using the FTR process and tool described in this paper. Furthermore, two different functional specifications were subject to the above treatment conditions; and two different groups of collaborators were selected (by convenience) to participate in the two treatments. We therefore accomplished a set of 2x2 experiments.

Eight employees with different professional expertise participated in the experiments. The participants were organized in teams of five and three persons. The team of five, constituted by three reviewers, a review leader and an author was subject to the experimental treatment. The team of three, consisting of three reviewers, one of them acting informally as leader, was subject to the control treatment. The facilitator role, which is required by the experimental treatment, was fulfilled by one of the researchers.

We selected the following evaluation criteria: (1) number of considered proposals; (2) number of considered arguments; and (3) number of changed positions toward consensus. The comparison was directed by the assumption that high quality means: more considered proposals; more registered arguments; and more changed positions toward consensus.

The functional specifications selected for the experiments were controlled to ensure they possessed the same level of complexity and quality.

5.1 Preliminary Evaluation Results

Comparing the control and experimental treatments, we observed that the experimental treatment resulted in an increased number of arguments and number of changed positions towards consensus. This may be a sign that the FTR tool supports the negotiation-collaboration and promotes higher levels than the traditional FTR process.

In the detailed presentation of the experimental results we will refer to the functional specifications as FE2950 and FE22520. First, it should be noted that FE2950 and FE22520 were rejected in both the control and experimental treatments.

Regarding the control treatment of FE2950, we had 4 proposals, 6 arguments and 2 changed positions. The experimental treatment of FE2950 resulted in 31 proposals, 15 arguments and 3 changed positions.

Regarding the control treatment of FE22520, we had 9 proposals, no arguments and no changed positions. The experimental treatment of FE22520 resulted in 23 proposals, 1 argument and 3 changed positions. The results from FE22520 show that the participants (and in particular the leader) took the immediate decision to reject the functional specifications, which explains the absence of arguments.

Comparing the control and experimental treatments, we may notice that the experimental treatment shows a higher number of proposals, arguments and position changes than the control treatment. This gives preliminary hints that the FTR tool stimulates the argumentation-collaboration.

The number of proposals was much higher in the experimental than the control treatment. This may not only be caused by the FTR tool. One influence to pounder is the team that participated in the experimental treatment is not only larger but also more diverse.

Apparently, the simplicity of the proposed process and the short training required before the evaluation are sufficient to accomplish the review with success. We noticed however that the arguments were not always used as such. For instance, several comments were inserted as if there were arguments. Comments such as "I agree with the item above" are not real arguments but appeared as such. This may impact the above comparisons. From a total of 78 registered arguments, only 47 (about 40 %) were actually identified as real arguments.

5.2 Questionnaires

The participants in the experimental treatment were requested to complete an open questionnaire about the FTR tool. The answers to the questionnaire seem to point, in a general way, that the tool supports the dynamics of the negotiation-collaboration process and promotes collaboration in formal technical reviews.

The main advantages pointed out by the participants were: (1) it was easy to learn; (2) had clear rules; (3) managed knowledge evenly; and (4) preserved the history of the argumentations.

Also, the support to asynchronous and geographically distributed meetings was identified as an advantage. Though the participants emphasized the face-to-face meetings ease the understanding between the persons and offers more possibilities for expressiveness.

It is important emphasize that the participants, in general, valued the capability to register all the arguments in an organized way. This seems to ease changing positions towards consensus and enriches the review as a whole.

One of the principal problems identified in the FTR process currently utilized by the target organization is that the review repeats itself several times without necessity, only because the review's recommendations seem to be unnoticed by the authors. The FTR tool was seen by the participants as a mechanism to overcome this problem.

Overall, the comments produced by the participants indicate that the solution presented in this paper is coherent to the desired objective: supporting collaboration and negotiation. The participants in the experiment indeed recommended the adoption of the FTR tool in their organization.

6 Conclusions

We developed a negotiation-collaboration process for FTR and a tool to support it. The research allowed us to understand the dynamic of the negotiation-collaboration. The observations and the results we obtained from the case study provided some insights about people's behavior in view of a somewhat contradictory process. The experimental results indicate the proposed process and tool is capable to support FTR.

The experiments also allowed us to identify some points that may constitute subject for future developments. Nevertheless, some challenges to continue this research include: (1) apply the FTR tool throughout the whole software engineering process; (2) develop more functionality, especially promoting argumentation; (3) improve the strategic visualization of the negotiation-collaboration process. And finally, we should complete the experimental evaluation of the FTR tool and process, adding new experiments, selecting different organizations, and covering more reviews cycles.

Acknowledgments

This work was partially supported by grants No. 479374/2007-4 and 567220/2008-7 from CNPq (Brazil).

References

1. Pressman, R.: Software Engineering a Practitioner's Approach. McGraw-Hill, Inc., New York (2000)
2. Antunes, P., Ramires, J., Respício, A.: Addressing the Conflicting Dimension of Groupware: A Case Study in Software Requirements Validation. Computing and Informatics 25, 523–546 (2006)
3. Fagan, M.: Design and Code Inspections to Reduce Errors in Program Development. IBM Systems Journal 15, 182–211 (1976)
4. Grünbacher, P., Halling, M., Biffl, S., Kitapci, H., Boehm, B.: Integrating Collaborative Processes and Quality Assurance Techniques: Experiences from Requirements Negotiation. Journal of Management Information Systems 20, 9–29 (2004)

5. Veenendaal, E. (ed.): Standard Glossary of Terms Used in Software Testing. International Software Testing Qualifications Board (2007)
6. Janis, I.: Groupthink: Psychological Studies of Policy Decisions and Fiascoes. Houghton Mifflin, Boston (1983)
7. Esser, J.: Alive and Well after 25 Years: A Review of Groupthink Research. Organizational Behavior and Human Decision Processes 73, 116–141 (1998)
8. Schwenk, C., Valacich, J.: Effects of Devil's Advocacy and Dialectical Inquiry on Individuals Versus Groups. Organizational Behavior and Human Decision Processes 59, 210–222 (1994)
9. Ramires, J., Antunes, P., Respício, A.: Software Requirements Negotiation Using the Software Quality Function Deployment. In: Fukś, H., Lukosch, S., Salgado, A.C. (eds.) CRIWG 2005. LNCS, vol. 3706, pp. 308–324. Springer, Heidelberg (2005)
10. Haag, S., Raja, M., Schkade, L.: Quality Function Deployment Usage in Software Development. Communications of the ACM 39, 41–49 (1996)
11. Gutwin, C., Greenberg, S.: The Effects of Workspace Awareness Support on the Usability of Real-Time Distributed Groupware. ACM Transactions on Computer-Human Interaction 6, 243–281 (1999)
12. Karacapilidis, N., Papadias, D.: Computer Supported Argumentation and Collaboration Decision Making: The Hermes System. Information Systems 26, 259–277 (2001)
13. Vetschera, R.: Preference Structures of Negotiators and Negotiation Outcomes. Group Decision and Negotiation 15, 111–125 (2006)
14. Jennings, N., Faratin, P., Lomuscio, A., Parsons, S., Sierra, C., Wooldridge, M.: Automated Negotiation: Prospects, Methods and Challenges. Group Decision and Negotiation 10, 199–215 (2001)
15. Dubrovsky, V., Kiesler, S., Sethna, B.: The Equalization Phenomenon: Status Effects in Computer-Mediated and Face-to-Face Decision-Making Groups. Human-Computer Interaction 6, 119–146 (1991)
16. Raiffa, H., Richardson, J., Metcalfe, D.: Negotiation Analysis. In: The Science and Art of Collaborative Decision Making. Harvard Business Press, Cambridge (2002)
17. Weingart, L., Bennett, R., Brett, J.: The Impact of Consideration of Issues and Motivational Orientation on Group Negotiation Process and Outcome. Journal of Applied Psychology 78, 504–517 (1993)
18. Fjermestad, J., Hiltz, S.: An Assessment of Group Support Systems Experimental Research: Methodology and Results. Journal of Management Information Systems 15, 7–149 (1999)
19. Borges, M.R., Pino, J.A., Fuller, D.A., Salgado, A.C.: Key issues in the design of an asynchronous system to support meeting preparation. Decision Support Systems 27, 269–287 (1999)

Generating User Stories in Groups

Cuong D. Nguyen, Erin Gallagher, Aaron Read, and Gert-Jan de Vreede

Center for Collaboration Science, University of Nebraska Omaha
{cdnguyen,egallagher,aread,gdevreede}@mail.unomaha.edu

Abstract. Communicating about system requirements with user stories is a distinctive feature of Agile Software Development methods. While user stories make system requirements intelligible to both customers and technical developers, they also create new challenges for the requirements elicitation process such as personal bias and requirements coverage. In this study we propose that when elicited from groups instead of individuals, the number of stories generated, the uniqueness and the comprehensiveness of the stories is likely to increase. A lab experiment design is delineated and partially completed. Future research will need to be conducted to determine conclusions.

Keywords: Requirements elicitation, user stories, group story telling.

1 Introduction

Software development remains a challenging process with nearly half of the projects considered late, over budget, and completed with fewer features than planned [1]. Poorly defined requirements are considered to be a leading factor in project failure [2]. Agile software development methodologies address difficulties of developing requirements resulting from rapidly changing customer needs by allowing a development team to respond quickly to changing requirements [3]. They encourage incremental releases, cooperation between customer and developers, simplicity (ease of learning), and adaptivity [4]. User stories are an integral part of several Agile methodologies including XP and Scrum [5, 6]. A user story is an account in the user's own words of a way that (s)he would like to use the software and enables the communication of software requirements between developers and customers without needing familiarity with a specific method of delivery or jargon [6]. While using stories as a means of gathering requirements has been shown to be beneficial in a number of studies [7-9], collecting system requirements in the form of user stories can also be problematic for several reasons: A customers' tacit knowledge may be partially hidden, stories are subject to multiple interpretations and personal bias, and the completeness of the set of stories is difficult to determine [8]. These problems may be addressed by collecting stories in groups. Group story telling can create an environment that supports evaluation of experience and promotes problem-solving [10]. This can help surface conflicts in goals among users and enable them to create shared understanding. Group story telling can also help elicit the tacit knowledge of participants with the richness of several different perspectives [11]. The purpose of our research is to understand to what extent groups will outperform individuals in generating user stories.

L. Carriço, N. Baloian, and B. Fonseca (Eds.): CRIWG 2009, LNCS 5784, pp. 357–364, 2009.
© Springer-Verlag Berlin Heidelberg 2009

A key criterion to assess the quality of a set of requirements is completeness with a minimal amount of conflicts and overlaps in requirements [12]. We therefore compare individual and group storytelling in terms of the level of completeness and lack of overlaps or conflicts in requirements.

In the following section we define stories and explain their use in requirements engineering. We then explain the design of our study. We present some initial results. The paper concludes with a discussion of the envisioned contributions and expected challenges regarding the further execution of the study.

2 Background

Stories are narrative retellings of personal experiences of phenomena or events [13]. Individual tells a story either in one-on-one situation or in a group to another individual (usually a researcher) using either free-association or through the use of an interview and/or prompts [14]. Further, stories can be topical, biographical, or autobiographical [15]. Stories can appear in various forms, including terse stories – a simplified and succinct retelling that leaves components such as plot or characters to the imagination of the individual hearing the story [16].

In requirements engineering, stories are used and structured in at least two different ways. First, stories may be used as a means of understanding the experiences and needs of users. They can capture the experiences of users with a current system as well as aid in capturing the desired attributes of a system [7]. Second, in Agile software development stories are used as a source of documentation of requirements specifications and structured in a way that helps stakeholders to easily relate essential details of software requirements for a new system. Stories must focus on the experience of the user and must be short enough to fit on an index card [6, 17]. These short stories are then used as conversation starters with developers who confirm the details of the story in acceptance tests on the software [17].

User stories play an important role in the requirements elicitation process in agile development[5]. The purpose of the requirements elicitation activity is to be able to describe the goals of the new system, with an understanding of the needs of the stakeholders and the constraints of the system [18]. Eliciting requirements in the form of user stories allows stakeholders to convey their needs in a way that is natural to them, allowing them to relate more tacit knowledge [7]. Documenting system requirements in the form of user stories allows customers to communicate desired features of a system without having to know a specific modeling language [19]. Details of stories are worked out through oral communication between the users and requirements engineers, thus avoiding errors of interpretation which may occur with written requirements [17]. However, stories consist of an individual's view of a system and may therefore make it difficult for requirements engineers to grasp a complete view of the system [8]. Valuable information might not be volunteered by users as they might assume it is already known to the requirements engineers, or forget abnormal cases [20].

Allowing users to generate stories in groups can alleviate some of the aforementioned shortcomings of collecting user requirements with stories individually. Story telling in groups allows a problem to be seen from multiple perspectives [11]. When users tell stories in groups, the knowledge of one user can be verified and expanded by another user, since the knowledge of one user helps to activate the knowledge of another group member [21]. When users meet in groups, they are able to evaluate one another's information and ask for clarification [10]. Nevertheless, working in group also creates new troubles. Users might not tell stories that touch sensitive and private matters or speak what they really think in public and therefore important issues might be ignored [22]. Traditional challenges in group collaboration like groupthink and participants' lack of focus can aggravate the effectiveness of the meeting.

With this understanding of the benefits of group storytelling and brainstorming techniques, we propose to test the following hypotheses:

H1a: Groups generate a larger quantity of stories than those working individually.
H1b: Groups generate more comprehensive stories than those working individually.
H1c:Groupsgenerate more unique stories than those working individually.

As telling story is also an idea generation or brainstorming process, it is interesting to know how different brainstoming techniques affect the group storytelling behavior. In this study, we consider examining two brainstorming techniques: brainstorming with prompts and brainstorming without prompts, or free brainstorming. In a brainstorming session with prompts, the requirements engineers will add some prompts, or questions or suggestions that can direct the users' thoughts to areas that they might ignore while in a free brainstorming session, people tell their stories freely without the intervention of the requirements engineers. While free brainstorming is traditional and has been applied in requirements gathering for a long time [23], brainstorming with prompts is a relatively new technique that was proposed by Santanen et al.[24] under the name "directed brainstorming". The technique was claimed to make a group produce more unique and straight to the point solutions for problem solving tasks than free brainstorming [24]. Brainstorming with prompts is also considered to reduce cognitive challenges like human memory constraints during a requirement elicitation process in [25]. Based on the findings of these previous studies, we propose the following hypotheses:

H2a: Groups using brainstorming with prompts generate a larger quantity of stories than those using brainstorming without prompts.
H2b: Groups using brainstorming with prompts generate more comprehensive stories than those using brainstorming without prompts.
H2c: Groups using brainstorming with prompts generate more unique stories than those using brainstorming without prompts.

3 Method

3.1 Participants

Forty-one participants enrolled in psychology courses at a Midwestern university provided useable data for this study. All participants electing to volunteer received extra credit in one course of their choice. Of these participants, 21 (53.7%) were female, 19 (46.3%) were male. Participants had an average age of 20.154 years ($SD = 2.16$, $Range = 18 - 27$).

3.2 Task

The task of this experiment is to provide system requirements in form of user stories about a system called Book Exchange. Book Exchange is a web-based system designed to allow students at a university to buy and sell text books at a reasonable price. The system does not provide payment services; it simply allows sellers to post items for sale, allowing potential buyers to search for their textbook offerings.

3.3 Independent Variables

- *Group Composition.* Participants will generate stories either individually or in a small group. In the individual treatment group, three individuals will generate user stories for a text book exchange system by themselves. In the group condition participants will be randomly assigned to a group of three individuals to generate user stories.
- *With prompts and without prompts brainstorming.* Participants will or will not be given some prompts in their session.

3.4 Dependent Variables

- *Quantity.* The number of requirements coded from user stories generated by subjects
- *Comprehensiveness.* The comprehensiveness of the requirements coded from user stories will be assessed in a manner similar to Pitts and Browne [25]. Comprehensiveness is determined by the number of predetermined categories covered by the requirements generated (breadth) as well as the number of requirements in each category (depth).
- *Uniqueness.* Story uniqueness is the extent to which stories generated do not overlap or duplicate the functionality of other stories generated.
- *Additional Measures.* Participants will complete additional measures assessing their perceptions and attitudes towards the story generation process, including demographic information, satisfaction with the process, understanding of the generated stories, and level of agreement regarding the generated stories.

3.5 Design

This study employs a 2x2 between-participants factorial design by varying the two independent variables discussed: group vs. individual and prompted vs. unprompted. An initial pilot study was conducted. Based on the results of a pilot study we came up with this final design the general procedure for all treatment groups is as follows. After providing a brief introduction explaining the purpose of the research, participants will provide consent and be randomly assigned to one of the experimental conditions. Participants will then complete identical demographic questionnaires and receive answers to any questions they may have. Each session will begin with standard script read to subjects providing a description of the project as well as delineating the task required. A sample script can be seen in Appendix A. Reflecting on the results of the pilot study we modified the scripted activities for each condition to make the groups more comparable across treatments. We also clarified the instructions for the activities in each treatment group to reduce misunderstanding about the desired outcome of the task. All conditions will last 40 minutes. Following completion of the story generation task, all individuals will complete additional measures and receive debriefing.

3.6 Analysis and Initial Results

Thus far, we have only analyzed results for quantity of stories for the 41 participants mentioned above. Quantity was determined by the number of on task, feature specific, requirements coded from user stories generated by subjects. The comprehensiveness of the requirements coded from user stories will be assessed in a manner similar to Pitts and Browne (2007) in future analysis. Inter-rater reliabilities for story/not a story ratings averaged 85.65%. Disagreements between coders were resolved by a third party to form the coding results used in the analysis.

The individual treatment groups generated on average 33.71 Stories with a standard deviation of 12.50 and 10.29 respectively. The participants in the group treatment generated on average 46.14 on task items and 41.57 stories, with a standard deviation of 10.81 and 12.35 respectively. For the difference between On Task Items between treatment groups, the two sample t-test statistic is $t=1.3032$, with 12 degrees of freedom. This is associated with a two-tailed p value of p value of .2170. This does not conclusively support Hypothesis 1a. The two sample t-statistic for the difference between the Stories in each group is $t=1.293$ with 12 degrees of freedom. This is associated with a two-tailed p value of .2203. Hypothesis 1b is also not supported with conclusive evidence.

4 Discussion

This research in progress describes how we have and will continue to explore whether or not generating user stories in groups will result in a significantly more numerous and more comprehensive set of user stories. Such understanding will provide insight for practitioners and researchers wishing to understand the benefits of group storytelling in the requirements elicitation setting. Initial findings show no conclusive difference between groups and individuals for quantity of stories. However, it is likely that there will be a stronger effect for comprehensiveness of stories, as groups may be more likely to generate a wider variety of stories. In terms of practical implementation, a positive result from the research will imply that instead of interviewing users individually, a practitioner may prefer to utilize groupware tools to let users tell stories in groups. From the academic perspective, this study contribute further understanding of the effectiveness of story telling technique in requirements gathering. Creating this research design is not without challenges.

References

1. Rubenstein, D.: Standish Group Report: There's Less Development Chaos Today. Software Development Times (2007),
   http://www.sdtimes.com/content/article.aspx?ArticleID=30247

2. Hofmann, H.F., Lehner, F.: Requirements Engineering as a Success Factor in Software Projects. IEEE Software 18(4), 58–66 (2001)
3. Highsmith, J., Cockburn, A.: Agile software development: the business of innovation. Computer 34(9), 120–127 (2001)
4. Abrahamsson, P., et al.: New directions on agile methods: a comparative analysis, pp. 244–254. IEEE Computer Society, Portland (2003)
5. Beck, K., Fowler, M.: Planning Extreme Programming. Addison-Wesley Longman Publishing Co., Inc., Amsterdam (2000)
6. Cohn, M.: User Stories Applied: For Agile Software Development. Addison Wesley Longman Publishing Co., Inc., Amsterdam (2004)
7. Alvarez, R., Urla, J.: Tell me a good story: using narrative analysis to examine information requirements interviews during an ERP implementation. SIGMIS Database 33(1), 38–52 (2002)
8. Sutcliffe, A.: Scenario-based requirements engineering. In: Proceedings of 11th IEEE International Requirements Engineering Conference, pp. 320–329 (2003)
9. Ima, M., Benyon, D.: How Stories Capture Interactions. IOS Press, Amsterdam (1999)
10. Banks-Wallace, J.: Emancipatory Potential of Storytelling in a Group. Journal of Nursing Scholarship 30(1), 17–22 (1998)
11. Valle, C., Prinz, W., Borges, M.: Generation of group storytelling in post-decision implementation process. In: The 7th International Conference on Computer Supported Cooperative Work in Design, pp. 361–367 (2002)
12. Grünbacher, P., et al.: Integrating Collaborative Processes and Quality Assurance Techniques: Experiences from Requirements Negotiation. Journal of Management Information Systems 20(4), 9–29 (2004)
13. Creswell, J.W.: Qualitative Inquiry and Research Design: Choosing among Five Approaches [with CD-ROM], 2nd edn. SAGE Publications, Thousand Oaks (2006)
14. Yanow, D.: Writing organizational tales: Four authors and their stories about culture. Introduction Organization Science 6, 225–226 (1995)
15. Merthens, D.M.: Research and evaluation in education and psychology: Integrating diversity with quantitative, qualitative, and mixed methods, 2nd edn. Sage Publications, Thousand Oaks (2005)
16. Boje, D.M.: The Storytelling Organization: A Study of Story Performance in an Office-Supply Firm. Administrative Science Quarterly 36(1), 106–126 (1991)
17. Jeffries, R.E., Anderson, A., Hendrickson, C.: Extreme Programming Installed. Addison-Wesley Longman Publishing Co., Inc., Amsterdam (2000)
18. Nuseibeh, B., Easterbrook, S.: Requirements engineering: a roadmap, pp. 35–46. ACM, Limerick (2000)
19. Davies, R.: The Power of Stories (2002)
20. Sutcliffe, A.: Scenario-based requirements engineering, in Requirements Engineering Conference, 2003. In: Proceedings of 11th IEEE International Requirements Engineering Conference (2003)
21. Leal, D.J.: The Power of Literary Peer-Group Discussions: How Children Collaboratively Negotiate Meaning. Reading Teacher 47(2), 114–120 (1993)
22. Goguen, J.A., Linde, C.: Techniques for requirements elicitation. In: Proceedings of IEEE International Symposium on Requirements Engineering (1993)
23. Maiden, N.A.M., Rugg, G.: ACRE: selecting methods for requirements acquisition. Software Engineering Journal 11(3), 183–192 (1996)

24. Santanen, E.L., Briggs, R.O., de Vreede, G.-J.: The cognitive network model of creativity: a new causal model of creativity and a new brainstorming technique. In: Proceedings of the 33rd Annual Hawaii International Conference on System Sciences (2000)
25. Pitts, M.G., Browne, G.J.: Improving requirements elicitation: an empirical investigation of procedural prompts. Information Systems Journal 17(1), 89–110 (2007)
26. João Pissarra, J.C.J.: Idea generation through computer-mediated communication: The effects of anonymity. Journal of Managerial Psychology 20(3/4), 275–291 (2005)

Appendix A: Experiment Scripts

Every subject in the groups will be given the same overview description of the system to be designed as follows:

The Book Exchange is a website which will be designed to allow students at this university to buy and sell text books at a reasonable price. The website will not provide payment services; it will simply allow sellers to post items for sale, allowing potential buyers to search for their textbook offerings. The website will also have features that facilitate a buyer's search for textbooks. For example, the website will have access to which textbooks are required for a given course.

In the short user story treatment, the subjects receive the following instructions. In the treatment where user stories are not required, the experiment participants are encouraged only to provide stories about software requirements.

- Provide as many user stories as possible. A user stories is a story that provide a feature that the system to be designed should have in your opinion. A recommended form for a user story is:
  "As a <type of user>, I want <some goal> so that <some reason>."
     e.g. "As a buyer, I want to be able to see the prices of all the books so that I can decide whether to buy the book or not."
- Organize the stories into a given set of categories.
- Continue to brainstorm user stories

Your stories should not be more than two sentences in length. You are NOT being asked to come up with a technical description of the website (i.e., it will use mySQL database for data storage). Instead we are asking you to describe what the website can do from the perspective of the website's users.

The experiment will be conducted in one hour long sessions and the experiment processes applied to each of four treatment groups are as follows:

- *Individual story telling – Unprompted* group: For this group, the session starts with letting the subjects complete a questionnaire. After that, the investigators give a short presentation about the purpose and the procedure of the session. Then, the subjects are trained on using GroupSystems. Next each subject is required to generate his/her user stories individually, i.e. no contact with other subjects is allowed, by typing them in a computer. After that, the individuals reorganizes their stories by putting them into the appropriate prioritization category: *critical, important,* and *if resources permit.* Finally, each member in the group will spend a portion of the time generating additional user stories.

- *Individual story telling –Prompted* group: For this group, the session starts with letting the subjects complete a questionnaire. After that, the investigators give a short presentation about the purpose and the working process of the session. Then, the subjects are trained on using GroupSystems. Next each subject is required to generate his/her user stories individually, i.e. no contact with other subjects is allowed, by typing them in a computer. As the group brainstorms, *prompts* will be displayed relating to the overarching goals of the system. After that, the individuals reorganizes their stories by putting them into the appropriate prioritization category: *critical*, *important*, and *if resources permit*. Finally, each member in the group will spend a portion of the time generating additional user stories.

- *Group story telling – Unprompted* group: For this group, the session starts with letting the subjects complete a questionnaire. The subjects are suggested to provide no more than two sentence length stories. Then the subjects are divided into groups of three people. After that, the investigators give a short presentation about the purpose and the working process of the session. Then, the subjects are trained on using GroupSystems. Next, each subject group is required to generate their user stories together by contributing user stories to the same electronic page or list at the same time. After that, the group reorganizes their stories by putting them into the appropriate prioritization category: *critical*, *important*, and *if resources permit*. Finally, each member in the group will spend a portion of the time generating additional user stories.

- *Group story telling – Prompted* group: For this group, the session starts with letting the subjects complete a questionnaire. The subjects are suggested to provide no more than two sentence length stories. Then the subjects are divided into groups of three people. After that, the investigators give a short presentation about the purpose and the working process of the session. Then, the subjects are trained on using GroupSystems. Next, each subject group is required to generate their user stories together by contributing user stories to the same electronic page or list at the same time. As the group brainstorms, *prompts* will be displayed relating to the overarching goals of the system. After that, the group reorganizes their stories by putting them into the appropriate prioritization category: *critical*, *important*, and *if resources permit*. Finally, each member in the group will spend a portion of the time generating additional user stories.

Author Index